BROWN & SHARPE
AND THE MEASURE
OF AMERICAN INDUSTRY

Brown & Sharpe and the Measure of American Industry

Making the Precision Machine Tools That Enabled Manufacturing, 1833–2001

Gerald M. Carbone *with the*
Rhode Island Historical Society

Foreword by Steven Lubar

McFarland & Company, Inc., Publishers
Jefferson, North Carolina

LIBRARY OF CONGRESS CATALOGUING DATA ARE AVAILABLE

Names: Carbone, Gerald M., author.
Title: Brown & Sharpe and the measure of American industry : making the precision machine tools that enabled manufacturing, 1833–2001 / Gerald M. Carbone with the Rhode Island Historical Society ; foreword by Steven Lubar.
Description: Jefferson, N.C. : McFarland & Company, Inc., Publishers, 2017 | Includes bibliographical references and index.
Identifiers: LCCN 2017005225 | ISBN 9781476669212 (softcover : acid free paper) ∞
Subjects: LCSH: Brown & Sharpe Manufacturing Company (Providence, R.I.)—History. | Machinery industry—United States—History. | Labor—United States—History. | Technological innovations—United States—History. | United States—Social policy. | Industrial revolution—United States—History.
Classification: LCC HD9705.U64 C37 2017 | DDC 338.7/6219020973—dc23
LC record available at https://lccn.loc.gov/2017005225

British Library cataloguing data are available

ISBN 978-1-4766-6921-2 (print)
ISBN 978-1-4766-2919-3 (ebook)

© 2017 Rhode Island Historical Society. All rights reserved

No part of this book may be reproduced or transmitted in any form or by any means, electronic or mechanical, including photocopying or recording, or by any information storage and retrieval system, without permission in writing from the publisher.

Front cover: *inset* the DigitCal, billed as the first hand tool to embrace microchip technology, introduced by Brown & Sharpe in 1978. The three-story "fire proof" factory designed by Frederick W. Howe in 1872 stands in the foreground, next to a foundry that fired for the first time in December 1880 (photographs courtesy Rhode Island Historical Society)

Manufactured in the United States of America

McFarland & Company, Inc., Publishers
Box 611, Jefferson, North Carolina 28640
www.mcfarlandpub.com

Table of Contents

Acknowledgments   vii
Foreword by Steven Lubar   1
Preface   5

PART ONE—AGE OF INVENTION

1. A Burning Curiosity   9
2. A Curiosity Shop   14
3. "The Genius of Its Maker"   27
4. "An Eye to Business"   34
5. Prosperity   41
6. Panic   51
7. The Age of Steel   64
8. Providence, Paris, Chicago   75
9. Spoke Wheels, Turning   86
10. Death and Succession   97

PART TWO—CENTRALIZATION

11. "Clean the Damned Place Out!"   109
12. When Peppermint Creams Meet Steel   121
13. "If I Had Known This Was Coming"   136
14. "Don't We Ever Play a Waltz?": The 1930s   148
15. World War II: Defense Workers Wanted   158
16. One War Ends, Another Begins   172

PART THREE—A PEOPLE'S CAPITALISM

17. Flying into the Jet Age   183
18. Retooling   193
19. An Industrial Eden   203

PART FOUR—MANAGERIAL CAPITALISM AND THE GLOBAL CORPORATION

20. Fenced In	215
21. The Longest Strike	228
22. Locked Out	244
Chapter Notes	253
Bibliography	260
Index	263

Acknowledgments

First, thanks to Phil West, a friend and author of *Secrets and Scandals*. Phil knew I was looking for work, and he knew that Henry Sharpe, Jr., and his wife, Peggy, wanted someone to write the story of Brown & Sharpe, the company that the Sharpe family was connected to for 153 years. Phil introduced us. Thanks also to Henry and Peggy Sharpe, who hired me with only one condition: "We do not want a vanity book." They wanted an unvarnished, objective look at the company's history—I hope I have delivered.

The Rhode Island Foundation and the Rhode Island Historical Society were instrumental in bringing this project to fruition. Thanks to RIHS Executive Director Dr. Morgan Grefe, who convened two panels of experts to critique early drafts of this story. The panelists offered valuable commentary, though did not necessarily condone the work—in fact one of them, industrial historian Pat Malone, emphatically told me this was *not* the story he would have written about Brown & Sharpe! Besides Dr. Malone, insightful readers included the late Rick Greenwood, Ted Widmer, Steve Lubar, Jaimie Warren, Steve Proctor, Dr. Grefe, Matt Burriesci, Ned Connors, and Rick Slaney.

The staff at the Rhode Island Historical Society's Mary Elizabeth Robinson Research Center embraced this project, even allowing my research assistant access when the library was closed for emergency repairs. About this research assistant, my sister, Jean Carbone: she is so much a part of this work that I offered her credit as co-author, which she humbly declined. But she was a colleague and generator of ideas, and much of this book is drawn from her work in the archives.

Numerous people gave their time to be interviewed for this story, and again without implying that they condone the final draft, I would like to thank some of them: Velia Constantino, a real Rosie the Riveter who worked at Brown & Sharpe during World War II; machinist Ed Green; engineers Merrick Leach, Arthur Parrillo, and Larry Webster; former apprentice Bill Nixon; Bruce MacGunnigle, East Greenwich town historian; labor historian Scott Molloy, and labor lawyer Marc Gursky; Dick and Chris Jocelyn from the company's human resources office; sales executive Rudy Bernegger; the children of Henry and Peggy Sharpe: Henry III, Douglas, and Sarah; Susan McGregor and her brother, Bill, who hail from generations of Brown & Sharpe workers, as did the late Maybury Fraser.

Traveling to archives, machine shops, and museums was educating and fun, thanks to: the late Fred Jaggi and the New England Wireless and Steam Museum; Droitcour

Manufacturing Company, which still uses dozens of Brown & Sharpe screw machines in its Warwick, Rhode Island plant; the Starrett Tool Company of Athol, Mass.; the staff and volunteers at the Charles River Museum of Industry and Innovation in Waltham, Mass.; welcoming staff at the Baker Library of the Harvard Business School; familiar faces at the Rhode Island State Archives, particularly Gwen Stern and Ken Carlson, and new faces at the University of Rhode Island archives; patient circulation staff at Brown's Rockefeller Library; and people at the American Precision Museum in Windsor, Vermont, particularly Ann Lawless and John Alexander.

Living near Brown University gave me access to roundtable workshops led by Professor Seth Rockman, where good historians unveiled books that they are currently writing. Seeing established historians such as Michael Zakim and Stephen Mihm in the early phases of wrestling new ideas into manuscripts was somehow bracing. Again, Zakim and Mihm have not even seen this work, much less endorsed it, but I single them out because Zakim is cited in it and Mihm writes about the standardization of things such as currency, weights, and measures, which is all relevant to the story of Brown & Sharpe.

Finally, there is my wife, Mary Preziosi, who not only contributed her technical expertise as a network administrator, but offered a listening ear, good feedback, and high tolerance as the manifold books, articles, and artifacts required to write this spilled out of my home office, onto the dining room table, and beyond. For three years this project and these people shaped my perspective and my life in unforgettable ways, and I am thankful for that.—G.M.C.

Foreword by Steven Lubar

On one of my first days working as a curator at the Smithsonian's National Museum of American History, I discovered the Brown & Sharpe micrometer No. 1. The museum was planning an exhibition on the American industrial revolution. I was visiting storerooms, looking for artifacts that would tell the story of the hard work and ingenuity that powered America's transition from agricultural colony to industrial juggernaut.

I knew a bit about micrometers. My father had been a sheet-metal worker at the Philadelphia Naval Aircraft Factory. I had briefly worked in a machine shop, as a gofer, and my father had given me his micrometer—a revered symbol of the metal worker's skill. With a twist of the wrist, the micrometer told him the size of a part, the thickness of a sheet of metal, to a thousandth of an inch. But it took that skilled twist of the wrist—it wasn't a digital readout that just presented a number. It required a feel for the tool gained only through use. The twist of the wrist was symbolic of the deep knowledge of tools, measurement, and materials that was embodied in the metal worker's everyday work.

I knew something about Brown & Sharpe, too. It was one of the most important American machine tool firms, and had been, for a very long time. Machine tools were the technological heart of the American economy. They were the tools used to make other tools. A skilled machinist using Brown & Sharpe's machine tools could make just about anything. As a historian of industrial technology, machine tools were central to my work, too. The micrometer was a fine item for the exhibition. It captured, in one small device, entrepreneurial energy, technological innovation, and the skill of workers, all essential elements of the industrial revolution, and of American society.

The Smithsonian had a large amount of other Brown & Sharpe material. The collections included dozens of other measuring tools: the firm got its start by creating—inventing and manufacturing—some of the most accurate, and affordable, rules, scales, and calipers for metal work of all kinds. There were several Willcox & Gibbs sewing machines, an early success for the firm, which adapted the Armory System of making interchangeable parts to produce them. Machine tools in the collection included a gear-cutting machine and two universal milling machines—these were the breakthrough tools that defined Brown & Sharpe in the late nineteenth-century, and that helped make possible the mass production that shaped America's twentieth-century history. The library had hundreds of catalogs from the company, telling the long history of the firm through the products it had sold. And on "permanent loan" to the museum was the symbolic heart of Brown & Sharpe: a linear dividing engine used to make precision measuring tools, built in 1852, a true national treasure.

Just as the micrometer would allow an object to tell the story of American technology and work, Brown & Sharpe would nicely tell the story of American industry. In the Brown & Sharpe collection were essential elements of the story of American industrial revolution. After all, Brown & Sharpe was at the center of that story, making the machines that made the looms, sewing machines, locomotives, automobiles, and weapons that made America modern.

I moved to Providence a dozen years ago, trading the Smithsonian for Brown University. Exploring the city, I found other traces of Brown & Sharpe. The firm had once been among Providence's largest employers and that history was still visible. Its original factory had been located just blocks from my office; that building was gone now. On the other side of downtown was the massive plant the firm had built starting in 1872, expanding it over the course of 70 years to include more than a dozen red brick factory buildings, spread over several blocks. It was once the largest tool manufacturing factory in the world.

That factory was mostly vacant, slowly being converted into offices and apartments. The neighborhood nearby was in decline. It had been home to many of the skilled workers at Brown & Sharpe, many of them immigrants who through their skills and hard work had been able to live in attractive houses within walking distance of the factory. Brown & Sharpe had moved to the suburbs in 1964, taking its skilled workforce with it. But even that new plant was mostly empty. A 16-year strike, along with a changing economy, management troubles, and international competition, had nearly finished off the firm. Brown & Sharpe was bought by a Swedish competitor in 2001.

I knew something of the machines, and the buildings, and the outline of the history. Behind them, of course, were the people, and that's a story that, until now, has not been so well known. The history of Brown & Sharpe turns out to be not only an important technological and economic story, but also a fascinating human story. Joseph Brown, the founder, was a skilled clockmaker-turned-machine-maker who invented new machines, and new ways to make things, as needed. Samuel Darling was an eccentric inventor from Maine, a one-time competitor who joined the firm and brought with him his prized dividing engine. The Sharpes—Lucian, his son Henry, and grandson Henry Jr.—guided the firm for more than a century, and shaped not only

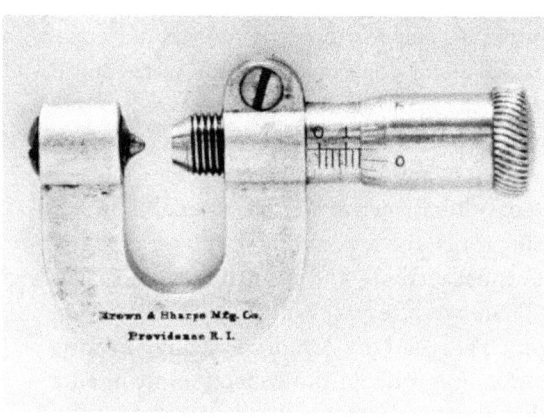

A one-inch "Pocket Sheet Metal Gauge" that Brown & Sharpe released in 1868, renaming it the "micrometer" in its 1877 catalogue. The micrometer made it possible for every mechanic to have a precise, portable tool for measuring metal thicknesses always at his disposal.

the company, but also the industry, and the city. And apprentices, skilled workers, union leaders, financiers, and family members.

Gerald Carbone's history of Brown & Sharpe tells these stories, bringing the people to life, putting them into the context of Rhode Island's and the nation's history, and the history of technology and the economy. Brown & Sharpe's story is the story of the American Industrial Revolution. But Carbone does much more than tell a dry story of machines and money, of innovative design and engineering, profit and loss. The real story here is the human one, encompassing more than a century-and-a-half of change. How did the owners and managers negotiate the ever-changing economy, rapid technological change, changing expectations about work and pay? How did the men and women who worked at the firm learn their skills and organize their work to produce and market a dazzling array of measuring devices, sewing machines, machine tools? How did the firm help shape the city, and the nation?

That micrometer in the Smithsonian storeroom helped me connect with that work, those lives, a remarkable set of stories of entrepreneurship, corporate management, and skilled work. Carbone tell those stories with verve. He brings the history of Brown & Sharpe to life. It turns out that it's not only a rousing good history, but also a story with lessons for us today. The tools may change, the technologies differ, but smart people, working hard, trying new things, inventing and innovating, organizing production, selling their products; those are universal, as much the story of today's innovative high-tech firms as Brown & Sharpe's story. This is a book to be read not only as history, but also to help understand the present, and the future.

Steve Lubar is a Brown University professor of American studies, history and history of art and architecture, and former chair of the Division of the History of Technology at the Smithsonian's National Museum of American History.

Preface

On a fall day in 1849, a young mechanic's apprentice strolled to the northern edge of Providence to tour the old army campgrounds where hive-like depressions still pocked the earth. Here in the waning days of the American Revolution, 4,000 French troops had dug in for temporary lodgings while awaiting transport ships to carry them home after a successful campaign at Yorktown.

"About 4000 encamped here in the winter of 1782," wrote the apprentice, 19-year-old Lucian Sharpe. "A great number of holes remain, over which they pitched their tents." Here was tangible evidence of the American Revolution, essentially a capitalist insurgency fought by rogue colonialists such as George Washington and Nathanael Greene, to stamp out the vestiges of British mercantilism and to appropriate as their own powers and property previously relegated to the aristocracy.[1] Washington felt that the king had no more right to reach into his pocket than did his neighbor. Greene never cared a whit about the troubles with England until a British revenue cutter seized his family company's ship for evading customs—then he became obsessed with the cause. The Revolution's leaders were as motivated by Adam Smith's recently published paean to capitalism, *The Wealth of Nations*, as they were by Thomas Paine's appeal to *Common Sense*.[2]

Lucian Sharpe's grandfather had fought with Washington's troops at the Siege of Boston, but 65 years after the war's end the American Revolution had not fulfilled its promise of a free and equal society for the Sharpe family. Lucian's father, Wilkes, born among the generation that Joyce Appleby says "inherited the revolution" had barely escaped debtor's prison by selling his livery stable at public auction in 1834—horses, tack, sleighs, and all—to pay business debts.[3] The sale had left Wilkes so destitute that for a time he had to send young Lucian out of state to live with relatives while Wilkes got back on his feet as a hostler, a kind of valet for horses ridden to a local hotel.

Joseph Brown, born in 1810, also struggled to find a toehold in the economy of the New Republic. At age 23 he went into business with his father, David, making clocks, watches, and mathematical and nautical instruments such as protractors and sextants using only hand tools. They worked together for eight years, enduring a fire that razed their shop and destroyed most of their tools, before David Brown left his son to fend for himself. Disgusted with a Rhode Island law that restricted voting to an elite group of wealthy property owners, and frustrated in his vain attempts to change it, David moved to Illinois Territory in 1841.

This book traces the development of the political economy of the United States from the days of the New Republic into the twenty-first century, as seen through the lens of a single Rhode Island metals manufacturer, Brown & Sharpe, that traced its roots to the Brown's shop in 1833. There was nothing in the Constitution that made capitalism's dominance of the American political economy inevitable; but once policy makers favored capitalism over other ways of organizing economies, the standardization of time, currency, and metrology did become inevitable by-products of capitalism. Standardization is necessary for assessing the relative value of commodities, a function of efficient capitalism; and Brown & Sharpe, co-founded by Joseph Brown with his apprentice, Lucian Sharpe, was a key driver of both industrial capitalism and its attendant standardization of currency, time-keeping, metrology, and manufactured goods, defining features of modernity as we live it.

I became interested in the Brown & Sharpe story largely by chance. A friend, Phil West, knew that Henry Sharpe, Jr., and his wife Peggy (both of whom appear in the latter chapters of this book) wanted to hire somebody to write the story of the company that the family had owned for more than 150 years. I was skeptical: I had been a union steward and contract negotiator for the Providence Newspaper Guild, and Brown & Sharpe was infamous in local labor circles as the company that had presided over a long and exceedingly bitter strike—arguably history's longest. But I agreed to meet with the Sharpes to discuss writing this story. They told me their family was not interested in a vanity book to clog up their coffee tables, they wanted a good, objective look at the company's history. After voting to sell the company in 2001, Henry Sharpe, Jr., had spent seven years organizing the firm's files into archives that he then donated to the Rhode Island Historical Society; he and Peggy wanted a writer to look through the 42 linear feet of archives, plus documents and objects, in the family's possession in order to tease out a narrative. We agreed that if there were serious problems in interpretation of the archival material, either one of us could walk away from the project. Happily, Henry Sharpe, Jr., and Peggy never second-guessed any of my conclusions while reading the draft. Peggy did correct a couple of errors I made in writing about her courtship with her husband. Naturally I wonder whether the Sharpe's role in underwriting this story affected my telling of it. I honestly do not believe that it did, though some of my union brethren may feel otherwise.

The principal manuscript sources for this work are housed in the voluminous Brown & Sharpe archive at the Rhode Island Historical Society. One of the reasons for writing this story was to showcase the wide span and deep detail of the archives so that historians may make future use of them. In the book's latter sections I also conducted interviews with people who worked at Brown & Sharpe from the 1940s into the 2000s. For secondary material I consulted more than a hundred books, journal articles, and newspaper stories. The most similar work to this one is Max Holland's *When the Machine Stopped*, which also examines the history of a machine tool company from its founding to its failure; a key difference between the two works is the time span: the company in Holland's story, Berg Tool, was founded in the 1940s, while Brown & Sharpe planted its roots more than a century before. To understand the causes of economic panics from 1837 to the Great Depression, I leaned heavily on Scott Nelson's *Nation of Deadbeats*, which I found to be a good, interesting, and accurate work.

The narrative of American labor history has long favored what Seth Rockman calls the "heroic hammer-wielding mechanics" such as Lucian Sharpe, men who emerge as the citizen workers on the front lines of the Industrial Revolution.[4] This book is another one of those heroic epics, especially so because it begins, in true Horatio Alger style, with a poor youth, Lucian Sharpe, who through dint of diligence and industriousness, lifts himself up from one who sells his labor into the class of people that buys the labor of others.

There is an inherent danger in telling yet another antiquarian, epic narrative about a heroic mechanic who achieves social mobility through his uncommon virtues, for in doing so we ignore the vast majority of people who, because of race, ethnicity, or gender never could have had the opportunity available to Lucian Sharpe, regardless of their virtue. In telling a Horatio Alger–type history, we emphasize the rare case of a worker transcending the circumstances of his birth, and largely exclude from the narrative the vast majority of the available labor pool, people who held little social capital: women, non–Protestants, American Indians, Asians, Hispanics, and African Americans, free and enslaved. Indeed, in the long history of Brown & Sharpe, African Americans are conspicuous by their absence—the company did not hire its first black worker until it brought on Jesse Bradley as a foundry worker in the 1920s; the machinists union itself, the International Association of Machinists, enforced segregation into the 1940s.

In writing this story I try to make no claims of inevitability, and work instead to introduce agents and their actors who make choices that privilege one course of action, one set of people, one systemic approach over others in the constant give-and-take between labor, capital, and government that continuously shapes a nation's political economy. The narrative follows an epic storyline, because the story of Brown & Sharpe is a multi-generational epic; I use three generations of the Sharpe family to examine actors and events from the company's founding in Providence by David and Joseph Brown in 1833, until its sale to a Swedish company in 2001. I present the story in four sections. The first part examines the actions, interests, and inventions of Lucian Sharpe and his partner, Joseph Brown from 1848 till 1899, as they create the tools and the market underpinnings necessary for industrial capitalism to thrive; part two traces the management of Henry Sharpe from 1900 until 1951, a period of progressivism, increasing centralization, and government-regulated capitalism; and part three follows Henry Sharpe, Jr., from the 1950s, which saw the rise of a "people's capitalism" that created a large middle class of white, male workers, into the late 1970s and 1980s with the advent of Reaganomics and the managerial capitalists who directed the course of global capitalism from the 1970s into the twenty-first century.[5]

Stories about poor men who beat long odds to strike it rich are popular because they are good stories, and my goal with this book is to write a good story. The old-fashioned epic history has its virtues, particularly when applied to the experiences of white male workers at the intersection of capitalism, standardization, and modernity. Examining the vicissitudes of this relatively privileged segment of the population across a century-and-a-half of United States history is revelatory. This is the segment, not only the mechanics but the clerks and bookkeepers that forged industrial capitalism; this is the demographic that most benefited from the postwar people's capitalism of the

1950s and 1960s, and this is the population that felt keen betrayal by the rise of the managerial capitalist and the push for a capitalism dominated by multi-national corporations independent of sovereign fealty that left the white working class man at Brown & Sharpe walking the picket line in the world's longest strike, a sometimes violent, ultimately impotent protest against the U.S. political economy at the dawn of the twenty-first century.

PART ONE: AGE OF INVENTION

1. A Burning Curiosity

> The two [Joseph Brown and Lucian Sharpe] made a team almost irresistible. There might be a dispute as to which did most for the Concern. The only conclusion *should be* that each was indispensible in laying the foundations deep and strong. Like man and wife they were of one flesh, and equally necessary.—*Thomas McFarlane, former Brown & Sharpe Superintendent*

Joseph Rogers Brown wanted to build a machine that could make rulers. In 1848, nearly half-way through the century, no one in North America had ever built such a machine. If an American wanted to buy a ruler, he or she had to import one from Europe or hire someone capable of painstakingly inscribing one by hand.

Brown stood a little over 5 feet tall, and his rumpled suit coats always looked a size too big. The two times he posed for pictures he wore a furrowed look of deep concentration, with dark, recessed eyes, bushy eyebrows, thin lips and a wide beard. His cramped workshop occupied a narrow slice of a riverfront warehouse that once stored teas, porcelains, and silks hauled back from China before textile factories displaced shipping as a primary generator of Rhode Island's wealth. Here in a dimly-lit space measuring 30 feet-by-60-feet, Brown obsessed over the mathematical and mechanical challenges posed by customers who asked him to fashion or repair umbrellas, milk cans, eye glasses, quart measures, iron chain, tweezers, machinery to blow bellows, trusses, abdominal supporters and the like.

A staple of Brown's business was building tower clocks for churches, factories, and the state house down in Newport, big clocks with meshing gears turned by the swinging action of 13-foot pendulums.[1] To keep his staff of 14 tinkering at their gas-lit benches Brown also took on almost any odd job, particularly those dealing with grinding, drilling, planing, or turning of metal—sharpening knives and scissors, polishing silverware, fashioning small tools for watchmakers and jewelers, repairing watches, music boxes, pumps, or fixing the mechanical lamps that pedestrians carried over Providence's sandy, mostly unlit streets.

Brown could become so lost in solving the challenges posed by his customer's needs that he had to be reminded to eat, and his solutions often bore the stamp of genius. The obvious solution to building a machine that could precisely mark, or graduate, rulers was to move the ruler blanks past the diamond-tipped engravers by using a "lead screw" that turned in place and smoothly passed the ruler blank through the machine from thread to thread. The problem was, no one had yet been able to make a "perfect screw" with precisely equal distances between the screw threads. Even tiny differences in thread spacing would show up as repeated imperfections in a ruler's graduation lines. Rather than tackle the problem of inventing the perfect screw, Brown tried moving the blank ruler past the diamond-tipped engravers by harnessing the potential energy of tightly wound steel bands attached to a round index plate grooved with 1,080 teeth. By gradually relaxing tension in the bands he could move the blanks at a constant, controlled pace, with fractional turns of the index plate setting the spacing of the ruler's graduation marks.

Brown reveled in problems like those posed in building a graduating machine, but other skills required for running a small business tended to elude him—writing, for example. His correspondence was competent, but inelegant. He liked to joke that he spelled words "*variously* so that a reader might have his choice."[2] By 1848, standardized spelling had been customary in the United States for a generation, thanks to Noah Webster's dictionary, but Brown still had trouble with certain words, writing dimentions for dimensions, misspelling proffits, opperation, simplyfied, and frequently using principals for principles.

Brown's bookkeeping was neat but unsophisticated, a pastiche of receipts pasted into a scrapbook and lists of jobs done and money owed or received. By 1848 he had been in business for 15 years, 8 of them with his father, the rest on his own, but he still tottered on the edge of profitability. People in Providence flocked to his shop as a kind of store, not to buy premade products but to have things made. Brown had the skills to make most anything, except for a profit.

Joseph Rodgers Brown, circa 1860, when he was about 50. In his younger days his beard was shorter, but in an earlier portrait he wore the same expression of deep concentration.

1. A Burning Curiosity

Lucian Sharpe stood a head taller than Joseph Brown, with jug ears, thick lips, a flat belly, long limbs, and the tireless energy of a teen. Only 18 years old in May 1848, Lucian possessed an insatiable curiosity about developing technologies—the locomotives that first chuffed into Providence when he was 5, the modern gas works that would soon begin lighting a single row of flickering lamps along the city's main street, the new telegraph poles and wires that stretched northward for as far as his eye could see. Lucian viewed this column of poles marching off toward Worcester not as an aesthetic blight, but as a tangible sign of nineteenth-century progress.

Lucian worked from sunup to sundown in the Providence Machine Company's big, Gothic mill, with four octagonal towers looming over each corner of the big, brick complex. Lucian's father had bound him out to the company's owner, Thomas Hill, as an apprentice mechanic making fly frames and roving frames for cotton mills. Lucian was not happy with the arrangement, not so much with his employment as with his employer. Hill, a pious Methodist with a thin white beard, kept his workers on the same schedule as the textile mills that he stocked with machinery: sunup till sundown, six days a week. On some of Lucian's early time cards he clocked in for more than 80 hours a week. He began missing work quite a bit, sometimes staying in his room to perform experiments with chemicals, beakers, and test tubes that he bought at a downtown store.

Owen Mason, a wealthy 37-year-old merchant who took an interest in mentoring some of the local teens, was aware of Lucian's unhappiness with his apprenticeship. Mason suggested that the J.R. Brown Company might be a good place to learn a trade; though the shop was small it always seemed busy. The shop ran on a 10-hour day—work began at 7 a.m. and ended at 6 p.m. with the noon hour off for "dinner," a more merciful schedule than the machine company's grind. As clouds drew across the sky on May 7, 1848, Lucian determined that he had had enough of Mister Hill and his machine company. "Did not go to work today," he declared in his journal. "Went down to Joseph R. Brown's in south Main Street to see about getting a place with him for I have made up my mind to leave where I am now."

Lucian lacked Brown's natural gift for mathematics, but the two men, a generation apart, one short, the other tall, shared one thing in common: both had been born to fathers who had built and lost small businesses. Brown's father, David, worked as a 14-year-old tavern boy before apprenticing first to a jeweler, then to a watchmaker before opening his own store. Right after Joseph's birth in 1810, David's small silverware and jewelry shop in Warren, Rhode Island went broke, forcing him to leave his family behind while he took to the road as an itinerant peddler and knife sharpener. For three years David Brown pushed a cart up and down the Connecticut Valley, selling his handmade silverware and sharpening cutlery on a foot-powered grindstone of his own invention.

When Lucian was born in 1830 his father, Wilkes Sharpe, owned a large livery stable with 18 horses, 8 sleighs, bells, surreys, carriages, and buffalo robes. But in the run-up to the Panic of 1837, Wilkes found himself unable to pay back debts. In order to avoid the squalor and humiliation of debtor's prison, he sold at public auction all of the horses, the stables, sleighs, carriages, robes, and bells—everything but the family's household furnishings. Things grew so dire that he sent 4-year-old Lucian to live with relatives, first for a few months in Boston, then for years at the family's ancestral farm in Putnam, Connecticut. Now in 1848, Wilkes worked as a Providence hostler, a valet for the horses at a local hotel.

A month after his first inquiry about working for Brown, Lucian again strode through the door of the J.R. Brown Company, dropping in after working much of a summer Saturday at Thomas Hill's plant. Brown, Lucian reported in his journal, "didn't know but there might be a place for me next fall." In July, Mason the mentor told Lucian the good news that Brown had indeed "concluded to give me a place in his shop," but this would require delicate negotiations between Mister Hill and Wilkes Sharpe, who had signed an apprenticeship agreement binding Lucian to Hill. The two men worked out the terms of Lucian's separation, and since Wilkes Sharpe could not afford it, Owen Mason agreed to pay the standard $50 apprenticeship fee to Joseph Brown so Lucian could begin working there in August.

With the promise of employment at Brown's shop in a couple of weeks, Lucian had a fortnight in August free to do as he pleased. He travelled by stage over dry, dusty roads to Pomfret, his grandfather's home and the place where had spent much of his childhood exile from Providence. He carried with him a piece of white phosphorous for his chemistry experiments, "thin as a pipe stem" and about a half-inch long. Lucian concealed the phosphorous in a small, corked bottle in his pocket. On contact with air, white phosphorous spontaneously ignites, producing a garlic-smelling smoke.

One morning as he descended the staircase in his grandfather's house, the bottle cork unstopped, exposing the phosphorous to air in his pocket. At first, Lucian felt a burning in one thigh. He dug into his pockets to empty them and "upon doing so … I blistered my hand badly." He slathered his burned hand and thigh in Dalley's Pain Extractor, billed as the Great Family Ointment. After the pain subsided "in an hour or two," Lucian dryly noted in his journal: "Don't think I shall be so careless as to carry such things in my pocket again." The pain lessened, but the infection had only begun.

On the morning of August 12, 1848, Lucian hauled some corn and oats to the Pomfret depot for milling, then took the eastbound stage for Providence, sharing a crowded coach with nine passengers. As soon as he disembarked he made a beeline for J.R. Brown Company to discuss his start date, but Brown was not in. "Think today I shall be able to go to work next week," Lucian wrote in his journal, but this was not to be.

The next morning the phosphorous burn on his leg felt "quite sore." He stayed home all day, and awoke the next with a headache; he took some pills for his infected leg that made him sicker. His mentor walked over to the Sharpe house in a flat, miasmic section near the wharves to deliver books, including *Handy Andy* for his convalescence. Lucian, an insatiable reader, laid awake till after midnight reading *Handy Andy*, "and didn't know but I should split my sides laughing."[3]

Throughout Lucian's convalescence his new job at J.R. Brown's was very much on his mind. As soon as his leg felt well enough to support a walk from his father's house "downstreet" to Brown's shop on South Main he made the trek, as if to reassure Brown that he could start soon. But Brown was not at the shop. A few days later he tried again, stopping first to see a Doctor Chapin who observed his sore and ordered a new ointment to slather on it. On September 12, 1848, Lucian Sharpe finally felt strong enough to write in his journal: "Went to work at Mr. Brown's this morning."

In his job book, Joseph Brown wrote: "Lucian Sharpe came to work for me this day as an apprentice."

These brief, declarative sentences marked the beginning of a synergistic partnership between an ingenious inventor and a brilliant businessman that would fundamentally change how people cut and shaped metal, catalyzing the inventions of the bicycle, the automobile, and the airplane.

2. A Curiosity Shop

> August 22, 1850: Mr. Brown's Graduating Machine went into operation today. By simply turning a crank the rule is accurately divided into any number of parts....
>
> January 4, 1851: Stormy day. Snow and rain by turns. At home writing all day. Began to keep my account of expenses, etc., in the form of a Double Entry.—*from Lucian Sharpe's Journal*

A travel writer from a Boston magazine toured Providence, and as he strolled along South Main Street his attention turned to "the friendly face of an old clock" keeping precise time in a shop window. The scene evoked in his mind the "Old Curiosity Shop" of an 1840s Charles Dickens serial.

"While we were standing on the sidewalk, opposite the establishment where the clock may be seen, a friend came up and very kindly volunteered to give us an introduction to one of the proprietors of this manufactory," wrote the writer, in an 1854 edition of *The Illustrated Waverly Magazine and Literary Repository*. The proprietor who agreed to guide the writer through this shop of curiosities was Lucian Sharpe, who had joined Joseph Brown as a full partner the year before at the age of 23. In just five years Lucian had risen from an apprentice paying fifty dollars for the privilege of learning the watchmaker's trade, to partner in a machine shop on the cutting edge of nineteenth-century technology.

Lucian wowed the writer with a behind-the-scenes tour of J. R. Brown, L. Sharpe, Clock and Watchmakers and Manufacturers of Small Machinery, a narrow shop, thirty-feet wide and sixty-feet deep. The closely packed shop sported a gas light at every one of the fourteen work benches, plus an assortment of clocks, watches, jewelers' tools, and "mathematical and nautical instruments." Of all the tools and machines that Lucian Sharpe showed to the writer from *Waverly's* one stood out: the hand-cranked linear dividing engine that J.R. Brown had built to line or "graduate" rulers and protractors quickly and with precision. "This ingenious contrivance," wrote the *Waverly's* writer, "appears to be endowed with intelligence; for it takes the rule of steel or wood in its metallic fingers, and marks it at the rate of nearly one hundred marks per minute, and the work is done is such a manner, as to defy improvement."

"We looked at this curious combination of metal with intense interest. It is an idea translated into iron; a good thought stereotyped in steel." Four Ivy League professors, three from Brown and one from Harvard, had also tromped through the shop

2. A Curiosity Shop

Joseph Brown's first linear dividing engine, which he built in 1850 to graduate rulers and other small measuring instruments. Originally powered by a hand crank, subsequent models worked by steam-powered drive belt. The index plate in the center was notched with 1,080 teeth, and fractional turns of the plate set the spacing of the graduation marks, depending on the gearing used. This No. 1 linear dividing engine, photographed here in 1944, is now housed in the American Precision Museum in Windsor, Vermont.

to study the ingenious invention of Joseph Brown, a man whose formal schooling ended in a local district school, schools that taught children of working men the Three R's and Republican morals of virtue, a good work ethic, deference to adults, and self-restraint.[1]

As a teen, Brown briefly apprenticed as a machinist at Walcott and Harris in Valley Falls, making cotton machinery, while his father started a shop making tower clocks across the river in Pawtucket. Brown left Walcott and Harris after a wage dispute and found work in nearby Central Falls, turning spindles for a throstle, a cotton-spinning machine, while continuing to help his father make clocks. The experiences of clock-making and hands-on machining gave Joseph Brown insight into the abstract challenges solved by the clock-maker, and the machinist's feel for precision.

To build his linear dividing engine, the first such machine built in America, Brown likely drew on his remembrances of a circular dividing engine that he and his father had imported before fire razed their first shop on South Main Street. Brown completed his first linear dividing engine in 1850 and then improved upon it, adding steam power instead of a hand crank in 1854, just after the *Waverly's* writer's visit. As an apprentice, Lucian had been required to fashion his own six-inch ruler, using an existing one as a template then painstakingly scribing dividing lines. Brown's newest linear dividing engine, powered by steam, could cut such rulers, graduated to 64ths of an inch, in boxwood or in steel within four minutes.

With the linear dividing engines, Joseph Brown was transforming his shop from specializing in horology—the measurement of time for factory owners, church-goers, and railroad magnates—to metrology, the science of precision measurement. The *Waverly's* writer enthused about the measuring devices sold in the little shop: "Steel Rules suitable for every division, diameter and circumference. Logarithmic slide rules, Bevel Protractors, Shrinkage Rules, Architectural Scales, Box-wood Rules, indeed every thing belonging to that branch of business are made here in the most perfect manner."

He was enamored by the silvered Vernier calipers, two-feet long with a sliding, finely graduated scale to measure inside and outside diameters to within thousandths of an inch. Brown sold only a few of these large bench calipers—they used a lot of steel, were expensive, and parts to be measured had to be brought to the bench where the caliper was mounted. A few years he later made a nine-inch, hand-held model that quickly caught on, bringing to the machine shop floor a small tool that measured to the thousandths of an inch at a price that the average machinist could afford, introducing the precision of the Vernier system to American machinists.[2] The dividing engine also churned out six-inch, steel pocket rulers that, like the Vernier calipers, became a staple tool for machinists everywhere.

The *Waverly's* writer expressed disappointment that Brown was not in the shop to explain his linear divider; however, he was also impressed by Brown's young partner. While he was visiting a watch arrived from France accompanied by a note in French, "which was readily understood by Mr. Sharpe." Lucian rose from apprentice to partner by making himself indispensable to Brown by acquiring the right blend of skills to complement Brown's mechanical genius. During his apprenticeship years he took weekly French lessons with a native French speaker named Monsieur Laeger, whose class also drew several young women of Lucian's acquaintance—the mercurial Martha Thayer, musical Lucy Metcalf, and the "sensible and intelligent" Louisa Dexter, who had just moved into the city from Smithfield to be near her brother.[3] Even in his apprenticeship, Lucian's French proved useful to Mr. Brown: one May evening the boss climbed down the hill from his Congdon Street home and crossed the river to the flats on the west side, where he sought his apprentice's help in translating an article describing a French machine for dividing circles and engine plates.

On Wednesday nights, Lucian took "drawing" or mechanical drafting from Mr. Wagner, a young German immigrant. The first thing he drew was a locomotive's steam engine, but he "did not succeed very well" due to a wobbly T-square and other bad drafting tools on loan from a friend, giving him a keen appreciation for good drafting tools. Friday nights Lucian busied himself learning geometry and algebra from a Brown University student named Bond. He liked geometry, but found the algebra puzzling.

In mid–December of 1850, Lucian and Owen Mason had held a frank discussion about Lucian's mathematical abilities and concluded that rather than trying to emulate Brown in mechanical design, he might be better suited to taking courses in business and accounting. After dropping algebra Lucian Sharpe took up bookkeeping at A.G. Scholfield's Academy on Westminster Street, a school that promised to "qualify any young man of ordinary abilities for the duties of the counting room, and all that relates to the accountant's desk." Lucian took an immediate liking to bookkeeping, writing after his third week of classes in October 1851, "I think I shall like book keeping very well." Within months he kept his personal accounts in double entry form—recording every transaction twice: when he logged a payout he also showed the asset purchased, and when he shipped goods he logged the money received in exchange.

Double entry bookkeeping was not a new technology—it had been around since the 1300s. But Colonial America, an agrarian, cash-strapped society, did not have much use for it. By the 1820s, that had changed. Industrialization brought rapid transactions and more of them that bookkeepers needed to explain. As sure as the clock-maker measured time and the metrologist precisely spliced distance, the bookkeeper set and measured value, providing the market with an underlying logic that made possible the market's existence. Historian Michael Zakim has observed:

> Bookkeeping thus proved capable of naturalizing and then taming the market, an essential contribution to the rise of industrial capitalism. ... It is little wonder then, then, that of all the competing forms of knowledge jockeying for position in an age yet to be crowned the age of capital, bookkeeping proved to be one of the most powerful of all.[4]

Schofield's Academy was just one of scores of such schools that sprung up in American cities in the early 1800s, giving young clerks and mechanics such as Lucian Sharpe the skills necessary to act as foot soldiers in the industrial revolution, employing an old technology in new ways to create a market society. Lucian's new skill was not lost on Brown, whose wife, Caroline, had died in 1851 at age 33, leaving him a widower with one daughter and no son to join him the way he had joined his own father in the family business.

Shortly after Lucian started bookkeeping classes, Brown began sending his young apprentice to call on customers to collect bills. The city streets that Lucian walked through on his dunning duties had changed greatly in the two decades since his birth. Population, largely driven by Irish immigration, had more than doubled to 41,513; a majority now lived, as did Lucian, on the flat, former marshlands on the west side of the river at the foot of College Hill. Eight cotton mills and two woolen mills employed more than 1,000 textile workers from sunup to sundown; Hill's machine shop, where Lucian had briefly apprenticed, supplied the textile factories with machines such as cotton combers, lap machines, worsted drawing, slubbing and roving frames. The Eagle Screw Company turned out screws on its patented machines to anchor the heavy machinery required to absorb the forces of the bigger steam engines driving them, engines made possible by George Corliss' recently patented shutoff valve. Telegraph wires from three companies now wove above the city; steam engines hauling railcars whistled in from north, south, and west, converging in a great circle of tracks looping behind an impressive, twin-towered depot built of brick.

Providence was quickly becoming a center for "technological convergence," a term coined by economic historian Nathan Rosenberg to explain the synergy between people who improved or invented new technologies and people who used them to increase production. When a shop such as the Corliss Steam Engine Works of Providence developed something new—a more efficient, more powerful steam engine—industry found new uses for it that, in turn, demanded new developments such as larger, stronger, more precise gears to convert the increased steam power into useful energy without snapping a cog. Better gearing created new possibilities for production, which created new challenges for machine designers such as Joseph Brown, who responded by patenting a new kind of gear cutter in 1855.[5]

Even as an apprentice Lucian Sharpe made it his business to understand the latest solutions to production problems. On his annual summer trips he went out of his way to visit the factories in whatever city or town he was in—a carding factory in a Connecticut village with "twenty carding machines all operated by one horse"; or a Fall River nail factory "with 100 nail machines cutting off the nails as fast as you could count." Lucian's tour of the Allen and Thurber pistol factory in the summer of 1849 turned out to be life-changing, for it exposed him to the Armory System of manufacturing.

The Allen and Thurber firm made Pepperbox Pistols, the revolver of choice for gold prospectors rushing west. Like many American small arms manufactures of the mid-nineteenth century, the Worcester firm used the Armory System of manufacture, developed after decades of expensive experimentation in the United States armories at Harper's Ferry, Virginia and in nearby Springfield, Massachusetts. The ultimate goal of the Armory System was to build weapons with interchangeable parts so that guns damaged on battlefields could be broken down and rebuilt into functioning arms. The pursuit of this goal resulted in important developments such as specialized "machine tools" to turn precise parts, and the division of specialized labor—rather than build an entire gun the way skilled gunsmiths did, each group of workers built only one component of a gun.

Machine tools played an important role in the Armory System, and it is important to define what they are: "a power-driven machine, not portable by hand, which, by the use of cutting or abrading tools, removes metal in the form of chips."[6] Essentially, they are heavy power tools used to cut or shape metal.

The national armories did not try to keep secret their methods for achieving interchangeable parts; on the contrary they encouraged private arms makers to embrace the Armory System. Machinists who once worked for the Springfield Armory circulated throughout New England to Hartford, Worcester, and Providence carrying the system's principles of interchangeability: (1) an ideal, model product for subsequent products to follow; (2) custom jigs and fixtures to firmly hold metal stock (movement is the enemy of precision;) (3) and gauges—measuring devices. If a part fit slipped through a gauge too quickly, it was too small; if would not pass through the measuring device at all, it was too large; but if it fit snugly into the gauge opening, it was precisely right to fit the model product. After observing the assembly of Allen and Thurber Pepperbox Pistols, Lucian wrote in his journal: "There's a great division of labor in making them, and the work is very well finished."

2. A Curiosity Shop

By the 1850s, the Harper's Ferry and Springfield armories could each churn out 30,000 well-made muskets per year, a figure that astounded the British Board of Ordnance. At the height of the Crimean War, Great Britain was buying arms from small shops employing highly skilled labor, which crimped production of much-needed guns. The board appointed a Select Committee on Small Arms to determine whether America's Armory System could quickly produce quality arms at an affordable rate. Board members had their doubts. Machine tools, after all, were prohibitively expensive to build—the United States government had spent lavishly for decades developing such machines—and the machines were dumber than oxen; many felt that machines could never replace the precise work of a skilled craftsman.

At a hearing in 1854, the Select Committee called on American arms manufacturer Samuel Colt to testify about production of his Colt revolver—a gun so similar to Allen and Thurber's Pepperbox model that he successfully demanded a settlement to avoid a patent suit. When asked whether machine tools could really produce accurate small arms Colt replied with American swagger: "There is nothing that cannot be produced by machinery."[7]

Besides making clocks, watches, and mathematical instruments, Brown & Sharpe worked on any kind of mechanical device in need of repair or development, forcing Lucian to keep current in the latest technology. In the early 1850s he drilled into glass a half-inch thick to repair an insulator for the Merchants Telegraph Company; repaired a telegraph register for Bain's Telegraph ("this confounded telegraph is running in my head the greatest part of the time") and helped a local inventor develop a railcar positioning system that automatically telegraphed the whereabouts of a train when it tripped a signal on arrival at a new station. He also tightened a helix for a House's Telegraph containing six miles of wire.

In working with so much wire, Lucian felt frustrated by all the different diameters of wire manufactured by different companies—it seemed as though no two manufacturers made wire the same way, which complicated repairs. To rectify this he set about devising a system to measure the thickness of wire by passing it through U-shaped grooves cut into the edge of a steel disc. Each groove was of a different width, and each served as a gauge—when a piece of wire fit snugly into a groove, that determined the wire's diameter. He presented his wheel of wire gauges to the Waterbury Brass Association, which adopted the grooved gauges as standard in 1857; eight of the leading American wire manufacturers agreed to fit their products to Lucian's gauges, creating the standard American Wire Gauge, also known as the Brown & Sharpe Standard.[8]

The same year he established the American Standard, Lucian Sharpe also logged a personal milestone: on June 25 he married the sensible and intelligent Louisa Dexter from his French class, whom he found to be "a pleasant talker." The young couple picked a tough time to take on the financial burdens of setting up house: the Panic of 1857 threatened to shutter the shop; at one point the company pared its workforce to Joseph Brown, Lucian Sharpe, and just six hands.[9]

Brown & Sharpe still earned much of its income through making clocks and repairing watches, mixing centuries-old techniques of hand craftsmanship with new technology, such as the pioneering gear-cutter Brown built to make gears for himself

THE AMERICAN
STANDARD WIRE GAUGE,

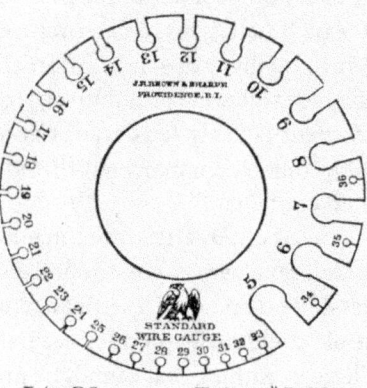

Adopted by the Brass Manuf'rs, January, 1858.

These Gauges are made from the best steel, and are tempered, adjusted, and warranted accurate.

☞ None genuine unless stamped as in the engraving, with our trade marks.

PRICES.

Round Gauges, sizes, 9 to 36, - - - $4.00
"   "   "   5 to 36, - - - 3.00
Angular Gauges, - - - - 8.00

Tables giving the weight of Iron, Steel, Copper, and Brass Wire, per lineal foot, and of Sheet Iron, Steel, Copper and Brass per square foot for each number of the Gauge, from 0000 to 40, are printed on another page.

5 to 36 - - Price, $3.00.

RESOLUTION OF THE BRASS MANUFACTURERS.

Whereas, it seems desirable that some steps be taken to arrive at a more complete uniformity in Wire Gauge, used by the Brass Makers; and whereas, J. R. Brown & Sharpe, of Providence, R. I., have, at considerable expense, prepared a Gauge with a new grade of sizes, a plan which is by us approved, therefore,

RESOLVED. That we will adopt said Gauge, and be governed by it in rolling our metals, and will use our exertions to have it come into general use as the Standard U. S. Gauge.

SIGNED,

BENEDICT & BURNHAM MF'G CO.,
  Chas. Benedict, Secretary.
BROWN & BROTHERS,
  Philo Brown, President.
SCOVILL MAN'FG CO.,
  S. M. Buckingham, Treasurer.
THOMAS MAN'FG CO.,
  S. Thomas, Jr., President.
NEW YORK & BROOKLYN BRASS CO.,
  John Davol, President.
NEW YORK BRASS & MAN'FG CO.,
  J. Hoppock, President.
JAS. G. MOFFETT.

WATERBURY BRASS CO.,
  L. W. Coe, Treasurer.
HOLMES, BOOTH & HAYDENS,
  J. C. Booth, Secretary.
BRISTOL BRASS AND CLOCK CO.,
  E. N. Welch, President.
WALLACE & SONS,
  Thomas Wallace, President.
ANSONIA BRASS & BATTERY CO.,
  J. H. Bartholomew, Agent.
BELLVILLE WIRE WORKS,
  G. DeWitt, Agent.
WALCOTTVILLE BRASS CO.,
  W. M. Hungerford, Pres't.

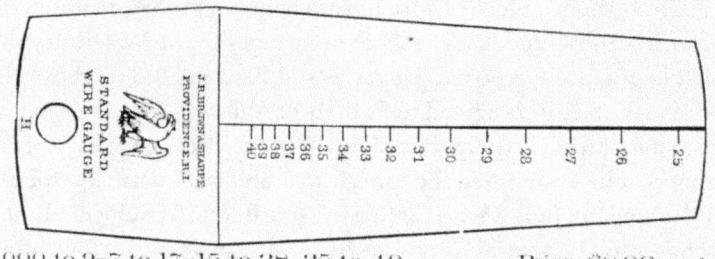

000 to 9—7 to 17—15 to 28—25 to 40 - - - Price, $8.00 each.

Frustrated by different wire thicknesses produced by various manufacturers, Lucian Sharpe invented a gauge to measure and standardize thicknesses in order to make ordering wire easier. Eight of the leading American wire manufacturers agreed to fit their products to Lucian's gauges, creating the standard American Wire Gauge, also known as the Brown & Sharpe Standard. Brown & Sharpe first included the gauge in its catalogue of 1858, when Lucian was 28.

and to sell to others in 1855. But the clock and watch making business was losing its luster. Even as early as 1850 Lucian was complaining, "It requires a *great deal* of patience to perform the mechanical part of the watch business correctly and a *great deal more* to deal with people who think that a watch never wants anything but cleaning and that we charge 3 or 4 dollars for nothing." Both he and Brown were looking for greater challenges, and in the late 1850s they found one that threatened to give them more than they could handle.

One day in 1855 an Appalachian Mountains carpenter spied a woodcut of a sewing machine advertised in a newspaper. He lived in an out-of-the-way place on the backside of the mountains in what would soon become West Virginia.[10] With lots of quiet time on his hands this jack-of-all-trades, James E. A. Gibbs, decided to make a close study of this woodcut to see how this contraption might work. The problem was the woodcut displayed only the top half of the Grover & Baker sewing machine; as he imagined its construction, Gibbs got it wrong. He assumed that the machine used one spool of thread, and figured that a hook on the underside of the machine must snag that thread and twist it into a chain stitch. In fact, like all sewing machines then in the early stages of production, the Grover & Baker machine used two spools of thread and laid down an entirely different kind of stitch.

Gibbs got a chance to see an actual sewing machine in use a few months later, when he travelled over the mountains to visit his father's farm down in rural Rockbridge County. Sitting in a tailor's shop in his old hometown, Gibbs spied a Singer sewing machine. He could not believe the complexity, heft, and exorbitant price of the machine. Compared to this, the one-spool design he had sketched out in his head was elegant in its simplicity.

On returning to his mountain home, Gibbs, a vegetarian with a long nose, intense blue eyes and a bushy black beard, spent evenings and foul-weather days building a prototype of his machine, using carpentry tools, a penknife, chisel, and sawed wood. He fashioned the hook from a sprig of mountain ivy; needles he cut from spare bits of metal. By April he had built a wooden model with enough integrity to convince a local sawmill owner to buy half-interest in the machine. With this seed money, Gibbs made his way to Washington to research patents, and found that no one had yet hit on his idea of using a single spool with a revolving hook to create a chain stitch.

Gibbs felt that his machine could be lighter, cheaper, and more efficient than anything on the

James E.A. Gibbs, an Appalachian jack of all trades, who serendipitously invented a single-spool sewing machine that laid down a chain stitch instead of a lock stitch. Lucian Sharpe described him as a "long-nosed vegetarian" with intense blue eyes. Gibbs lived in a part of Virginia that became West Virginia when the federal government divided the state during the Civil War.

market. To negotiate the complex world of patent law he travelled up to Philadelphia for a consult with James Willcox, a hardware merchant who built patent models. Willcox liked Gibbs' idea, and gave him space in his back office in the Masonic Hall. Willcox also put his son, Charles, on the project. Though only 17 years old, Charlie was mechanically inclined, and eventually improved the machine with a couple of patents of his own. The Willcoxes and Gibbs built a crude model that successfully won them a patent for a single-spool sewing machine. They refined the design; then they faced a challenge: who could manufacture such a machine in sufficient quantities to succeed in an already competitive, cutthroat market?

One of the businesses in which James Willcox held an interest had hired the New England Butt Company of Providence to make iron castings for its knife cleaning machines. Willcox asked his Providence contacts for a likely manufacturer of the Willcox and Gibbs Sewing Machine, and they put him touch with William Angell, head of the Eagle Screw Machine Company. Despite the recent Panic of 1857, both Eagle Screw and New England Butt Company had all the work they could handle manufacturing their core products: metal screws and, for New England Butt, braiding machines for textile companies. Angell could not take on a new product line, but he was very impressed by the work coming out of Brown and Sharpe, the small shop that backed up to the river on South Main. Joseph Brown had followed up the invention of his famous linear dividing engine with a vastly improved gear-cutter his shop unveiled in 1855. Brown's newest machine precisely cut cogs into steel discs as large as 37 inches in diameter, making it possible for factories to install gears strong enough to take advantage of recent improvements in the steam engine. Brown took a pardonable pride in his gear-cutting machine, giving it a "place of honor" in a room by itself where a machinist covered it every night to protect its precision.[11] Given Brown's track record, which included repairing sewing machines and making a few needles for Cushing & Shearman, Angell thought that Brown and Sharpe might be up for the challenge of manufacturing a new kind of sewing machine; he told them about Willcox's quest for a manufacturer to bring a crude patent model to market.

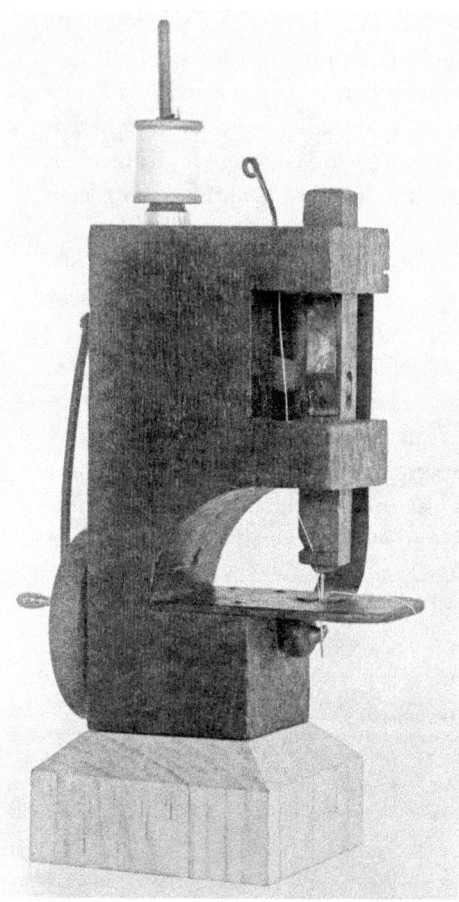

James E.A. Gibbs' prototype of his single-spool sewing machine. As a carpenter, Gibbs cut the major pieces from wood, using a saw, chisel, and pen knife. He fashioned the hook from a sprig of mountain ivy, and cut his needles from spare bits of metal.

2. A Curiosity Shop

In 1858, Lucian Sharpe took a train from Providence to Philadelphia to discuss business with the senior Willcox. From the start, Lucian planned to manufacture machines with interchangeable parts by using the Armory or American System that he had seen nine years earlier at the Pepperbox revolver factory. This was risky business that would require an initial outlay of capital to build the custom model, tools, gauges, fixtures, and machines; if unsuccessful, a venture like this could swamp a small shop like Brown & Sharpe. But the rewards could well be worth the risk.

For inventors on both sides of the Atlantic, a viable sewing machine was something of a Holy Grail. After the invention of the Arkwright water-powered spinning frame and the power loom in the late 1700s, the nineteenth-century was relatively awash in textiles; women were no longer chained to the spinning wheel and the loom, but somebody had to take all those factory-produced textiles and stitch them into clothing. Well into the 1800s the age-old technique of sewing still reigned: hand-stitching two pieces of fabric together with a common needle and a piece of unspooled thread. Ready-made clothing had not won widespread acceptance; shop tailors and seamstresses working out of their homes or in machine-less factories tailored most clothes. A good sewing machine had the potential to become a staple in almost every house and in clothing factories, huge and potentially lucrative markets.

The Willcox & Gibbs Sewing Machine that Brown & Sharpe manufactured through the American System, using interchangeable parts. First made in 1858, the machine sold for half the price of a Singer, and quickly became popular with those tailors and seamstresses open to the idea of using sewing machines. Brown & Sharpe manufactured sewing machines for 96 years, ending production in 1954.

The dream of perfecting a sewing machine is what brought George Corliss, the steam engine genius, to Providence. Corliss had obtained the third sewing machine patent issued in America, in 1843, for an invention that stitched together two pieces of shoe leather. A year after winning his patent he travelled from his home in upstate New York to Providence in search of funding for his sewing machine, but he failed to find it. After his failure he took a job as a draftsman. While working at that he turned his mind to improving the stationary steam engine, a century-old technology widely viewed as a backup for water power before Corliss developed a system of valves to harness the steam engine's previously wasted power.

In the 15 years between Corliss's shoe-stitch patent and the patent won by Gibbs for his single-spool machine in 1858, the U.S. Patent Office had issued more than 100 sewing machine patents. The field was so crowded that the four biggest sewing machine makers pooled their patents and agreed to stop suing each other for infringements, which made it much harder for smaller companies to break into the market. The potential for making a fortune was still there, but competition was intense.

Before signing an agreement to manufacture machines for Willcox, Lucian completed a due diligence investigation into the man's and character and finances. Reports compiled by an investigations agency and through other businessmen ran the gamut from an unsigned reference's penciled scrawl of "not reliable, Is free to give his note, but not so willing to pay it," to the New England Butt Company's glowing, italicized recommendation: "a man of *Sterling integrity* and *uprightness of character*." Giving weight to the latter, Lucian Sharpe mortgaged his company's future. He agreed to tool up to produce the Willcox and Gibbs Sewing Machine on the American System.

In late February 1858, the teenage Charlie Willcox lugged a hand-built sewing machine into Brown and Sharpe's small shop. This model machine acted as the template for every part that Brown and Sharpe would make for new machines. Brown and Sharpe agreed to make one dozen machines. Work commenced in mid–March—New England Butt Company cast some iron bases for the machines, and the young Willcox joined a few Brown and Sharpe machinists in fashioning tools and jigs to make the moving parts. Three months later they were still making tools, jigs, and fixtures without having produced a single machine. Lucian apologized to James Willcox for the delay, writing that "tools proved to be three to four times as expensive as was contemplated by us at commencement though they will doubtless be cheap in the end if many machines are manufactured."

The elder Willcox remained calm. He upped the agreement to 100 machines—50 of the model size, and fifty built from a new, smaller model. The new size doubled the time and expense of tooling up for production. Brown and Sharpe bought new screw machines and refitted its lathes to boost their metal-cutting capacity. By August the firm still had not produced a single sewing machine. Lucian wrote to Willcox "we cannot but feel nervous at the continued postponement."

By mid–September, the cost of making or "pushing" tools had run 10 times higher than anticipated. But, Lucian assured Willcox, "By the experience we now acquired we now hope to turn out nearly perfect machines at the outset." Six weeks after that, workers began assembling the parts cut from steel for the first 50 machines. The Willcox and Gibbs machine contained fewer parts than competitors' models, and with tools and machines in place they could now be made and assembled for much less than a machine built of handcrafted parts. Completed, the machines hit the market at half the price of a standard Singer sewing machine, $50 as opposed to $100, with Brown & Sharpe receiving $8 for each machine.

In some quarters sewing machines met with the same skepticism that had dogged machine-made rifles in Britain. Tailors and seamstresses feared the competition, and doubts persisted about a machine's ability to perform as well as the human hand.[12] But pleased with the success of their first models the partners signed a contract that summer in which Lucian committed Brown & Sharpe to building a minimum of 6,000 machines within 10 months, packed in crates and delivered to the railroad depot or wharf. He also agreed to expand their shop, machinery, and tools to be able to churn out 10,000 machines if necessary. The little 30-by-60 foot shop that evoked the Curiosity Shop of a Dickens serial now spread throughout two of the three floors of the barn-shaped building that housed it; the firm also took over the second story above the adjacent marble works on South Main. To move stock throughout the building, Brown installed a steam-powered elevator. It was a busy time for Brown, who had remarried, this time to Jane Francis, a woman 12 years younger. Every work day she stopped by the shop at noon to fetch him home for lunch, and most days he was bent over his drafting board working so intently that he had forgotten the hour and was surprised by her arrival.

After Brown & Sharpe sank all that money into tooling up and expanding the business, Willcox fell behind in payments to the firm. In October 1859 the shop made 500 machines, but received no money from Willcox. "[W]e were obliged to borrow money to meet our Pay Day," Lucian wrote to Willcox in late November, a time when Lucian was feeling the strain of impending fatherhood; Louisa was 8 months pregnant with their first child. "We are straining every nerve to make both ends meet."

In addition to tensions over money the two partnerships were at loggerheads over improvements to the machines—Willcox and Gibbs continuously tinkered with their sewing machines, making improvements that were good for sales but terrible for production. Each new part meant new tools, expenses, and delays. From the outset a culture clash between Gibbs, a Virginian, and Rhode Island Yankees at Brown & Sharpe tested the partnership. Gibbs—an eccentric Appalachian Mountain man described by Lucian as "the long-nosed vegetarian"—had come to Rhode Island to help assemble the first machines, bringing with him what one mechanic called "the most enormous spittoon ever made. It was always full of 'cuds' and was the bain" of the foreman's life.

The job of keeping together this uneasy alliance fell on Lucian. He dunned Willcox for payments, negotiated change orders, and sounded out Gibbs on his political views as the nation drifted toward civil war. Weeks after John Brown's raid on the Harper's Ferry arms depot Lucian wrote to Gibbs down in Virginia: "What are you doing and when shall you make us another visit. We abolitionist's (sic.) should not dare to come and see you now."

When Gibbs sojourned to New York in April 1860, Lucian teased: "Glad to hear that you have arrived in the land of *freedom* once more...." But with the election of Abraham Lincoln in November, Gibbs's stance on secession became more entrenched, a worrisome development. Brown & Sharpe faced the prospect of seeing a major partner, the chief inventor of its primary product, joining in a rebellion against their common government.

On December 21, the telegraph brought news that South Carolina had voted for secession. Lucian laid out his thoughts on the matter for Gibbs: "Your favour of the 5th was duly rec'd and was interesting as showing the rapid progress of disunion sentiments in your own mind as well as in Virginia. ... Notwithstanding all this talk and flurry I cannot think that any serious consequences will come [,] from the fact that the only safety for slavery is in the union.[13]" A month later, Lucian was still sanguine about secession, writing to Gibbs: "My new idea is that this *rebellion* for it deserves no other name, will not last long after the inauguration of Lincoln...." In a post-script he added: "You had better bring your family on and settle here where the people are sane for they are 'stark, staring mad' at the south."

By mid-summer of 1861, Willcox's tardiness in sending cash for its machines became a recurring theme; it appeared as if the anonymous reference who observed that Willcox "is free to give his note, but not so willing to pay it" was correct. Willcox owed more than $15,000, much of it secured by the 1,500 sewing machines Brown & Sharpe held in storage. The younger Willcox worked in the Providence shop and in late July, shortly after the birth of a second Sharpe daughter, Lucian gave him an earful. He told Charlie that when his father "commenced the enterprise he was not worth the first cent if his debts had been paid" and he was "indebted to our mechanical skills also."[14]

James Willcox was travelling in France when Lucian wrote him a scathing letter on July 25—four days after the Battle of Bull Run—imploring him to come home from France. Until Lucian heard from him, the firm would not do "any more work of any kind on your machines. Given the affairs of Bull's Run the prospects of business look worse than ever...."[15] With Virginia's secession that April, Gibbs had donned the gray uniform of the Army of Northern Virginia, his whereabouts unknown to the firm. The partnership was unraveling; civil war threatened the market for household sewing machines; so Joseph Brown and Lucian Sharpe turned their sites to an industry that promised to be profitable: Guns.

3. "The Genius of Its Maker"

> Here the "knee and column" type of milling machine which had been struggling towards birth for so many years finally emerges in a form of such classic "rightness" and simplicity that at first glance the machine does not proclaim the genius of its maker.—*L.T.C. Rolt, author,* A Short History of Machine Tools

The timing of James Willcox's letter to Lucian Sharpe could not have been worse. It reached the office of Brown & Sharpe in late July 1861, while the nation was still staggered by reports of the carnage at the Battle of Bull Run. Rhode Island's dead included two officers known to Lucian, Colonel John Slocum and Major Sullivan Ballou. Nothing in Lucian's life had ever upset him as much as the news of this war, and he was in no mood for nonsense as he cracked open the envelope that Willcox had sealed in Paris nearly a month before.

In the enclosed letter, Willcox again promised that he would send money soon; he also admonished Lucian against taking Brown & Sharpe into the gun business, writing: "Your neglect of my and your interest in the [sewing machine] business may operate much against you when *rifles go out of fashion* which I trust may be sooner than either of us expect."

With the nation now in an open civil war, it looked as though rifles would be very much in fashion for some time to come. And Lucian had already been taking steps to get into that business. With 1,500 sewing machines in storage and a nearly equal number needing only assembly, the company needed something to keep idle hands busy. He had heard that a rival manufacturer had turned their shop into a firearms factory, "and if the state of things is going to last long it would be the best thing for us to do the same," he told E.P. Hatch, the business agent for Willcox & Gibbs.

> It is a matter I have been thinking of ever since these troubles began but it is impossible to foresee what is best to do. As a general rule it is best to stick to one branch of business and we have always intended to stick to the sewing machine business ... but perhaps it may be good policy for some time to come for manufacturers to make two kinds of goods, one for war and the other for peace.

For advice in the gun-making business Lucian turned to William Angell, the screw manufacturer who had introduced him to Willcox and Gibbs. "I have always felt desirous of obtaining some article to manufacture the sales of which we could control ourselves," he explained to Angell. Clocks and watches were out—the firm had recently sold that business, only retaining contracts to build "watch clocks" that tracked the progress of watchmen as they made the rounds of factories.[1] Revenue from the precision tool business—calipers and various kinds of rules—was dampened by tough competition from a firm up in Maine, Darling & Schwartz. Its head, Samuel Darling, had perfected his own linear dividing engine, and seemed to be the one man who could match Joseph Brown in the field of precision rules.

So right now the best article to manufacture appeared to be rifles, specifically breech loading rifles, guns that loaded from the back end of the barrel near the stock, doing away with the cumbersome business of ramming powder and ball down the front end of a musket barrel. The U.S. Army had conducted tests of breech loading rifles at West Point, and Lucian sought to obtain results of that test without tipping off potential competitors that Brown & Sharpe had an interest in guns.

"General Burnside or Col. Slocum would know all about it but I do not like to write to them as I prefer that they should not know that we are taking any interest in arms," he wrote before learning of Slocum's death. "…From what I can learn Sharp's Rifle is the best in use and Burnside's would rank next if well made, are these conclusions correct? If so then it would only remain to decide whether our friend Mr. Howe's 'improvement' is superior to those."

The "Mister Howe" who laid claim to an improved breech loading rifle was Frederick W. Howe, a machinist of some repute. Lucian had known of Howe for a couple of years and had already consulted with him on methods for making sewing machine castings. Howe learned his trade at Gay, Silver & Company in Massachusetts before moving to the Robbins & Lawrence Company at Windsor, Vermont, which made arms and the machinery for making them. In Windsor, Howe devised a machine for milling the spiral grooves on the inside of rifle barrels. Howe learned more about the gunmaking business when he went to work for the Savage Arms Company, but his stint at Savage did not end well—he was in fact suing his former employer. Still, Howe had proven experience in the arms industry, so his claim of making improvements in breech loading weapons sounded plausible.

"Mr. Howe proposes to be here early this week to show us his improvements which he has put on a pistol for that purpose, but before we take any steps in the matter we propose to obtain answers to the following questions as near as possible," Lucian told Angell, before laying out a list of a dozen questions that revealed the analytical bent of his mind. Is Howe's invention a decided improvement on all other breech loading fire arms and in what particulars? Does Howe's invention impinge on any patents? What will be the probable cost of a full size model and experiments in getting it up and how long would it take to complete it? How much capital will have been invested before any returns are realized?

"Of course it will be impossible to answer all the above definitively but I propose to investigate the matter as thoroughly as possible should the improvement appear of sufficient importance to warrant it," he told Angell. "Perhaps there is some person well qualified to judge of improvements in arms who might be consulted confidentially as to its value. If it is in your way to learn without trouble please do so."

3. "The Genius of Its Maker"

There was one problem with Brown & Sharpe's attempts to woo Howe into working with them: They likely could not afford him. Good machinists were much in demand, and those considered to be among the best commanded a lot of money. As anticipated, Howe did come to Providence in mid-July to visit the small shop of Brown & Sharpe. But he also toured Providence Tool, the biggest shop in the city. Not to be confused with Providence Machine where Lucian briefly apprenticed, Providence Tool drew its initial capital from a host of Providence investors and the Borden family of Fall River. The company, which manufactured a rival line of sewing machines, had recently moved into a sprawling brick complex covering more than five acres with boiler rooms, a distinctive bell tower, and a drop forge next to the Corliss Steam Engine Works, about a mile north of downtown.

Providence Tool enticed Howe with an attractive package that included a one-fifth ownership, and he signed on there as a superintendent. Howe's decision to work with a bigger shop caused no animosity—it was strictly business. The government was not interested in his breech loader anyway; the Army hired Providence Tool primarily to make the Springfield Model 61 rifled percussion musket, a traditional muzzle loading gun. The initial contract called for thousands of guns per month, and it almost swamped the Providence Tool company; it could not keep up. A piece of the musket called the percussion nipple caused a bottleneck in production. To fire a musket ball down the barrel, the Springfield rifle used a percussion cap—a small metal cylinder holding "shock sensitive" material that would ignite when struck by the gun's hammer. The cap rested on a protrusion or nipple with a hole in it so the spark of the tiny explosion could pass through the nipple and into the gun's barrel to ignite the gunpowder rammed in behind the musket ball.

Manufacturing the percussion nipple presented two challenges—shaping the nipple, and then drilling a small hole in it. Howe had plenty of experience with making percussion-fired muskets from his years at Robbins and Lawrence, but with a war on he now needed to build guns at a far faster clip. So he turned to Joseph Brown. First he asked Brown to make a turret lathe to assist in shaping the nipple. A turret lathe allowed a machinist to switch metal-cutting tools by rotating the turret, a revolving block that gripped multiple tools, allowing the machinist to click in the tool he wanted when he wanted it.[2] Brown essentially copied a turret lathe that Howe had already designed, and then improved upon it by adding features that made the cutting tool far more versatile—a self-revolving turret, and a feed stock feature that enabled machinists to feed in fresh metal stock without shutting down the machine. Brown's new turret lathe was not a revolutionary design as much as it was a vast improvement on Howe's existing design. The sale of this machine-powered cutting tool marked Brown & Sharpe's first entry in the business of making machine tools for use outside of the firm's own production needs. Brown & Sharpe was not the first company to make a machine tool for sale to someone else, but it was an early entry into the machine tool field.

The turret lathe solved the problem of quickly shaping gunlocks, but Howe was still puzzling about how to drill all of those tiny holes in each percussion nipple. Twist drills were the way to go, but making a twist drill bit consumed a lot of time as it required hand filing a spiral groove in a cylindrical piece of steel over and over and over to sharpen and deepen it. Howe watched with frustration as his best, highest-paid machinists gripped thin rat-tail files and meticulously filed spiral grooves into tool-steel wire in order to make twist drills. Howe raised this problem of making drill bits with Brown, who had felt a similar impatience with the process while drilling holes for his Willcox and Gibbs sewing machines.

Brown tackled the problem by radically redesigning the milling machine, the basic machine used to cut metal with a toothed, rotating blade. He added to it a "knee and column" that essentially allowed a machinist to raise and lower the metal to be cut. On top of the adjustable knee protruding from the machine he added a slide that moved left and right, in and out. The universal miller could cut on an x, y, z axis. Writing more than a century later L.T.C. Rolt, an M.I.T. industrial historian, described Brown's work as "a machine of such classic 'rightness' and simplicity that at first glance the machine does not proclaim the genius of its maker."[3] Brown's universal milling machine had no problem raising and lowering cylindrical steel stock while advancing the cutter to cut deep grooves or "flutes" into it, quickly making all the twist drills his and Howe's machinists could want; within a decade the now familiar twist drill became ubiquitous.

Brown & Sharpe delivered their first universal milling machine to Providence Tool on March 14, 1862. Its obvious potential made it a popular item. Within days, Burnside Rifle submitted an order for a universal miller "to be done in two months if possible or sooner." Starr Arms Company of Yonkers quickly followed suit. The partners converted the empty third floor of their shop above Tingley's Marble Works and went to work building machine tools—the self-feeding turret lathe, or screw machine; tapping machines to cut screw threads inside of cylinders; and the universal miller. By year's end they had built and sold 10 of the revolutionary milling machines, a pace exceeded in 1863.[4] Brown & Sharpe never did get into the business of making rifles, but the Civil War acted as a catalyst for developing turret lathe improvements and inventing the universal milling machine, the partnership's first entries into the nascent machine tool business. The war spurred what Lucian Sharpe had been seeking all along: an article to manufacture "the sales of which we could control ourselves."

James Willcox eventually made good on his note, so Brown & Sharpe did not suspend production of his sewing machines; but true to Lucian Sharpe's fears, news of the 1861 Battle of Bull Run brought sewing machine sales to almost a dead halt.[5] The firm took up the slack by building the new universal milling machines, turret lathes, a tapping machine, calipers, standard rules, and quite a few "watch clocks" that helped factory owners in Lawrence and Fall River ensure that night watchmen made their appointed rounds; but none of these machines, rules, and tools approached the sewing machine in terms of profitability. For Brown & Sharpe the business climate of the Civil War looked unpromising. Then the unexpected happened. In October 1862, monthly sales of sewing machines leaped from 129 to 332. They never looked back. By June of 1863 the company was crating and shipping more than 700 units per month.[6] The flurry of business caught the attention of the *Providence Daily Journal*, which reported in December:

"On South Main Street just below Crawford Street, stands a large, substantial and unsightly building.... But this building unsightly as it is in appearance, is the 'local habitation' of a machine business that has grown from small beginnings to a fame not confined to this side of the Atlantic.... The large sale and increasing demand for the Willcox & Gibbs machine affords pretty conclusive evidence that the public appreciate its merits and are satisfied with its utility."[7]

Though Lucian Sharpe did not see it coming, the Civil War spurred popular acceptance of the sewing machine in American culture. Before the war, sewing machines faced the same kinds of suspicions that machine-made rifles had met before Britain's Select Committee on Small Arms: machines could not possibly craft as well as the human hand; they were expensive; and, as British gunsmiths feared that machine-tooled arms would harm their business, American tailors and seamstresses saw sewing machines as a threat to theirs.[8]

By 1862, tailors and seamstresses were awash in work, far more than they could handle. That summer President Lincoln called for 600,000 new soldiers, who needed hats, boots, trousers, tents—all manner of equipage that required sewing. In war, production time was of the essence, and even the most Luddite of tailors could see that sewing machines demonstrably saved time. In 1861, machine manufacturer Wheeler & Wilson published the results of a study that showed sewing machines cut the time for making a man's shirt from 14 hours and 26 minutes by hand, to 1 hour and 16 minutes.[9]

The No. 1 Universal Milling Machine that Joseph Brown invented in 1862, in order to help the Providence Tool Co. cut spiral drill bits so it could more quickly make rifles for Union forces in the Civil War. Brown added the adjustable "knee and column" in front of the machine that allowed a machinist to raise and lower the metal to be cut. On top of the adjustable knee he added a slide that moved left and right, in and out. The universal miller could cut on an x, y, z axis, giving it a flexibility that went far beyond cutting drill bits. In this photograph there is no cutting tool in the arbor. Brown's later invention of the formed tooth cutter made it economical to use the Universal Milling Machine in gear-cutting.

Sewing boomed into big business: In the course of the war tent companies did twice as much business with the government than privately held arms makers such as Providence Tool.[10]

The huge government contracts for these supplies favored contractors with lots of capital to buy material and hire sub-contractors to shape it into the desired form. Sub-contractors hired mostly women as sewers and stitchers, and many of them used sewing machines both in factories and in home production.[11] Though the single-thread machines were light and designed for home use, *Scientific American* found the Willcox

& Gibbs machines durable enough for factory use, writing in 1863: "they will stand a great deal of hard usage without injury. Sewing machine operators are not generally the most careful persons in the world (where they do not own their machines) and manufacturers of clothing &c., will find these durable in this as well as other respects."[12]

Between 1861 and 1864, annual sales of the Willcox & Gibbs machine more than quadrupled, soaring to 8,669. Through the Civil War, the sewing machine won widespread acceptance in American households and factories. Its use diversified the workforce by putting women to work in the war effort, and gave the Union a technological edge that enabled it to equip many more troops than the Confederate States, in less time and at lower costs. Milling machines for making gun parts and sewing machines used to stitch clothing employed demographically different work forces, but both played key parts in equipping Union soldiers.

Lucian Sharpe began the war believing that it may be good policy to make two kinds of goods, one for war and the other for peace; by the war's waning days his firm was producing four kinds of goods—precision instruments, small tools, machine tools, and sewing machines—that were profitable both in war and in peace. Brown & Sharpe had never sold a machine tool before selling a turret screw machine in 1861; by December of 1864 it had sold 145 machine tools, the bulk of them to New England armaments makers.[13]

In 1864, the fertile mind of Joseph Brown delivered another invention: a patented milling cutter that expanded the possibilities of the universal milling machine as a tool for cutting gear teeth. Elegant inventions often look obvious in retrospect, and Brown's formed tooth milling cutter was no exception.[14] It looked like a thick wheel with wide, curved teeth shaped like cresting waves. The cross-section of each tooth wore evenly, allowing a machinist to easily sharpen it. Before Brown's formed tooth milling cutter, teeth wore unevenly making it more cost efficient to discard a barely worn, expensive cutter rather than waste a man's time trying to evenly sharpen its uneven teeth. Like the Mona Lisa's smile and the "Ode to Joy" theme in Beethoven's Ninth, Brown's milling cutter transformed a tour de force, in his case the universal milling machine, from an elegant execution of an idea into a classic work for the ages.

One Fall day in 1865, a bearded, weather-beaten man wandered into the Willcox & Gibbs showroom at 658 Broadway in New York City, sales headquarters for the sewing machine company. James A. E. Gibbs, inventor of the single-stitch sewing machine, had emerged from the war. When the father-son team of James and Charles Willcox caught sight of him, they might have looked as if they had seen a ghost. They had not heard from Gibbs since the previous April, when he had managed to smuggle a brief letter to Lucian Sharpe, saying: "I hope I still have many personal friends in the north notwithstanding the '[impossible] conflict' between our *countries*."[15]

One-in-five Confederate soldiers died in the war, and Gibbs had had his close calls. He briefly led a cavalry unit before typhoid fever cut him low; he recovered from that and rejoined the army as a lieutenant in the Ordnance Department overseeing the manufacture of saltpetre, a key gunpowder ingredient, at a processing plant near his Rockbridge County home. Yankees under the command of Major-General David Hunter sniffed out the gunpowder works, destroyed them, and razed Gibbs' farmstead. In June of 1864, Gibbs joined 5,500 of his comrades to fight Hunter's forces at the Battle of Piedmont, a bloody fight that resulted in 1,500 Confederate soldiers killed, captured, or missing.[16]

Now here he stood in the flesh, broke, with no farm, and four daughters to feed back home in what was now the new State of West Virginia. With the cessation of hostilities he had come north, a Confederate veteran, a survivor, to see what had happened to the partnership that bore his name. The Willcoxes broke the news: During four years of war, they had sold more than 20,000 machines. Hearing no word from him, they had not known whether he was even alive, so they had taken his share of the profits and placed it in an escrow account. He was now worth more than $10,000, about what a laborer would earn in 30 years, a lifetime's accrual of wealth.

4. "An Eye to Business"

> The American system not only insures accuracy, but is economical, and to this system, and to the fact that we are constantly improving the design of our machines, we chiefly attribute our success in being constantly considered superior to our imitators.... —*Brown & Sharpe advertising circular*

On a quiet Sunday night in April 1865, Lucian Sharpe strode up the steep pitch of College Hill on his way home to his wife and their daughters, now numbering three. Suddenly the boom of a musket—followed by a general clamor in the city below him, caught his attention. Lucian pivoted back downhill and made for the offices of the *Providence Journal,* where an official telegraph had just arrived announcing that Lee had surrendered to Grant. After four years the Civil War, which had killed one out of every fifty Americans, was over.

From church steeples and mill towers all across the city came the clamor of bells. Candles began flickering in house windows, a spontaneous celebration of illumination. Crowds collected in the streets, gathering round bonfires in Market Square, and near the train depot.

"Every one now seems to think there will be no more fighting," Lucian wrote the next day to his aunt, Lucinda Potter. "I hope not but I glory in the fact that we have *whipped* them out. Now if we can only catch Jeff Davis we will hang him on a '[sour] apple tree.' Or anything else which comes handiest, and then we shall have peace and quietness."

"I have some news to tell you," Lucian continued. "I am going to England in about a month principally for my health but partly to see the world."[1]

Lucian had already booked passage to London aboard the steamer *Scotia*. Even before the official cessation of war he could see the end coming and figured, correctly, that the summer would bring a lull in business. A good friend, the noted book-dealer Sidney Rider, planned to sail for Europe on a buying trip, and now seemed like the best time for Lucian to go: he needed a break from business; as the staggered nation got its bearings, trade was bound to be dull; and if business rebounded the way he hoped, Brown & Sharpe would be building a new shop that would inevitably consume much of his time.

Lucian explained to E.P. Hatch, the business agent for Willcox & Gibbs:

I shall of course have an eye to business that in so far as looking about among machine works and manufactories is concerned and I have an idea that we are making some tools which can be sold over there.... My health however is the main reason. [I am] not exactly sick but feel as though I should be in a year or two longer if I kept on the present routine.

Before he sailed, Lucian scurried to make innumerable arrangements. He tried, and failed, to convince his aunt to stay with aging parents during his absence; he asked customers, such as the Springfield Armory, for letters recommending Brown & Sharpe's universal milling and screw machines to present to arms makers overseas; and he asked industrialists who had been to Europe for tips on which machine shops and factories he should see.

"My principal object in writing now is to ask whether there are any revolving head screw machines in operation in England," he wrote to A.C. Hobbs of Bridgeport; "if not would it in your opinion *pay* to patent and introduce them there? Also the Universal Milling Machine? Do you think our steel rules and machinist tools could be sold there to advantage?"

Lucian Sharpe circa 1860, when he was about 30. Months after the Civil War ended in 1865, he traveled to Europe to develop new, overseas markets, and to recover from symptoms of an illness that may have been Bright's Disease, a chronic kidney inflammation that pained him throughout his life.

By the time he sailed on May 4, Lucian's itinerary was freighted with appointments to tour machine shops in England and on the Continent. The passenger list had also swelled to include, besides him and Rider, the young Charlie Willcox, who looked to drum up business for the sewing machines; a Providence jeweler last name of Owen, who planned to buy wares in France; and an older tag-along named James Brown, no relation but a good friend of Joseph Brown's who owned a textile machinery company in Pawtucket. Joseph Brown remained home to mind business while Lucian spent the summer away hoping to restore his health, see the world, and open new markets.

The steamship *Scotia*, carrying 465 passengers and crew, cleaved through icy waters off of Newfoundland, running through the fog at a good clip. Then the lookout cried: "Ice ahead!" Peering into the chill fog, Lucian Sharpe spied the iceberg looming a short distance away. The *Scotia's* thumping steam engine stopped. The ship's bow slowly swung in an arc as the captain twisted the ship's wheel to avoid what Lucian described as "an ugly lump of ice about ten feet high." The ship "just shot past without touching," he wrote from his stateroom to his wife, Louisa, whose fear of dying at sea had prevented her from accompanying him on his first trip to Europe.

Despite this close call, Lucian teased Louisa for the fear that had kept her stateside. With advances in maritime technology a transatlantic crossing was now "rather safer than staying at home," he wrote her. "The Cunard line has now been running 25 years, and their vessels have made say 2500 trips without losing a single passenger."

He landed in Liverpool after a six-day crossing, and immediately plunged into learning all he could about the city's commerce, geography, architecture. He found the port city dingy, with quarrelsome harridans peddling fish, tall warehouses darkening the maze-like streets, and coal smoke hanging in the perpetual fog.

"Queer old place," he wrote to Louisa, "though much of it is new comparatively, and it is increasing very rapidly on account of the extensive commerce, one-third of which is with the United States." During the war, England had done its best to divide the United States in two, in an attempt to weaken its growing influence. Even with the fighting ended, Lucian saw lingering signs of British support for the Confederacy: a "Rebel flag" and songs lauding Stonewall Jackson in a theater; repeated vandalism of George Washington's bust in a museum. "Most everybody here hopes we won't hang Jeff Davis," Lucian reported to Louisa. Despite British support of the South the Union had survived intact, and the nation was now an emergent international power.

On his first carriage ride into the English countryside, Lucian enthused about the "yellows, whites, and purples of laburnum, hawthorn, and rhododendron" in full bloom. But the more he saw of England, the less he liked it. "Thank God I wasn't born in England," he groused to Louisa. "I hate their climate, the people, and everything but their roast beef, mutton, landscapes, and steamships!" To be fair, he also admired their trains.

Lucian rode the rail from London to Birmingham, 113 miles in just 3 hours. In the United States trains stopped every 20 to 25 miles to replenish the steam engine with water, but the Birmingham train chuffed along for 83 miles before stopping thanks to an ingenious solution to refilling the tanks: every so often, half-mile long troughs full of water appeared between the rails. "When the engineer wants he lowers a tube pointed forward into one of these troughs, and the motion of the train drives the water into the tank or boiler," Lucian observed.

Lucian toured the Birmingham Small Arms Company, a factory built by a recently formed consortium of gunsmiths resolved to learn the American System of manufacturing guns with machinery. The smiths banded together in response to the Board of Ordnance's decision to equip the government armory at Enfield with American made machine tools, threatening to make them obsolete if they did not master new methods. So gunsmiths formed a trade association, built a plant on Armoury Road, and equipped it with machine tools made in America in order to learn how to use them.

"Birmingham is a great manufacturing place for guns, jewelry, and iron and brasswork of various kinds," Lucian wrote Louisa. "It has a forest of chimneys and the atmosphere is filled with the smoke from the soft pit coal which they burn. Here Mr. [James] Brown and myself visited a gun manufactory on the American plan...."

The Birmingham Small Arms Factory was one of only a few factories in England making anything on the American Plan. The others were the Enfield Armory which, Lucian adjudged, "has run down since the Americans left it" before the Civil War, and the Hobbs Sock Factory, "where the system is only partially introduced. (I am writing this mechanical dissertation for Mr. Brown's and Mr. Howe's benefit," Lucian parenthetically told Louisa. "Please read it to them.)"

Three weeks into Lucian's trip, he declared it a partial success: away from the daily stresses of the shop, his health was indeed improving. He wrote to Louisa: "I feel quite well and have got browned up, and they say I look better than before." But, he penned in the same letter of June 7, "I do not think I shall do much in a business way," at least not in England.

The problem was, English machine shops were not sophisticated enough to employ the latest American technology, a shocking reversal of fortune. The English had virtually invented the machine shop. At the dawn of the nineteenth-century, almost every significant invention in the metal shaping field was of English design, from John Wilkinson's cylinder boring mill that made possible James Watt's steam engine; to Henry Maudslay's screw-cutting lathe that reliably created screw threads accurate enough to use in dividing engines[2]; to Matthew Murray's planing machine that made it possible to machine a true plane surface on metal. But after pioneering the machine tool industry, the English grew complacent. One twentieth-century historian, L.T.C. Rolt, posited that English complacency may have been rooted in notions of class. In England "class distinctions were such that all who engaged in trade, whether business men or technicians, were held inferior to those who followed the learned professions," whereas Americans did not mind getting their hands oily if the work made money.[3]

Whatever the reasons, Lucian Sharpe saw evidence of English complacency, and it bothered him. The English, he wrote, "think themselves the center of the world." In Birmingham, Lucian toured C & W Harwood's sewing machine factory, and found it far inferior to Brown & Sharpe. "They make [sewing machines] for Thomas in London," he reported to Joseph Brown through Louisa. "They were very polite to us, and showed us over their works. Gave them some of our circulars but don't expect to sell them or anyone else in England any tools as they don't know the use of them, and are not in a way to learn."

The slums of Great Britain also shocked Lucian Sharpe. After touring Golden Lane, one of London's most dangerous quarters, he concluded that "The poor of London are to be pitied as the slaves of America ever were." Manchester turned out to be "a dirty, miserable place. We have nothing so bad in America, and never shall have. It makes me feel bad to see the poor miserable women and children about the streets." He took a train trip from Warwick to Wolverhampton, 13 miles through "the black country" of the West Midlands, a coal mining region. From horizon to horizon the dark landscape looked "a perfect cinder heap.... Looking across the country as far as the smoke will permit you, you see nothing but a forest of chimneys, the fire from the tops of blast furnaces, and the red-tiled roofs of the miserable houses. These various structures stand at all angles, the ground having sunk under many of them."

Not everything in the British Isles was bad; in places he saw signs that Great Britain had not entirely lost its edge in metal working. At Sheffield he toured the manufacturing plant of Joseph Rodgers & Sons Cutlery. Louisa owned an old pair of Rodgers scissors that she loved, and though he found the prices high, Lucian bought her a new pair. "Afterwards we went over their establishment and saw the process of manufacture." At Leeds he called at the Fairbairn, Kennedy, & Naylors machine works, where he obtained a photograph of a machine for making cutters for gears to show to Joseph Brown. In Scotland he found "The working classes live more like hogs than men. The streets are in good order, but the yards of both houses and manufactures are very muddy and filthy." Outside Glasgow blast furnaces "lighted up the country around and with the smoke and flame gave the idea that we were in the neighborhood of the infernal regions." But here, too, he saw evidence of England's early domination of the machine tools industry.

England still held its edge in forging iron and steel, which was Brown & Sharpe's bread and butter: the firm made sewing machines from it, and tools and machines for working it. Iron is the fourth most widely distributed element on Earth, but typically it is bound up in other rock and chemically bonded with other elements, so that it cannot be worked. To make ore workable, ironmasters needed to force oxygen to break its chemical bond with iron. From the early 1600s till the mid–1860s, English ironmasters led the way in smelting iron from ore: Abraham Darby perfected "coking"—removing impurities from coal through controlled burning in order to use the cleaned "coke" as an efficient fuel for smelting. (Uncoked coal imparted sulfur or phosphorous to the iron, making the metal friable when hot or brittle when cold.) John Wilkinson, the most famous of the eighteenth-century ironmasters, adapted James Watt's steam engine to run a reciprocating pump for blowing air through a blast furnace, freeing the furnace from reliance on water powered bellows.

Early nineteenth-century improvements in furnace design—making them circular instead of square, increasing pressure of the blast, adding iron oxide to the mix, tripled weekly production rates of iron while omitting the need for further refining it to make a good wrought iron. British inventors also developed slitting mills for cutting iron into manageable pieces, and rolling mills for flattening it into iron plates. At Napier's Steam Engine Works along the River Clyde, Lucian saw shipbuilders constructing iron frigates for the Turkish Navy. An upright planer planed a true surface an impressive 12 feet square.

On the summer solstice, Lucian finally crossed the English Channel to set foot on the continent in France. He stayed in Hotel du Louvre, with 600 rooms facing a glass-ceilinged courtyard. He wrote Louisa:

> In England none of us felt at home, but here we do feel so. The people are so different from the English. They act and look more like Americans. The atmosphere, too, is more like ours. In fact I like this place decidedly and I want you to study up on your French for I am going to take you over here sometime if I am ever rich enough.

4. "An Eye to Business"

By mid–July, Lucian had met the primary goal of his trip: he felt well enough to climb a mountain in the Rhone-Alps of southwestern France. For more than three hours Lucian climbed while "Mister Owen," the jeweler, rode a mule, toward the summit of Brevent Mountain.

Lucian wrote to Louisa: "Just before we reached the top I took a handful of snow from a bank which we crossed and on reaching the top I put it in a little brandy from Mr. Owen's flask and it made quite a refreshing draught...."

The next day he crossed a mile-wide glacier that flowed from Mont Blanc, rising against the horizon more than 50 miles away. Outside Chamonix he visited a grotto cut into a glacier replete with a "refreshment saloon" where he took refuge from the mid-summer's sun. "The grotto in the ice extends about 180 feet into the glacier which is as clear and solid as anything Col. Carpenter [a Providence ice dealer] ever furnished."

In the village of Chamonix, Lucian noted with patriotic pride, the "American, French and English flags floated from the staffs" of a local hotel. From there he rode in a carriage north, through the Arne Valley to St. Martin, then onward for Geneva through "a valley with nearly perpendicular walls of rock from 1 to 2,000 ft. high."

In Geneva he visited the shops of Patek Phillipe & Co., a noted manufacturer of fine Swiss watches. "Brown bought one for his wife," he teased Louisa. "We thought it fortunate for us that our wives were supplied." Before Lucian left Mister Patek told him, while lifting his hands above his head, "The Americans are 'vaary' high now."

Back in Paris by late July, Lucian visited "The Conservatory of Arts and Trades," housed in the buildings of an old convent, where he saw models of French machinery, standard measures used in government business ("many of them of extreme delicacy";) and a large collection of Philosophical instruments. He spied an "electrical machine" with a glass plate 5 feet in diameter; inspected a room full of clocks, watches, and watch tools. The Conservatory exhibited machine shops in a former Catholic church. One room held 50 glass cases containing miniature models of machine shops "with all of the tools and articles" used in French manufacture. He told Louisa: "Mr. Brown would be delighted with this place."

On August 1, Lucian visited several French machine shops "but find that they are not up to us in America." The next day he visited the largest sewing machine manufacturer in Paris, a firm that made Singer sewing machines for sale in Europe. An agent said he expected to visit the United States soon to buy tools. Lucian then called on another sewing machine manufactory "belonging to Mr. Goodwin who has some of our screw machines and wants more."

Lucian had restored his health. With orders for Goodwin's plant and a tip that Singer's Paris division was in the market for tools, the business objective of his trip was also bearing fruit. As August dragged through its first week, he felt like the time had come to go home. He had received word from Providence that his mother was seriously ill. He came down with a toothache, and had to seek out an American doctor to fill it. He liked France, particularly Paris, though the Catholic tradition of displaying public images of Christ and the crucifixion, even on street corners, repulsed his staid, Unitarian sensibilities. And he missed his wife, writing to her "I shall show my appreciation of your merits more than heretofore if I reach home again in safety." He booked passage home on the steamer Asia out of Liverpool, though it would not sail till September 2. He wrote to Louisa: "I dread the three weeks I have to stay in England, that land of fog, drizzle, and what is worse, *Englishmen*."

While he waited to sail, one piece of news buoyed his spirits: "Atlantic cable gone up!" he gushed to Louisa. Just 17 years after the young Lucian Sharpe had written in his journal "the idea of telegraphing across the Atlantic is absurd," a second attempt had been made to lay cable from Europe to North America. An earlier attempt in 1858 had failed, but not before successfully, albeit slowly, transmitting messages between Ireland and Newfoundland. This second attempt failed before it relayed a single message; the cable snapped from the shipboard spool that was paying it out and sunk to the ocean floor.

But the next year, 1866, one ship successfully laid a working cable between the two continents; a second snagged the broken cable from the sea floor and established another connection between Ireland and Newfoundland. News that used to be a week at sea now spread from continent to continent at the speed of 8 words per minute. Steamships had dramatically reduced the hazards of sea travel, reliably ferrying people and goods across the Atlantic. Commerce and communication between the United States and Europe stood on the cusp of a boom, and from Lucian Sharpe's observations the United States had gained a technological edge. Through his summer's sojourn, Lucian had restored his health and positioned Brown & Sharpe to take advantage of newly accessible markets.

5. Prosperity

> He was an inventor with all the peculiarities of that class of man.
> —*Supt. Thomas McFarlane on Samuel Darling*

In the fall of 1865, fresh from a salubrious sojourn through Europe, Lucian Sharpe settled back into the rhythms of the machine shop that bore his name. He and J.R. Brown now employed 300 men working 60-hour weeks in a "low dingy place" of small rooms crowded with work benches lit by gas light. The rooms hummed with the sounds of machines cutting metal and leather drive belts slapping overhead.[1]

Brown & Sharpe now pursued three types of business, but most of the work still went into making sewing machines for the Willcox & Gibbs Company. The company also built and sold machine tools—turret screw machines, screw machines, small hand lathes and universal millers—but this was a nascent industry, decades away from morphing into the "large, independent sector of the economy" it later became.[2] Most factories made fairly simple products such as textiles that did not need expensive, relatively sophisticated equipment such as Brown's universal milling machine.

For Lucian Sharpe the company's third line of business, precision tools, looked most promising. Reconstruction had begun in post-war America. Stimulated by government aid, crews were laying out thousands of miles of track—34,000 miles of new track between 1865 and the panic year of 1873.[3] Railroad construction created new markets for precision tools both in the field—plumb bobs (Brown & Sharpe made them in bronze or iron) for surveyors, and in locomotive factories.[4]

From Lucian's standpoint, two obstacles blocked Brown & Sharpe's path to prosperity in the precision tools line. The first, ironically, was the success of the sewing machine. Brown & Sharpe devoted so much of its manufacturing capacity to making machines for Willcox & Gibbs that it lacked labor and space to expand its line of precision tools. As Lucian told E.P. Hatch, business agent for the Willcox & Gibbs Co.: "A large field seems to be opening to us in the tool line but we cannot think of extending much in that direction as long as your work is on our hands."

The second obstacle proved to be formidable competition already afoot in the precision tools field, in the form of a firm called Darling and Schwarz. Samuel Darling, the mechanical genius behind this firm, seemed to be among the few men in America who could match J.R. Brown's ingenuity in metrology, the science of measurement. Darling grew up on a farm in Bangor, Maine, near the logging country of the north woods. He left the farm in the mid–1840s, apprenticed in a machine shop, and eventually joined forces with Michael Schwarz, a Bangor saw maker who advertised as "The Only Saw Manufacturer in the State."⁵

In 1852, just a couple of years after Joseph Brown built his first linear dividing engine, Samuel Darling built his own machine to precisely mark lines on measuring tools. Like Brown's machine, it could make rulers, protractors, and T-squares quickly and with remarkable precision. The firm used flexible saw stock for its rules and squares, evidence of its roots in Maine logging country.⁶ With their own graduating machine in place, Darling and Schwarz soon branched out from saw manufacturing into the field of precision tools and instruments.

J.R. Brown and Lucian Sharpe were painfully aware of Darling and Schwarz's emergence in the precision tools field. In 1856 the Franklin Institute of the State of Pennsylvania awarded Brown & Sharpe a bronze medal for machinists tools. The silver medal for a "Reward of Skill and Ingenuity" went to Darling & Schwarz.⁷

By the end of the Civil War, Darling and Schwarz were outselling Brown & Sharpe in this field. With their sewing machine contract and nascent machine tool line, Brown & Sharpe owned a bigger company, but they were playing second fiddle in the small tools line. Lucian decided that the best way to meet this challenge was to merge the two companies.

In early 1866, Lucian hired R.G. Dun Commercial Agency, forerunner of Dun and Bradstreet, to compile a confidential report on Samuel Darling, who had bought Schwarz's interest in the partnership. The agency reported:

Samuel Darling, a native of Bangor, Maine, who grew up on a farm, apprenticed in a machine shop, and invented his own method of graduating rulers, Vernier calipers, protractors, and other measuring devices. By the 1860s, Darling's company was out-selling Brown & Sharpe in the precision instruments line. Lucian decided that the best way to meet this challenge was to merge the two companies, so in 1866 he gave Darling $15,000 to form the firm of Darling, Brown & Sharpe to make small precision tools. Darling, an ingenious if eccentric inventor who patented the rotary pencil sharpener, sold his interest to Lucian in 1892, and the company dropped Darling's name from the line in 1897.

> Samuel Darling of the firm of Darling & Schwartz is considered as a very honest and honorable man. Very easy to "get along with," one who knows him well says, he would prefer "to give you ten dollars than to dispute with you about two." Is quite industrious and persevering. Has a very considerable inventive faculty.

Dun identified one flaw: a perseverance that bordered on obsessiveness, or, as the report put it: "He has not always expended his time and energies in the pursuit of objects of sufficient intrinsic importance to justify the devotion he bestows upon whatsoever he aims at."⁸

Most of the confidential report on Darling turned out to be true: he was honest, persevering, and inventive; but his eccentricities, and they were many, made him not so "easy to get along with."

William Viall, a longtime Brown & Sharpe worker whose father joined the firm a few years before the merger, recalled, "To meet him, he was the kindliest sort of a man, gentle and fulfilling the word gentleman in its fullest meaning. A Swedenborgian in his religion, he firmly believed it, and he practiced it in his every day life."

Darling believed strongly in the New Church or "Swedenborgianism," which held that love and charity are as important as faith. He believed, as Viall put it, "that only the best should be offered to God by man." He tried to live up to this ideal; he expected others to aspire to this lofty goal, and was often disappointed when he felt they had fallen short.

Merger negotiations proved difficult; though Lucian was willing to credit Darling with the lion's share of the precision tool company's assets, he felt that Darling overvalued his firm's contributions to the potential partnership. After a couple of months of haggling, Lucian and Brown boarded a Bangor-bound train and on arrival in late May gave Darling a check for $15,000. The firm of Darling, Brown and Sharpe was born.

Darling was very secretive about his linear graduating machine, but he did allow his insatiably curious partners to inspect it. Though the machine graduated instruments in the most precise fashion, it looked nothing like the "ingenious contrivance" with "metallic fingers" that Brown had built. Brown built his machine of good steel and even silver. Darling not only used saw stock in making his scales and squares—he had used it "quite freely" for parts of his graduating machine.[9] Yet the machine worked beautifully, cutting hair-thin lines into saw stock with great precision.

The Darling, Brown and Sharpe partnership succeeded from day one. After seven months the firm, which remained separate from J.R. Brown and Sharpe, returned a yield on investment of 30 percent. "We are well pleased to hear that you are satisfied with the result of the [first] seven months business," Lucian wrote to Darling in February 1867. He simultaneously complimented Darling while urging him to do more: "Perhaps you are disposed to be too modest about your share of the result.... We are as anxious as you to have a shop where more can be accomplished but meanwhile let us do, both here and at Bangor, *all we can* with our present facilities...."

To make a good impression on the new partnership's customers, Lucian wanted to stock the company's shelves with finished tools. With tools on hand the company could quickly fill orders, shutting new startup companies out of the market. Lucian advised Darling not to worry should "dull times" leave the company with storerooms full of tools. Rather than viewing that scenario as a catastrophe, Lucian welcomed it as an opportunity that would free workers to "consider what there is new that we ought to make."

The sewing machine business boomed in 1866; sales of the Willcox & Gibbs machine more than doubled to 16,408, and continued to be strong well into the following year. Businesswise, everything looked good for J.R. Brown and Lucian Sharpe; but on the medical front, things were less rosy. Both men lost parents in 1866—Lucian's mother, Sally Chaffee Sharpe had died that May; Brown's father, David, followed in August—and now Brown, at age 56, felt the effects of overwork. With the businesses running smoothly and Brown in need of a rest, the two partners decided to make a trip overseas.

"Mr. Brown has agreed to accompany me to Europe," Lucian informed Darling in August of 1867. "He has not been well for some months past and I am glad he has concluded to go although I regret that both of us will be away at the same time. We have however made such arrangements that we think the business will go smoothly."

The two partners, a generation apart, sailed for France with the goal of displaying their machine tools and precision instruments at the Paris International Exhibition. While strolling through the exhibits, they saw a nifty hand tool developed by a French mechanic, Jean Laurent Palmer, who had patented it as a "screw caliper." The device looked like a C-clamp, with a cylinder that rotated around a smaller, cylindrical handle to widen or narrow the gap in the C. By matching lines cut into the rotating cylinder with those cut into the stationary handle around which it rotated, a user could determine the precise width of the gap in the C. This device was perfect for measuring the width of a piece of sheet metal or wire placed in the open side of the caliper.

Palmer's screw caliper reminded Brown of a tool that that he had studied earlier that year at the request of the Bridgeport Brass Company. One of the Brass Company's customers had returned a lot of sheet brass, claiming that it was not of the advertised thickness. The sheets fit the gauge that the Brass Company used but did not square with the customer's gauge. A third party determined that neither gauge matched the one it used. To address the problem, the Bridgeport Brass Company devised a tool that looked a lot like Palmer's, then brought a set of six of them to Brown & Sharpe to see if would make a lot of them. The Bridgeport device was much smaller than Palmer's; in fact it was so small that there was no room on the cylindrical handle to cut in numbers that would show how wide the gap was in the C. The device could precisely measure to ten-thousandths of an inch, but without numbers it could only act as a gauge, not as a flexible measuring device. In Lucian's words the gauge "was too troublesome to read to be salable" so he turned down the offer to manufacture it.

The Palmer tool, on the other hand, displayed numerically and with the same precision as the Bridgeport device exactly how many millimeters the gap was. When a user closed it around a wire or piece of sheet steel, the device displayed at a glance the thickness of the thing being measured. Palmer's tool seemed to Brown and Sharpe to be too big, an unwieldy thing for a mechanic to carry around with him in a machine shop. Brown and Lucian talked it over and, in Lucian's words, concluded:

French mechanic Jean Laurent Palmer patented this "screw caliper" that Joseph Brown saw at the Paris International Exhibition in 1867, giving him the idea for a smaller, more rounded tool that became known to mechanics throughout the world as the micrometer.

5. Prosperity

"As a gauge was wanted for measuring sheet metal, we adopted Palmer's plan of division [which displayed numerically] and the Bridgeport man's size of gauge" with a few improvements to compensate for the wear of moving parts with use.[10]

On their return to Providence, the partners began manufacturing "the Pocket Sheet Metal Gauge" which dramatically improved machine shop precision by making it possible for every mechanic to have a precise, portable tool for measuring metal thicknesses always at his disposal. The firm began selling it in 1868, and renamed it the "micrometer" in its 1877 catalogue. By 1887, Brown & Sharpe advertised 76 different versions of micrometers that specialized in measuring the widths of screw threads, electric wire, thickness of bar stock, the diameter of rods, bullet caliber, paper, and other materials in English or metric measure.

In every way but one, the trip to the Paris Exhibition turned out to be a great success. The trip spawned Brown's idea for melding Palmer's screw calipers with Laws' sheet metal gauge to make the micrometer; the firm won silver medals for its universal milling and turret screw machines; and foreign companies placed 11 orders for machine tools in 1867, a small number but a significant 26 percent of the company's overall machine tool sales for that year.

The one goal that the trip failed to achieve was restoring J.R. Brown to full health.

Ten days shy of his 58th birthday Joseph R. Brown suffered a "slight paralytic stroke" while at work in the shop. Lucian Sharpe shared the news in a letter to Samuel Darling, written on New Year's Day, 1868:

> He recovered in a few minutes and rode home where he has remained ever since. He seems pretty well now and is anxious to get out but the doctor says he has overworked his brain and must give it rest. I am inclined to think it will be best for him to lay by for several months. The trouble is that he has not only been very much occupied at the shop but he is always busy at home evenings on similar matters.[11]

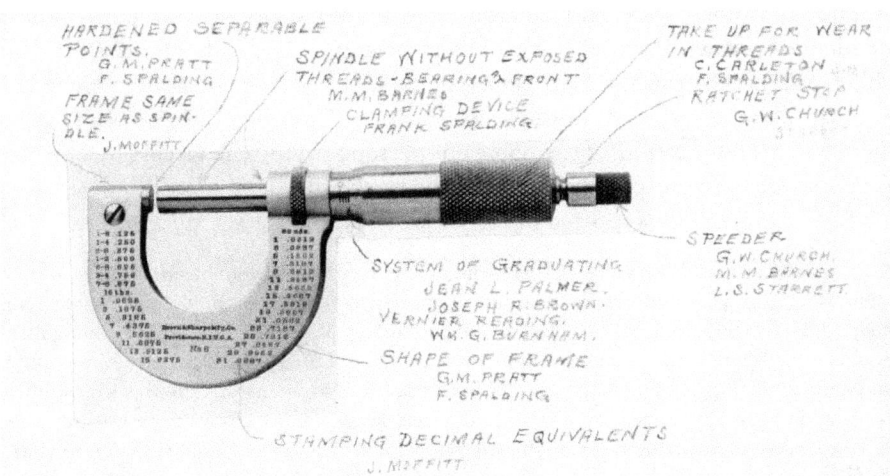

Within 20 years of its release by Brown & Sharpe, the micrometer evolved into a ubiquitous tool with a variety of shapes and functions, with features added to it by inventors such as Brown & Sharpe's Frank Spalding, who patented improvements in the shape of the frame and in the tool's measuring points. Between 1883 and 1919, Spalding took out 30 patents related to small tools.

Brown's doctor advised him to ease up on his business commitments; as Lucian put it in a letter to Hatch: "his physician tells him he must give up for the present at least if he desires to live." With a wife and a 23-year-old unmarried daughter to consider, Brown heeded his doctor's advice. He agreed to sell 60 percent of his stock in both companies to two new partners: shop superintendent Thomas McFarlane, and Frederick W. Howe, formerly of the Providence Tool Company.

McFarlane had begun work for the company as a bench hand making 15 cents an hour in 1856, and had quickly risen to superintendent with a say in hiring and firing, fixing wages and job prices, negotiating with the foundry that cast the company's sewing machines, and ordering stock. McFarlane was a prickly sort, often given, in his own words, to "a sense of depression." He had already tendered his resignation twice, but each time J.R. Brown and Lucian Sharpe ignored it.

Howe came to Brown and Sharpe with a sterling reputation as a mechanic and an inventor; Lucian hoped Howe could take Brown's place as ingenious inventor, though some of Howe's new co-workers quickly concluded that his mechanical skills were not as keen as his mind. McFarlane later observed of Howe: "As I viewed him he was cold, slow, unattractive, close mouthed and generally gave no signs of his inner feelings.... [H]e really *was* an impediment in the manufacture from the first...."[12]

With the stock sale, Lucian Sharpe, erstwhile apprentice, became the largest holder of J.R. Brown & Sharpe shares controlling 50 percent of them; Brown and Howe each held three-sixteenths, while McFarlane owned one-eighth. Lucian also owned a quarter of Darling, Brown, and Sharpe, making him a very wealthy 38-year-old man.

In the post–Civil War boom, both firms were making great profits.[13] To codify the new ownership structure, J.R. Brown and Sharpe incorporated in June 1868 as Brown & Sharpe Manufacturing Company, chartered, "For the purpose of manufacturing machinery, and working in iron and other materials...." Even in its first year the firm paid out handsome dividends to its four incorporators: Brown, Sharpe, Howe, and McFarland. Lucian's share of the company brought dividends of $12,000, about $200,000 in current value, and 12 times the salary of an average laborer. He held his wealth close to the vest, so as not to arouse jealousy. To Brown, Lucian wrote in July: "There was no publication of the names of persons paying the largest income taxes this year so we have escaped the notice of the public."[14]

To Darling, Lucian reported in early 1868:

> We do not know how much you tell other people about the profits of your business but we would impress upon you the importance of keeping it entirely to yourself, if you do not we are afraid that it may give both you and ourselves trouble in the way of competition. There is not a man in this establishment that knows how much the profits of DB&S or JRB&S are. Some of the accountants may guess but they have no direct means of knowing.... There is much more we would like to say but perhaps a word to the wise is sufficient. We think if no one but your wife knows what you made last year you will find your account in keeping the matter to yourself.[15]

Lucian's fears of competition were not without foundation. Darling, Brown & Sharpe had already staved off one attempt to pirate their graduating machines, made by a man who called himself Henry Flather and Henry Floan. He had written by one name to Darling, asking where he might buy a good graduating machine, and under the other name had come into Brown & Sharpe with a request to see their shop.

"He was shown about accordingly," Lucian told Darling, "but asked particularly to see the graduating machine...." Though Lucian had shown the machine to a writer from *Waverly's* decades before, the company now zealously guarded that machine and its two improved models. A man named Tom Bull operated the linear dividing machines in a small room petitioned off from the rest of the shop. Only Bull, Brown, and the superintendent could access that room, in the southwest corner of a ramshackle building, overlooking the river.

The building itself belonged to the Tingley Marble Company, whose work in gravestones and monuments boomed in the war years. Over three decades, J.R. Brown & Sharpe had grown like a wisteria vine around and above Tingley's street-level showroom for marble work and gravestones. Brown and Sharpe had built a three-story addition on the rear of the building, and added a second-floor atop Tingley's flat-roofed store. Now machinery cramped all floors and additions, and all tenants needed more space—Tingley's for monument carving, the newly incorporated Brown & Sharpe Manufacturing for expansion of its sewing machine and machine tools lines; and in May of 1868, a schooner arrived in the Providence River from Maine, carrying all of Samuel Darling's tools and machines.

The schooner moored at the foot of Elm Street, where Darling, Brown & Sharpe rented a riverfront building. Had a spy glimpsed Darling's graduating machine as it moved from ship to shop, it might not have mattered. Looking at it, there was no clear way to intuit how it ran. The machine built of saw stock was so quirky that its operation baffled even the few Brown & Sharpe mechanics who got to see it.

Darling's old grinding machines, used for putting straight edges on rulers and squares, also failed to impress the Brown & Sharpe men, who did their grinding on lathes that J.R. Brown had fitted with small grinding wheels. William Viall observed that Darling's grinders used grindstones as tall as a man, 5 or 6 feet in diameter with grinding faces 8 to 10 inches across: "Any mechanic, to have seen these machines and applied ordinary tests for good running machines, would have declared that any approximate degree of accuracy would be impossible," Viall wrote.[16]

Besides the machines, Darling also brought with him a crew of Maine machinists second-to-none in operating their quirky, old machines. Calvin Weld moved down from Bangor with his family to run the grinding machines with the most perfect precision. Steam power turned the wheels while the feeds guiding the metal to be ground, both longitudinally and transversely (left and right) were controlled by Weld's deft touch. "It meant that the operator knew just when and where to make the extra pass, and this he did day and day out," Viall wrote of Weld.

A Maine machinist named John Hall, and Samuel Darling himself ran the graduating machine, turning out rules and scales of the highest standards, better in many cases than Brown & Sharpe had been able to do.

Viall recalled:

In looking back at the work that came out of that shop, and at the tools that were used to do the work with, it seems hardly creditable that the results were obtained. In this establishment it was

Samuel Darling's wet surface grinding machine that he hauled down from Maine in 1868 to use at Darling, Brown, and Sharpe in Providence for grinding straight edges for rulers, squares, and other precision tools. Charles Weld, whose father moved from Maine to run the machine, continued to operate it until the mid–1930s.

one of the last cases where the skilled touch of the operator played a greater part in the product than the mechanical means that could be employed for the work.

At Darling, Brown & Sharpe, the science of building precision tools still depended upon the artistry of the craftsmen.

By the Spring of 1868, making machines that cut metal had become the poor step-child at Brown & Sharpe; machine tools took a back seat to Willcox & Gibbs' sewing machines and to Darling, Brown & Sharpe's precision instruments. Ironically, the firm that was among the first in adapting the American system outside of an armory did not itself employ that system when building machine tools. The American system required space to hold pre-cut parts before assembly, and the continued demands of sewing machine production was eating up all the room in the ramshackle shop on South Main. As a result, Brown & Sharpe built its machine tools in small lots of six at a time.

A long-time machinist named C.H. Phillips recalled a superintendent's response to a suggestion that they double machine tool lots to 12: "It would not be possible on account of the *immense amount* of stock we should have to carry."[17] Machinists planed all the nuts, bolt heads, and large parts one at a time, without the use of gauges, then filed pieces to fit. Such steel or iron that was pre-cut to length and diameter piled up on cramped shelves, making it difficult to retrieve. And yet, the machines sold steadily. Spurred by orders from Europe, mostly France, machine tool sales of 1868 kept a pace that nearly matched the high-water mark of the war years, when the company sold 67 machines to U.S. companies.

About the time that Darling's schooner unloaded in May, J.R. Brown, his wife, and daughter steamed for Europe in hopes of restoring his health. Lucian wrote to Brown frequently, keeping him abreast of business developments, deaths, gossip, and current events such as the progress of laying track for the trans-continental railroad. He sincerely inquired about Brown's health, which was not good. Lucian encouraged him to remain in France throughout the remainder of the year:

"We are getting along quite well without you, better than I expected in fact," he wrote to Brown in November. He reported that Howe had not made much progress developing the new grinding machine that Brown had left drawings for, but that was because he was so busy trying to keep pace with the company's orders. When Brown wrote on December 30 that his head still did not feel quite right, Lucian suggested that he stay put and live well: "I believe we made about $89,000 beside DB&S [Darling, Brown & Sharpe] business. ... So I think you need not starve yourself or wear your old clothes too long!"

By March of 1869, Lucian could report: "Business is very good with us. Machine Tools seem to go off lively and we expect to make 1500 sewing machines per month commencing in May." Things were going so well that Lucian and his wife, Louisa, decided to leave their daughters with a nurse and join the Browns on an overseas adventure. Charles Willcox's mother and sister planned to come too, along with one of Louisa's sisters.

"So I shall have 5 women under my charge on shipboard," Lucian joked to Brown. "Quite a harem isn't it. John B. Chace used to say that you couldn't have too much of a good thing when speaking of his large wife! Whether the same will apply to number as well as size remains to be seen." The Sharpes returned in September of 1869, but Brown remained in France, convalescing. "Found all our children and friends well," Lucian reported to Brown that Fall. "Business is good with us and seems likely to continue so."

In January, Lucian moved the company's machine tool work from the cramped quarters on South Main to the third floor of the New England Butt Company's building across the river on Pearl Street. Between those two shops, Darling's operation, warehouse space for sewing machines, and space leased for castings, et cetera, Brown and Sharpe rented "fourteen different places (as I remember it)," recalled Phillips, the old hand who began there in the 1850s.

Robert Dunlap, who worked at Brown & Sharpe from 1864 to 1869, recalled: "During these years Mr. Sharpe attended to the outside business around town driving to one place and another to get castings, etc. At this time it was a common sight for him to have one or more of his daughters [now numbering four] with him on the shop team when driving around town."[18]

The city Lucian rode through, flanked by his daughters, was growing and had grown to the 21st largest municipality in the United States, with a population of nearly 70,000. Entire mill villages had sprouted within the city's borders—Wanskuck around the Wanskuck Mills, and Olneyville, springing up around multiple factories, most notably the twin Atlantic Mills with their ribbed, elongated domes.

As a hostler's son, Lucian felt it important to present a good-looking team of horses to the public. So Brown & Sharpe developed an ingenious pair of horse hair clippers, sharp cutlery as large as hedge trimmers and operated with two hands, to ensure a nice, trim cut on the coats of the shop's horses. The business end of the cutters looked something like two garden rakes with sharp-edged tines. One of the "rakes" remained stationary, while the other, activated by a squeeze of the handle, slid across the stationary tines, a motion that captured and cut the horsehair. As he clopped through the lively city's cobblestone streets with his four daughters and well-trimmed horses, Lucian envisioned the façade of another factory rising against the city's skyline west of the river, one large enough to hold all of the firm's scattered businesses.

6. Panic

> I have often thought about him, and the remarkable work and great results which he accomplished, and have really grieved over it that he has never quite met with the recognition from the mechanical and engineering world which he so richly deserves.—*Henry Leland, founder of the Cadillac Motor Co., on Joseph R. Brown*

While he dreamed of raising a new factory on the west side of the river, the reality of working with too little space scattered through too many buildings continued to plague Lucian Sharpe. Sewing machine production flowed smoothly until it funneled into the grinding room, where machinists ground metal parts such as needle bars, foot bars, and shafts to precisely fit the model machine.

Grinding was the last step in production, the phase where a machinist could remove the smallest imperfections to shape a part just right. To complete this step, Brown had modified a lathe by adding a wheel of solid emery to the carriage, and attaching a "longitudal feed" that drew a piece of metal stock into and away from contact with the spinning emery wheel for fine grinding.

Grinding metal on a lathe took a lot of time, and an artisan's skill. This penultimate phase of production created a bottleneck—parts could be cast, turned, and cut in rapid secession, but they all piled up at the grinding machines awaiting the final touch of a skilled machinist, who ground by feel. A good grinder knew "where to put one's foot on the floor, or when and where to rest a hand on the machine to get the desired results." A grinder at Brown & Sharpe, Thomas Goodrum, earned $7 a day, an enormous sum for a shop worker of the late 1860s. People in downtown Providence knew Goodrum as "the Silk Hat machinist" because he strolled to work wearing a tall silk hat.[1]

Inventors constantly tried to invent a wheel that could grind metal faster and last longer than the standard wooden wheel coated with glue and sprinkled with emery. They won patents for wheels employing concrete lime; zinc oxide; granite and kaolin (a kind of clay) and mica—all without improving the process. Brown & Sharpe tried grinding with solid emery and with corundum, a hard abrasive which in its red form is prized as ruby. English dealers imported it from India, made a paste of two-thirds corundum and one-third gum resin, shaped it into a wheel that dried and hardened; then they bore a hot shaft through the center to melt a hole for mounting on a grinding lathe. Grinding wheels cost money, and they wore out after even light grinding. As the firm's business agent, Lucian always kept a keen eye for new sources of grinding material, particularly the corundum. In the two years that Joseph Brown convalesced abroad, Lucian often reminded his partner to keep abreast of the latest European news on corundum and emery.

"Mr. J. Morton Poole was here a few days since," Lucian wrote to Brown in June of 1868. "He has confidence that corundum exists in North Carolina and is going to investigate the matter. [Mineralogists did discover corundum in North Carolina, but not until 1872.] If you can find out where it can be bought in London I wish you would."[2]

Like many metal-working firms, Brown & Sharpe experimented with the different combinations of adhesives, shellacs, and glues for grinding. Lucian presented new mixtures to Goodrum, the Silk Hat Machinist so he could test them to "see how much difference it is going to make to us in dollars and cents."[3]

Brown, whose father had once roamed the countryside with a portable grindstone, believed that the answer to faster, cheaper grinding lay not in the wheel, but in the grinding lathe itself—he felt he could make a better machine for the job. Shortly after his stroke he dictated plans for a grinding machine to a draftsman, but went overseas to convalesce before he could pursue its construction. In the meantime Lucian sought new sources for corundum, tried out different combinations of glues and abrasives with Goodrum, and watched with frustration as machined stock piled up in the grinding room.

In April 1869, a "large black-whiskered man" entered the main shop of Brown & Sharpe, looking for employment. Oscar Beale arrived from his native State of Maine, where he had apprenticed as a machinist making farm tools, worked for a Portland maker of steam engines, and labored in the Portsmouth Navy Yard in Kittery. He quit the Navy Yard job after Congress passed a law cutting the work day for federal workers from ten hours to eight, and cut his pay commensurately.

Beale arrived in Providence from Kittery a few days shy of his 26th birthday. He had given up his dream of going to college, after a headmaster at the Towle School told him: "You have a good trade, there is no other substance that is wrought so much as iron. You never need to lack for employment."[4] Like the young Lucian Sharpe, Beale was insatiably curious; he studied geometry on his own, and in Kittery he had taken private lessons in drafting which he paid for by teaching algebra to the draftsman.

At Brown & Sharpe he sought out the foreman, E.H. Parks, to ask whether he needed any help. Parks said that he did not. Before Beale left he dropped a comment about toothed gearing that proved to Parks he knew something about the complex mathematical challenges posed by gear construction. Parks sent the young Mainer down to see the "man in the linen coat," Thomas McFarlane, the irritable superintendent who often teetered on the verge of tendering his resignation. McFarlane hired Beale on the spot, and sent him over to the Butt Company on Pearl Street to work on the machine tools.

Brown & Sharpe, along with their ancillary company Darling, Brown & Sharpe, now employed more than 300 men. They had Goodrum and his silk hat on the grinding machine, Maine craftsmen turning out rulers and scales on the graduating machines, and some of the most recognizable names in the network of New England machinists on their staff in F. W. Howe and Samuel Darling. Occasionally Charles Willcox or James E.A. Gibbs dropped into Providence from New York or Virginia to work out some new improvement to the sewing machines.

McFarlane, who could rarely resist tempering praise with criticism, wrote of Gibbs: "Mr. Gibbs was a typical, tall Southerner with rather rude habits and form, a clever, quiet, common sort of a man, whose success, if attained at all, was by accident rather than by genius."

Of the younger Willcox, McFarlane observed: "He was one of the best young men I have ever known, a superior mechanic, always a gentleman to all with whom he had dealings, patient and efficient, and of good judgment. Somewhat slow as to production of results, but sure when a thing was finally passed upon."

The job of yoking all of this ingenious, often eccentric, talent to pull in the same direction fell to Lucian Sharpe, who had his hands full. Darling in particular kept Lucian busy, as he often devoted his energies to ideas that had nothing to do with Brown & Sharpe's core products—a better system for ventilating railroad passenger cars; a new and improved ink stand; a cranked, cylindrical pencil sharpener. Darling's tendency to go off on tangents that did not always make the best business sense made him a difficult, but well-meaning, business partner.

"You say you can't help inventing!" Lucian wrote to Darling in a jesting letter soon after the firms had merged.

> There are some people who say they can't help stealing! For such people there are houses provided where they can be boarded and lodged at the public expense and thereby kept from the exercise of their proclivities. Perhaps it would be a good plan to have institutions to keep people in who cannot help inventing and thus prevent them from spending their savings on patent agents....

In a sense, Brown & Sharpe was an institution for people who could not help inventing. Brown himself was so obsessed with ideas and inventions that even after his stroke he could not keep away from the business. After a two-year hiatus overseas he returned in 1870. McFarlane remembered of Brown:

> On his return he attempted to take up again his branch of work, but his physical powers did not respond, and thereafter until his death, he was a semi-invalid, but greatly helpful, all the same in the business. His presence in the establishment was always a stimulus to all who came into contact with him.... He craved discussion and suggestions on mechanics. He used often say to me—I must talk to somebody if it is only a post, provided it could say "Yes" or "No."[5]

An institution for people who could not help inventing required its own brick-and-mortar building to house them all: Brown, Darling and his crew of Maine craftsmen; Goodrum, the Silk Hat Machinist; Thomas Bull, master of the graduating machines; bewhiskered Oscar Beale, and the famous Frederick Howe. So in 1870, Lucian Sharpe bought land west of the confluence of the Woonasquatucket and Mohassuck Rivers in order to build a new, cutting-edge factory for Brown & Sharpe.

One day in the early 1870s, Richmond Viall, a Brown & Sharpe foreman in his late 30s, packed his wife and two sons into a carriage for a ride into the country west of their Providence home. The carriage rolled down the slope of Sheldon Street, crossed the briny Providence River at the Great Bridge and trundled past the bustling Union Railroad Depot.

Out beyond the railroad's switchyards Viall showed his sons a big hole in the ground. This, he told them, was the cellar for the new plant of the Brown and Sharpe Manufacturing Company. To William Viall, a boy of about 10, the building site seemed "out of the world, situated as it was behind the old State Prison ... with the low, unattractive wooden building of the Harris-Corliss Engine Company beside it, and a sandbank behind it."

Promenade Street, a straight, westward-running road, separated the factory site from the Woonasquatucket River, which flowed a straight course here, channeled between two stone walls before mingling with the Mohassuck's waters in the fetid cove basin. The Woonasquatucket raced at the foot of sparsely-settled Smith's Hill, which rose steeply to the north, behind the factory site.

For most workers the new factory would mean long walks as the horse-drawn streetcars came no nearer than a half-mile to this outpost on the city's west side. After laborers sunk the foundation the factory's brick façade began to rise along Promenade Street—three stories of brick, granite, and glass covering a steel frame, the first steel-frame factory in the United States.

While building the new plant, Lucian Sharpe simultaneously oversaw construction of a new summer house on Nayatt Point, a small peninsula jutting into Narragansett Bay. With his brother-in-law, Louis Dexter, he had bought a large lot in an exclusive enclave eight miles south of the city, where they were building twin Italianate style mansions replete with large lawns seeded with two-and-one-half bushels of "Central Park Lawn Mixture," the very seed used in sewing New York's new park.

Lucian attended every detail of his Nayatt house, right down to the rugs. For the factory, he put Frederick W. Howe in charge of construction and design, a job Howe tackled with a passion bordering on obsession. A machinist named George Potter recalled that in everything that Howe did, from designing a machine to building a factory, he was "very particular and wanted to keep changing things over." In making a new machine, "He would have a few castings gotten out and then see a chance to make an improvement and have them all broken up again. Some of the men thought he would never get started on account of all those changes."

One evening a man stopping at the City Hotel on Broad Street said to Potter's father, "What has become of Fred Howe?"

Mr. Potter replied, "He is putting up a fire-proof building for Brown & Sharpe."

"Good God," the man said, "he will ruin them!"

"He was a man whose ideals were very high," William Viall recalled of Howe, his father's colleague.... "The building was built on a day-payment plan, and if things did not exactly suit Mr. Howe's idea of what they should be the work was torn down and replaced."

"Sometimes he would ride into the stable in his buggy at night remaining sitting therein asleep from sheer exhaustion," McFarlane recalled of the construction years.

For parts of three years the building rose, in fits and starts, brick-by-brick, until a new brick-red feature stood against the city's western skyline. Howe set the building with its long front facing south, toward the river, so the sun flooded the floors with natural light. He inserted tall windows with sashes of 20 panes over 20 panes, which workers could lift open in summer to capture prevailing southwesterly winds. The factory Howe designed was a marvel for its time—66,000 square feet of manufacturing space built of brick over steel with granite ledges; its design was believed to be "fire proof" or as fireproof as the age would yet permit. He installed a steam-powered elevator and a system of overhead cranes to move heavy stock through the building, a first in American factory design. The new factory cost $300,000 to build, about equal to each year's gross annual revenue during those three years of construction.

In September of 1872, the Willcox & Gibbs sewing machine division packed up and moved from the Tingley Monuments building westward, out beyond the rail yards, the cove, and the State Prison. Before leaving the old shop, the men posed for a picture that shows them spilling out of an over-stuffed building like pillow wadding. Men crowd the windows, clog the doors, and stand in places a half-dozen deep along the sidewalk. Close observation shows three men wearing tall silk hats: the principals, Joseph Brown and Lucian Sharpe—and Thomas Goodrum.

The factory on Promenade Street proved "airy, well-lighted," William Viall recalled, with systems for "heat and ventilation that were unheard of." Oscar Beale and

In September 1872, the sewing machine division moved from the crowded building on South Main Street, where Lucian Sharpe began his apprenticeship, to a new brick and steel-framed factory on the west side of Providence. Before leaving the old shop the company's 300 hands posed for a picture that shows men and boys spilling out of an over-stuffed building, crowding the windows, clogging doors, and standing in places a half-dozen deep along the sidewalk. Looking straight down from where the sign reads J.R. Brown, close observation shows three men wearing tall silk hats: the principals, Joseph Brown and Lucian Sharpe, and to their right "The Silk Hat Machinist," Thomas Goodrum.

the machine tools crew moved in after the sewing machine crew, emptying out their third-floor rooms over the Butt Company, leaving only Samuel Darling and his crew of Mainers to move their old machinery into the new fireproof building on Promenade Street.

In September of 1872 the Manchaug Mill Co. failed, costing Louis Dexter most of his money. Lucian Sharpe felt the failure of his brother-in-law's textile mill as a one-two punch: Manchaug owed Brown & Sharpe "considerable money"; and Lucian had to start paying his broke brother-in-law's bills. But for Brown & Sharpe, business continued to be good: for the third year in a row, Lucian received $16,000 in dividends, exclusive of his Darling, Brown & Sharpe earnings (by comparison Thomas McFarlane reported an annual wage of $1,400 just after the Civil War.) Despite his brother-in-law's failing, Lucian felt wealthy enough to build a new house for his family—now numbering five with the birth of his first son, Lucian Jr. in 1871—and a sixth on the way. He built a big duplex with a slate mansard roof on the exclusive East Side, a block away from Brown University's main green. When the house stood ready to welcome the Sharpes in the Fall of 1873, Louisa had delivered child number six, a second son, Henry.

In mid–September, the Philadelphia bank of Jay Cooke & Co. failed. The firm's failure was both a symptom and a cause of the Panic of 1873. Like many investors, Cooke & Co. had over-subscribed in the construction of western railroads. The bank overestimated returns it expected from selling Northern Pacific Railroad bonds; the Northern Pacific line proved too rural to draw significant investment. Cooke could not sell enough bonds to meet its own obligations; when word got out that Cooke & Co. was running short on cash, a run on the bank ensued.

Like the Panics of 1837 and 1857, the Panic of 1873 came as the result of multiple causes, at home and abroad; but its effects were depressingly familiar: cash shortages, shuttered mills, high unemployment. The Panic hit Rhode Island particularly hard when the A.&W. Sprague concern foreclosed, leaving $11 million in unpaid debts. A.&W. Sprague was not only Rhode Island's largest business, it was one of the largest concerns in the country. Sprague's chief business was textiles—it owned 25 percent of all the spindles churning out fabric in Rhode Island—and when it closed, 12,000 people lost their jobs.

Despite the worsening panic, 1873 had proven to be a smooth year for Lucian Sharpe, a 43-year-old man with a new house, summer house, and factory, plus a wife, and six children. In May there had been some troubling business with the Howe Machine Company of Bridgeport. The firm (named for sewing machine inventor Elias Howe) had tried to wrest from Brown & Sharpe the contract to build Willcox & Gibbs sewing machines. A protest from Lucian had quashed that, though Howe did win a lucrative contract to build a key piece, the iron tables, for the sewing machines.

Less than two weeks after Cooke & Co. closed its doors on panicked investors, Lucian received a letter from New York, bearing news that McFarlane compared to a "thunder clap … out of a clear sky." James Willcox had sold his interest in the sewing machine company to an Alden B. Stockwell, a Wall Street speculator and son-in-law of the late Elias Howe.[6] Stockwell and his brother now owned the Howe Machine Co.

of Bridgeport, the firm that had taken a piece of the Willcox & Gibbs business from Brown & Sharpe. If Stockwell moved all of the Willcox & Gibbs work to Bridgeport, it would spell the end of Brown & Sharpe, which was still largely dependent on the sewing machine contract.

Stockwell was a formidable character—a red-headed man from Cleveland who had worked as a riverboat clerk out west before marrying Howe's daughter and using his new found to wealth to speculate on Wall Street, where he earned the sobriquet of "The Napoleon of the Street." He became an officer of the New York Yacht Club, and was said to be worth millions.[7]

The day after he received word of Stockwell's takeover of the elder Willcox's stock, Lucian arrived in New York to meet with the Willcox & Gibbs firm's treasurer, J. Parmly, who told a tale of tension and intrigue long simmering inside the company, threatening its very survival. In explaining the whole affair, Parmly wrote: "Stockwell had apparently closed his eyes and jumped boldly into the control of a thing [Willcox & Gibbs] he knew little or nothing about. I think he had … spent but a few moments over the trial balances, when he launched nearly a half million dollars into the thing."

Though profitable, Willcox & Gibbs could not justify a $500,000 investment. According to Parmly, Stockwell quickly became frustrated with his returns. "In his disappointment and mortification, he lost the little self-control he had, and all hands came in for the exercise of his spite." To get out from under his investment, Stockwell wanted to liquidate all of the company's assets, and remove sewing machine production from Brown & Sharpe to the Howe Machine Company's Bridgeport factory.

Stockwell called in considerable loans that he had made to Willcox & Gibbs, which the firm could not pay. Parmly told Lucian: "In his distress for money, and fury at his misfortune, [Stockwell] determined to 'hedge' by 'gobbling' everything with any value owned by the company," including the horses and wagons, a London sales office, and a British plant that manufactured sewing machine needles.

All summer Parmly had used his position as treasurer and trustee to act as a "spy," maintaining Stockwell's confidence while scheming with a group of stockholders to thwart plans to liquidate the company. "Stockwell endeavored to tear the organization to pieces—while we tried to hold it together," Parmly recounted for Lucian. "His orders were received but ignored, and various excuses given by us, to calm him…. An open eruption had been avoided with Stockwell, up to the Panic of Sept. 20/73. Then it came in force!"

With runs on banks, cash became scarce; those banks with money to loan charged exorbitant interest. Willcox & Gibbs had a few thousand dollars on hand; Stockwell wanted it. Parmly, the treasurer, would not sign it over. "I was left alone to face this 'tiger-man,' and refuse him the balance of our bank account."

A Willcox & Gibbs trustee named Gray then used the company's close ties with Brown & Sharpe to draw money on account of the Brown & Sharpe Manufacturing Company. That's when Parmly wrote to Lucian. "Mr. Sharpe came immediately, and I unburdened myself to him," Parmly noted.

Lucian returned from New York to a hugely pregnant wife, and an airy, well-lighted factory that stood mostly empty. William Viall recalled: "With such depression on the country, and with a bank account extremely exhausted through building operations, it looked as though [Brown & Sharpe] had undertaken far more than they could by any possible means take care of.

"They discussed very seriously renting out parts of the building," but could find few takers in the crashing economy of 1873.

From his corner office in the new brick factory, Lucian Sharpe began hatching plans to keep Willcox & Gibbs out of Stockwell's hands. He ruled a sheet of paper into three columns titled "Sure Vote," "Philadelphia Vote," and "Outside Vote." The Sure Vote column listed those stockholders whom he knew were with him; the Outsiders were sure to support Stockwell, while the Philadelphia vote was anybody's guess. "Sure Voters" included the young Charlie Willcox, and this "Combination" held more shares of stock than the Outsiders—but not enough for a controlling vote. To thwart Stockwell's plans to liquidate the company, the Combination would need to snap up more Willcox & Gibbs stock whenever it became available, a risky investment in a time of panic.

In December 1873, the company was scheduled to issue new stock. Stockwell knew what the Insiders were up to, and tried to destroy blank stock certificates to block the new issue. Parmly locked the blank certificates and transfer books in a safe to protect them from being destroyed by Stockwell. Lucian hopped a train for New York, where he, Parmly, and Willcox crossed the East River to Brooklyn to consult with Gray. Around midnight they returned to Parmly's office, opened the safe, and issued new shares which they sold to others in the Combination, to Brown & Sharpe, and to themselves.

Meanwhile, "Stockwell was doing his best to secure stock enough to control the election. It was a very anxious time for all, as the chances of success in securing stock seemed to favor first one side and then the other," Parmly wrote.

On January 8, 1874, Lucian filed suit in New York, seeking to restrain Stockwell from interfering in the upcoming company election. Stockwell convinced a judge to postpone the election, then "redoubled his efforts to purchase sufficient stock to control the elections.... He offered fabulous prices for the little stock needed to give him the majority," Parmly wrote.

Then Charlie Willcox brought to Lucian's attention a quirk in the company's bylaws: trustees could expand the board at anytime, not necessarily at an election of the stockholders. So in March the board added two new trustees, both drawn from the combination: a man named Southworth, and Lucian Sharpe. "This settled Stockwell," Parmly recounted. "He saw all hope of success ... was gone. The case was hung up in the courts *forever*, with his foes inside and he out."

But troubles within Willcox & Gibbs were not over yet—Stockwell still owned a good deal of the company. In April, Lucian wrote Parmly a confidential memo, counseling patience. "At the present and for some time to come it seems to me best to shut our eyes to [many] things that could not be overlooked when we are settled.... I would urge you to use every means in your power to obtain any information you can affecting our interests. Let us hope for the best but be prepared for the worse."

In July the Howe Company, left to Stockwell by his father-in-law, shut down for two months because it was bleeding money in the Panic; the company was months behind in paying its workers. Seeing that the Howe Company was short of cash, Willcox & Gibbs bought out the contract it had held with Howe to cast iron tables for 50,000 sewing machines, and moved the work to Brown & Sharpe.

By mid-year the Insiders were in, the Outsiders were out, and work that had gone to Bridgeport had come back to Providence. But the sewing machine company's future, and by extension the future of Brown & Sharpe, was still uncertain. Under court order, stockholders still could not meet; and Stockwell still held a significant share of company stock.

Inside the new Providence factory, conditions continued to be tense. Lucian and the plant superintendent, Frederick W. Howe, disagreed on almost everything. Lucian had handpicked Howe to serve as Brown's likely successor, but they had never really pulled together from the first.

McFarlane was not the only worker who grumbled about Howe being over-rated as a machinist. In machine shops, men judge each other on how well they handle tools and machinery—a man who treats tools carelessly loses status in the eyes of his peers, because he will not be able to work with precision with dinged-up tools.

Robert Dunlap, an apprentice and journeyman for five years, recalled the day that a machine designed to count sewing machine needles developed a hitch in dispensing the needles. "Mr. Howe, to ease it up, put powdered emery into the bearings and joints," Dunlap recalled. Emery, an abrasive used in grinding, would wreak havoc on the machine's moving parts. Seeing Howe shake emery into the bearings sent "a cold shiver" down Dunlap's back, "lowering Mr. Howe's standing as a mechanic" by his lights. "Mr. Howe was of an inventive turn of mind," Dunlap recalled to a company historian, "but was very particular, especially as to appearance and was inclined to change a thing over and over ... making the work very expensive...."

Lucian and Howe frequently argued over cosmetics vs. costs. "Mr. Howe claimed that appearance and finish were important elements," Dunlop recalled, "while Mr. Sharpe was inclined to ask 'Will it sell for any more?' ... These two men taking opposite views along such lines gradually grew apart and misunderstandings developed."

The "misunderstandings" developed into an open rift. Howe, as one of only five stockholders in the company, began insisting on a right to inspect the company's books. And in 1874, the books were not good: Brown & Sharpe lost $49,000, and shareholders received no dividend. Lucian's company was losing money; Willcox & Gibbs stockholders still wrestled over the fate of his largest customer; and his superintendent was openly challenging his authority.

Given the circumstances, Lucian's tone was amazingly polite as he answered his cousin George's request for money:

> [B]usiness in all branches is quite dull now and has been ever since the panic. It will be impossible for me to comply with your request to lend you money as requested as I have used all my spare funds in the purchase of a house in the city last fall.... My family are all well. I have six children now, four girls and two boys.

Lucian also rejected Howe's proposal to sell back his share of stock to Brown & Sharpe at $700 a share—more than twice the $300 Lucian had paid that year to buy back some of McFarlane's shares.

The Panic of 1873 lingered into the late 1870s, even longer in England; and 1875 brought another year of losses for Brown & Sharpe. Again the company paid no dividends. Howe became a constant presence on the office, pressuring Lucian to see the books. The day after Valentine's Day, Lucian wrote:

> Mr. Howe has been here for five days from two to four hours per day.... This afternoon he came again and after warming his back before the fire for twenty minutes commenced to ask questions about amount of sales, how much we paid for Willcox & Gibbs Sewing Machine Cos. Stock &c all of which I answered correctly as far as I could remember.

Howe objected to Lucian's strategy of using Brown & Sharpe cash to buy the Willcox & Gibbs stock while wrestling control from Stockwell. "He then asked to see the Ledgers and Journals to which I objected as he had seen copies of them in the Trial Balances and that it was an annoyance to have him round here." The session ended with Lucian telling Howe to stop "coming round here and annoying us." Howe then filed a lawsuit against Brown & Sharpe.

Despite business troubles, creative energy still crackled through the factory on Promenade Street. The crew from Maine—John Hall on the graduating machine, and Calvin Weld with his sons running their lathe grinders, precisely graduated the small tools in a cluttered space on the first floor of the North Wing. "The room was never open to visitors," recalled William Viall," and had a stranger gone into that room he would never have believed it possible that any good thing, to say nothing of any accurate thing, could ever have come out of it."

On the second floor, in the northwest corner, James A.E. Gibbs, a "tall and straight" tobacco-chewing southerner, obsessively worked at improving the stitch laid down by his sewing machine. "He was working on a lockstitch machine that would buzz along rapidly for a few seconds and then break the thread," recalled Henry Corp, a machinist. "His patience was indefatigable. He threaded up and started again thousands of times a day."

Charlie, the younger Willcox, took over the job from Gibbs but the lockstitch "behaved no better for him than it had for Mr. Gibbs," Corp wrote. Literally thousands of times per day the machine whirred to life for a couple of seconds, then—snap, the thread broke. Then one day the needle chugged along for a full minute. Willcox emerged, smiling.

When Lucian wrote Darling in 1868 that "dull times" were a good time for developing new products, he had meant what he said. While Willcox and Gibbs busily perfected a lockstitch sewing machine, Joseph Brown resurrected his earlier drawings of a Universal Grinding Machine, and brought the plans to fruition. By early 1876, a prototype of the new grinding machine stood on the second floor at the top of the stairs. Corp recalled, "It was Mr. Brown's pet machine at the time."

Brown grew fond of showing the machine off to visitors, scorning the elevator and clanging up the iron stairway to show it off. While Corp ground a piece of metal on the machine, Brown began explaining in complex detail an improvement he had in mind for the prototype grinder. Corp "was so embarrassed at having all these ideas fired at me in his Gatling gun manner that I comprehended nothing of what he talked and I told him so frankly."

Brown replied: "Never mind. I cannot *think* ideas. I have to *talk* them and it has cleared my mind to talk with you. If I cannot get a man to talk to I sometimes talk to a post."

A few days after delivering his "Gatling gun" monologue about grinding to Henry Corp, Joseph Brown passed through the shop. His salt-and-pepper beard stretched nearly to his navel as he peered through gold-rimmed spectacles, shaking hands with workmen he met along the way. Brown planned to steam out to the Isles of Shoals, rocky islands off the coast of New Hampshire, to escape a weeks-long heat wave that fouled Providence with a noxious air.

Some blamed the enervating heat on a shift of the Gulf Stream; others pointed to "masses of burning magnesium" on the surface of the sun. "Speculation has extended itself in endeavoring to find out a reason for the almost unexampled heated term," the *Providence Journal* reported on July 24, 1876. "That it is now the minimum period for sun spots ... is one of the favorite theories." The *Journal* concluded: "The fact is the most scientific observers know very little concerning these great meteorological problems."

The heat wave's cause remained a mystery, but its effects were certain: The Moshasauck River, slowed to a trickle as it passed the long façade of Brown & Sharpe, emitted a stench "exceedingly foul and often very offensive." The city's death rate rose to 71 in a week, a number "remarkably large," the *Journal* reported, "greater, perhaps, than in any single week since 1872."

The Isles of Shoals, surrounded by the cooling waters of the Atlantic, may have offered Brown some respite; but it could not shelter him from man's ultimate fate. On July 23, while vacationing at the Shoals, Joseph Brown died. The next morning, a Monday, word of Brown's death filtered into the plant. "It came as a shock to us all," Corp remembered.

On Wednesday, the shop closed for the day. Workmen filed up streets freshly scrubbed by a cleansing rain to College Hill, where Brown's body reposed in his house on Congden Street. At noon an Episcopal minister performed funeral rites. A cortege carried Brown's coffin out to the garden cemetery on Swan Point. Gravediggers laid him next to his first wife, Caroline, and their toddler son, Walter. Their grave was marked by a chest-high, sandstone pyramid.

Brown willed his Congden Street house to his widow, Jane F. Brown; his house on Broadway, a wide thoroughfare graced with ornate Victorian houses, he gave to his daughter, Lyra Nickerson.

To the new Rhode Island Hospital, built during the Civil War, Brown left $8,000 for "two free beds in said institution one of which shall be called and named after me, the said Joseph R. Brown, and the other shall be named and called after my wife, the said Jane F. Brown."

An institution that had not even been chartered yet received even more: the Providence Public Library. Brown, a man with little formal education who spelled abysmally, bequeathed $10,000 to charter and build a library.

A couple of days after Brown's death, Brown & Sharpe publicly unveiled the new Universal Grinding Machine, before listing it in the company's catalogue of 1877. Though he never lived to see it, Brown's Universal Grinding Machine proved revolutionary,

Company executive Henry Corp poses with the original No. 1 Universal Grinding Machine, which he operated upon joining the company in his teens. Though Brown had made sketches of the machine as early as 1862, hence the patent date on the machine's door, he did not make a model until just before his death in 1876. Henry Leland, founder of the Cadillac Motor Company, wrote: "If all these [grinding] machines should be suddenly taken away, it is hard to imagine what the result would be. It would be impossible to make any more hardened work for the best parts of our machinery."

perhaps having an even greater effect on the base metals industry than his Universal Miller. The grinding wheel could shave hardened steel, a job beyond the power of grinding lathes. Instead of the wheel moving along the piece to be ground, the metal stock moved across a stationary, spinning wheel. Brown protected guideways from metal shavings to ensure smooth passage of the stock; to keep friction from heating metal he cooled the stock with a spray of water that flowed away into a bucket in the rear of the bed.

The Universal Grinder made it possible to grind hardened parts. Prior to its invention, precise parts such as gears had to be ground on a lathe then hardened, which distorted their shapes. With a grinder, machinists could cut even hardened parts to precision. With it, more people could grind metal to more precise tolerances and better surface finishes than ever before. At first, machinists used it to solve immediate problems: breaking the bottleneck at the needle-grinding station in sewing machine production; increasing production of the tools ground in the first floor warren of Darling, Brown, & Sharpe. Precision grinding of tool steel improved the quality of existing cutting tools such as drills, reamers, gear cutters, and even bettered milling cutters so that they could get the most out of Brown's milling machines.

Within a few years designers were applying the Universal Grinding Machine and heavier versions of it to solve problems such as grinding large batches of ball bearings and the circular raceways that contain them; batch production of bearings made possible bicycle tires, spawning the 1890s Bicycle Craze that quickly morphed into development of the automobile.[8] Henry M. Leland, who left Brown & Sharpe in the 1890s to help found the Cadillac Motor Co., wrote in the early 1900s:

> If all these [grinding] machines should be suddenly taken away, it is hard to imagine what the result would be. It would be impossible to make any more hardened work for the best parts of our machinery.... This in my judgment is one of the most remarkable inventions, and too much cannot be said of its praise, or in acknowledgement of Mr. Brown's perseverance, wonderful initiative and genius.

With his linear dividing engines that precisely graduated Vernier calipers in the 1850s, Brown brought new standards of precision to American machine shops. His Universal Milling Machine, along with his formed gear cutter, revolutionized the milling, or cutting, of metal. His micrometer, an ingeniously blended hybrid of existing measuring devices, put an easy-to-read precision tool in the palms of most machinists. And just before his death, Joseph Brown brought to fruition a Universal Grinder that allowed designers to take advantage of new steel alloys to invent technologies as yet unimagined.

In the *Providence Journal*, his passing commanded an obituary of just five sentences:

> The funeral of Joseph R. Brown, senior partner of the Brown & Sharpe Mfg. Co., of this city, was solemnized at his late residence, No. 119 Congdon St., at 12 o'clock yesterday. Rev. Daniel Goodwin, of Dedham, Mass., officiated in the funeral services, which were in accordance with the ritual ceremonies of the Episcopal Church.
>
> The funeral was attended by a large number of the relatives and friends of the deceased, and by the employees of the Company in a body. The works were closed for the day in token of respect, and to give the men an opportunity to attend the funeral services of their highly esteemed employer and friend.
>
> The remains were taken to Swan Point Cemetery and interred in the family burial ground.

7. The Age of Steel

The deadly heat wave of 1876 cast a heavy pall over the Centennial Exposition in Philadelphia, the nation's first World's Fair. The fair opened well enough on May 10, when President Grant and Brazilian Emperor Dom Pedro II pushed buttons to start the 30-foot flywheel of the mammoth Centennial Engine, built in Providence by George Corliss. The engine stood more than four stories tall. Its twin cylinders, each 40 inches in diameter with a 10-foot stroke, chuffed nosily along, powering most of the exhibits at the Centennial Exposition through a system of spinning shafts running more than a mile long.

Rhode Island was well represented at the Exposition. To draw attention to its yeast powder the Rumford Chemical Company of East Providence drew fairgoers to its exhibit with the enticing smells of freshly baking bread. The chemical company touted its Rumford Yeast Powder as "a combination of materials for making bread of all varieties." John Hope represented his Rhode Island firm of Hope & Co., exhibiting a "pantograph engraving machine" which used a camera and nitric acid to etch calico patterns into textiles. The National Rubber Company of Bristol showed the process of making rubber shoes with "the curious cakes of gum from South America."[1]

Brown & Sharpe a "company of world-wide renown for the manufacture of instruments of precision and of special machinery" exhibited in Section B of Machinery Hall. The firm displayed an array of small tools and a few of its largest machine tools—a gear cutting index, the new grinding machine, and a screw machine on which "one man may turn as many screws as three to five men on as many engine lathes," according to an exposition guide.

In another section of Machinery Hall, E. Remington and Sons showed off their new typewriter, a novelty that could not seem to gain market traction. Mark Twain referred to it as a "curiosity breeding little joker."

Over in the Main Exhibition Building, Alexander Graham Bell stole the show. He transmitted his voice through the gallery via a wire a hundred yards long, reciting Hamlet's soliloquy, "To be, or not to be—that is question," to the delight of the Emperor Dom Pedro.[2]

The intense heat of June and July depressed fair attendance, but by fall, more than 100,000 people strolled the fairgrounds every day, pushing attendance beyond 10 million, nearly 20 percent of America's population. When the fair closed a group of Rhode Island women—mostly the wives of mill owners who had raised money for the

Rhode Island exhibit—discovered that they had $1,676 left over. They decided to use the surplus as seed money to form the Rhode Island School of Design, in order to teach artists to apply "the principles of art to the requirements of trade and manufacture."

By the end of an eventful year in which the firm weathered Brown's death and introduced its universal grinder, Brown & Sharpe had eked out a surplus—a profit of $10,928; yet for the third consecutive year the company did not pay its small cadre of stockholders any dividends as it struggled to right itself from the move to the new factory, the recent Panic, and Stockwell's "grab" for Willcox & Gibbs.[3]

In the spring of 1877, Lucian Sharpe settled a lawsuit filed against the company by buying out the stock of his former superintendent and fellow stockholder Frederick W. Howe for $43,000. Then came a break in the Willcox & Gibbs situation which for four years had hung over the company like a Damoclean Sword. In mid–August, a Wall Street brokerage firm called a loan it had extended to A.B. Stockwell, demanding that Stockwell pay them $9,500 cash or they would seize the collateral that he had posted to obtain the loan: 1,000 shares of Willcox & Gibbs stock. Faced with a demand for $9,500, Stockwell had no choice: he had to sell his 1,000 shares of stock.

The Combination, a group of Willcox & Gibbs stockholders seeking to wrest control of the company from Stockwell, pounced. With an infusion of cash from Brown & Sharpe, the Combination snatched up the last of Stockwell's stock at $10 a share. "Well it is a relief to get this matter off our mind," Lucian wrote to Parmly on the morning of the sale. "It must have been a bitter pill to Stockwell to sell this stock to us." Less than four years before, Lucian had feared losing Willcox & Gibbs's business to the Howe Company in Bridgeport. Now he emerged from the fray as a Willcox & Gibbs trustee controlling 17 percent of the sewing machine company's shares, making him the firm's largest stockholder.

In his reminisces, Thomas McFarlane, a Brown & Sharpe superintendent who never really liked Lucian, concluded that the Combination *"by careful, patient manipulation got control of the Sewing Machine Company out of Stockwell's hands,* and once more saw clear sailing and hopeful signs; and then naturally Mr. Sharpe's spirits rose from depression to elation, and *sometimes almost to oppression,* that somewhat taxed his associates."[4]

The problem with McFarlane was that he too taxed his associates—the man always seemed to have some sort of drama going on at work: Someone overheard him talking down John Cranshaw who worked in the front office. (McFarlane claimed he was just trying to keep a good floor worker from desiring a front-office job.) After Brown's death, McFarlane tried to sell a two-volume mechanical dictionary that Brown had owned, contrary to the wishes of Brown's widow, Jane. (McFarlane said he was helping to "make a sale for the Brown estate.") Lucian, who had already bought all of McFarlane's stock, tired of the drama.

In February 1878, Lucian fired McFarlane, replacing him as shop superintendent with Richmond Viall, who in 15 years had risen from screw machine operator to superintendent of the sewing machine department. Viall seemed an unusual choice for the powerful role of factory superintendent. He was a farmer's son, raised on nearby Bullock's Point. Viall had almost no formal education, though he loved to read, and had read the Bible and the works of Shakespeare so closely that he often quoted them to make a point. Other than an apprenticeship with a jeweler he had had no formal training before coming to Brown & Sharpe as a $1.50-a-day laborer; his son, William, recalled that Viall's promotion to superintendent of the whole works touched off "a great furor through the shop.... Sneering remarks were made about the 'soft solder jeweler' placed in charge of the manufacture of machinery."[5]

While offering the factory superintendent's post to Viall, Lucian gave him this advice: "I have but two things to tell you: First, never keep about you men who are not in sympathy with and working with you; secondly, never take a job that you do not feel assured you are going to make money on."

Viall's promotion opened a vacancy as sewing machine superintendent, which Lucian filled with Henry Leland, a 34-year-old tool builder who had been in charge of the screw machine department. Leland accepted the job only after Lucian agreed that he could institute a piecework system, and set the pay rates for each job.[6]

Upon hearing of Viall's promotion some machinists resigned, starting their own businesses or finding work elsewhere. One skilled machinist who remained became "somewhat offish" to Viall, a surliness that, unchecked, could bring "a slow seeping out of poison" through the factory.

Richmond Viall, a farmer's son fond of quoting Shakespeare and the Bible, went to work for Brown & Sharpe as a 19-year-old screw machine hand in 1863, and worked his way up to factory superintendent. With Henry Leland, he introduced scientific management into Brown & Sharpe, and built the company's apprenticeship program into a nationally recognized model.

Viall called the man into his office and said, "Now I want to know, are you going to stay with me and work with me, or are you going to take yourself out?"

The man replied that he would think it over.

"You will think it over while you are sitting in that chair," Viall said, "and your answer will be given while you are in this room."

The machinist remained with the company.[7]

Others fell into line as Richmond Viall and Henry Leland ushered in a new era on the factory's floors.

With the Stockwell and Howe situations settled in 1877, for the first time in three years Lucian felt comfortable in paying dividends to stockholders, who now numbered four: himself, Darling, and Brown's two heirs, wife Jane and daughter Lyra.

"It is hardly necessary to say that it is very desirable to keep the matter of these dividends to ourselves both in number and amount," Lucian wrote to Brown's daughter, Lyra Nickerson in the summer of 1878. "I would be glad if *no one* in Providence knew anything about them. I would much prefer the business would appear to be an unprofitable one than other wise as the contrary impression will only tend to raise taxes and make parties, formally *in* but now *out,* uncomfortable and talkative."[8]

Shortly after Joseph Brown died, Lyra Nickerson came down with an illness that made it hard for her to walk. She set off to Europe on a steamer to convalesce, encouraged by Lucian in the summer of 1878 to live "in comfort with all the luxuries you desire." By Thanksgiving she wrote of improvement in her gait as she prepared to embark on an adventure to Africa. Lucian continued to encourage her spending, writing in May of 1879: "you are not living up to your income by quite a sum...."

By September he happily reported: "Business is improving with us and there is a much better feeling throughout the country." Each of the company's three branches earned about $20,000 more in 1879 than it had the previous year. In the machine tools line, this represented a doubling of business.

The growth in the small tools line received a boost from Brown & Sharpe's latest invention: barbershop hair clippers, which quickly became a staple tool in American barbershops. Success has many fathers, and there are various versions of who invented the popular Brown & Sharpe clipper.

As Henry Corp recalled it, young Brown & Sharpe workers began using the firm's large horse hair clippers to give each other "pineapple cuts," a style that was all the rage. "A barber down town, seeing these [pineapple cuts], wanted to know what tool it was done with, and afterwards came here, and bought a horse clipper, suggesting that a clipper could be made to use with one hand," Corp recalled. "He thought that by holding it way up close to the blades, he could use it with one hand, and a comb with the other hand." A supervisor named Cyrus Carleton, who had left Brown & Sharpe but returned with a salary boost after Brown's death, then patented a smaller version of the horsehair clippers that barbers could use to snip hair with just one hand.

But J.H. Drury, who left Brown & Sharpe to work for the Union Twist Drill Co., remembered the story differently: "I think it was in the summer of 1877, as a boy in the office, I took in horse clippers brought in by various barbers to be sharpened and collecting at that time $1.15 from each. One day, after taking in a clipper brought in by a colored gentleman, I was collecting the money when Mister [Richmond] Viall came through the offices."

Viall asked Drury why he was sharpening horse clippers in mid-summer, after horses had been clipped for season, and Drury replied that a black barber was using them on clients. Viall remarked that that must be hard work, because the barber would have to hold the clippers with two hands. He suggested developing a prototype clipper to be held with one hand. "I can see him now illustrating the movement of the shears," Drury said.

In Drury's telling, Viall borrowed the "colored gentleman's" clippers, and came back after some time with the model for a smaller clipper. Drury took it to the Narragansett Hotel Barber Shop to ask the head barber his opinion.

"He immediately got the rest of his small army of barbers about him and said that he would need a dozen of them...," Drury recalled. He then took the model to another barbershop up over the old New England Market, where barbers again expressed enthusiasm. Drury recalled, "Mr. Sharpe attempted to discourage it by saying that this 'pineapple' clip was only a fad and that it wouldn't last long...." But Carleton and another mechanic named George Noble persisted in bringing the hair clippers to market.

One of the earliest accounts of the invention came through a March 1889 issue of *American Machinist* magazine, which credited the clippers to Osceola Cook, a black barber who was a mountain of a man, weighing in at nearly 450 pounds and said to possess enormous strength. "Mr. Cook was a colored man and kept a barber and bootblacking establishment, and was a prominent local character," the *American Machinist* reported shortly after Cook's death. "He was a man of considerable mechanical ingenuity, having invented, besides the hairclippers, a shampooing machine, which worked well." *American Machinist* quickly retracted that story, likely at the request of Brown & Sharpe, which feared a claim on its patented clippers.

The first batch of hair clippers, told variously as numbering 300 or 500 units, quickly sold out, as did subsequent lots. Most versions of the hair clippers story share common threads: "the pineapple cut" spawned them, a black barber had something to do with them, Carleton perfected them, and they met lasting success: Brown & Sharpe made and sharpened barber clippers, including an electric version, for the next 78 years.

Lucian Sharpe spent some of his dividends on the latest technology—telephones. In 1879 he installed phones in his houses in Providence and at Nayatt Point, ran a line into the factory, and put one into Mrs. Brown's house. In summers he could call the factory from his Nayatt home and conduct business without having to make the eight-mile ride. "It is a great convenience," he enthused in a December letter to Lyra. He also laid out money in "concreting" the driveway at his summer home, and to enroll the boys, Lucian and Henry, in the open-air kindergarten that had just opened on the East Side.

With the company's gross income again crossing the $300,000 mark in 1880, Lucian began to grow nervous—the firm had not made that kind of money since just before the move. At age 50, Lucian had seen three Panics in his lifetime: 1837, 1857, and 1873. With the lessons of the latest one fresh in his mind, he began counseling caution to Lyra: "I have every reason to think that we are going to have a good business for sometime to come [but] I would ask you to consider whether $8000 a year is not more than is prudent for you to spend," at least until she diversified investments. He repeated his warning at the close of the year, 1880. The firm had just built a new foundry, so dividends for the next year would at best be small.

"I hope you do not regret buying so many beautiful things to take home but I cannot look so favorable upon such expenses as you do," Lucian wrote. "...I am afraid you have been so long away living among lords and nabobs as to quite forget the economical lessons inculcated by your father." Nevertheless, he conceded in a letter to Lyra in January 1881: "There never has been such a demand for the productions of machine shops as exists now. How long it will last no one knows but the general opinion seems to be that there will be good business for two or three years."[9]

The new foundry presented new challenges to Brown & Sharpe, a company with extensive experience in cutting metals, but none in casting them. The one-story wooden building ran for about 90 yards along Promenade Street and was nearly 20 yards wide. The men who first fired its furnaces in December of 1880 learned about forging, hardening, and tempering through trial and error. After firing one foundry foreman, Richmond Viall hired Elijah Lyon for the job. Lyon had no foundry experience—before his promotion he worked on the third floor making sewing machine needles—but Viall saw him as a "good fellow" who always told the truth, even if it meant admitting a mistake. One co-worker recalled that Lyon had some "convivial habits" that "took him into some rather trying positions" outside the shop. But Viall was willing to overlook Lyon's out-of-work shenanigans. He "knew how to get along" with workers, his colleague recalled, so they performed well for him.[10]

The science of metallurgy had changed dramatically in the latter half of the nineteenth-century. Henry Bessemer, an Englishman who had made a fortune developing a mechanized method of shaving bronze into powder to make sparkly paints, used some of his money to experiment with new methods of steel production. He determined that smelters could break oxygen's bond with iron by blowing air through molten pig iron, creating a nearly pure steel. High phosphorous levels in many iron ores stymied Bessemer's process—the phosphorous remained as phosphoric acid, sullying the steel—but a junior clerk in a police court in the London docks, Sidney Glichrist Thomas, solved that problem by adding limestone to the furnace lining to neutralize the phosphoric acid. William Siemens of the famous Siemens family, centered in Germany but inventing all over the world, developed an "open hearth" furnace that captured hot gases escaping from the smelting process and forced them back into the furnace heating ores to higher temperatures that cooked off impurities and created steel as good as Bessemer's.

Besides finding new ways to manage carbon levels and remove impurities from iron to make inexpensive steel, metallurgists such as Robert Mushet discovered that they could add desirable elements to molten steel, such as tungsten, to create alloys. Mushet's tungsten steel did not soften like carbon steel when a cutting tool heated it, allowing machinists to make full use of faster, heavier machine tools, such as Brown & Sharpe's grinding machine. The upshot of these advances was a new "Age of Steel." Until 1860, wrought iron had been king; by century's end it had been replaced by carbon or alloyed steel. Cast iron still had a place as a cheap, easily moldable product, used to cast items such as sewing machine bodies, but purer wrought iron was virtually squeezed out of production.

A new "Age of Steel" boded well for a base metals company such as Brown & Sharpe, which built the precision and machine tools that measured, cut, grooved, and ground it. But cutting steel was an entirely different operation than heating it for forging, hardening, or tempering. In making cast or pig iron from ore, foundry workers had to maintain a uniform heat throughout the furnace or the steel would crack like pudding skin. To make cutting tools workers forged high carbon steel then treated it with intense heat to give it different properties of hardness and softness. Heating steel until it glowed red then quenching it in a bath hardened the steel, but made it brittle.

Brown & Sharpe foundry workers in the 1890s demonstrating a pour of molten metal into a ladle attached to an overhead chain hoist for transport to the foundry floor, where workers will pour the metal into molds.

7. The Age of Steel

The three-story "fire proof" factory designed by Frederick W. Howe in 1872 stands in the foreground, next to a foundry that fired for the first time in December 1880. With its own foundry, Brown & Sharpe was able to increase its product line to manufacture a wider array of products, including barber clippers, foundry rattlers, and cast iron latrines for factories.

Reheating hardened steel then letting it gradually cool tempered it, retaining hardness but removing the brittleness. Each step was fraught with difficulty. In tempering, workers closely watched the steel's color to determine when to stop heating it: steel honed for use as cutters in a lathe was finished when it turned straw yellow; stouter milling cutters, taps, dies, and reamers needed to turn brown yellow, a little hotter than straw yellow; steel for twist drills needed to turn a light purple, at about 530 degrees. Away from the windows foundries were generally kept dark so workers could see the subtle variations in the glowing color of heated metal. Air was hot near the furnace, but not otherwise heated, so that on winter days workers sweated by fires of coke, charcoal, or gas, and felt their trickling sweat freeze beyond the furnace's circle of heat. Newly cast pieces of iron or steel noisily banged around in "foundry rattlers," revolving drums filled with metal castings and rotated so as to remove molding sands while smoothing the surfaces of the new castings.

Workers schooled in the craft of heat treating often cooled hardened steels in baths of either sperm whale or linseed oil. A 1900s treatise on hardening advised: "In using the fish oils, some means of ventilating that will carry away the disagreeable odor should be employed."[11] With coke fires burning, molten steel flowing, whale oil hissing, and castings rattling, foundries were loud, hot, dusty, dark, and dangerous places. And after a brief learning curve, Brown & Sharpe's was immensely successful, allowing the firm to make its own castings and multiple foundry products to sell to others.

People swarmed the decks of the steamship *Servia*, bustling about in a flurry of final good-byes before the ship's bells rang, signaling its imminent departure from New York to Liverpool. Mary Sharpe, at 23 the eldest of Lucian and Louisa's six children, had never seen such a collection of flowers as the aromatic array left aboard by the disembarking visitors. As she descended from the promenade deck on April 4, 1883, Mary felt dismayed by the two flights of stairs carrying her down to the ship's main level; but upon reaching her family's rooms she was "much pleased" with the rooms, finding them "convenient and quite large for staterooms."[12]

The *Servia* was a marvel of its era, the world's second-largest ship at 515 feet, and the first transatlantic vessel framed and plated with steel. By employing steel, created in quantity through the process developed by Siemens, designers at a Glasgow shipyard built a ship hundreds of tons lighter than a similar vessel built of iron would have weighed. The *Servia's* public spaces sprouted Edison's new electric light bulbs for illumination, the first ship of the Cunard Line to use electric lighting.

Mary was engaged to marry Zechariah Chafee, a Providence man whose family owned the Builder's Iron Foundry, and ran a bank that became the Union Trust Company. Mary's future father-in-law was also named Zechariah, but his son never went by Zechariah Junior. The elder Zechariah Chafee was then embroiled in a bit of a public scandal: After the Panic of 1873, creditors of the collapsed A.&W. Sprague Company had appointed Chafee as a bankruptcy trustee to straighten out the company's finances. He bought some of the debt for less than what the estate paid to retire it; when he emerged from those proceedings as a wealthy man, some felt as if he had diverted too great a share of the assets to himself. The Rhode Island Supreme Court censured him for speculating in the debts of a trust he controlled, but he never was convicted of wrongdoing. Legal issues surrounding the bankruptcy dragged on into 1927, long after the senior Chafee's death in 1889. The case involved some of Rhode Island's most prominent families besides the Chafees, including the Aldrichs and Rockefellers; when the Rhode Island courts finally disposed of the case in the 1920s, judges ordered all of the estate records to be burned.[13]

Before her marriage into the Chafee family, Mary wanted to take a grand tour of Europe with her parents and five younger siblings. In their travels the four Sharpe women delighted in flouting Old World customs concerning the proper behavior of women. In Amsterdam they piled onto an omnibus, an enclosed, horse drawn bus, and chose a perch atop the roof for better views of the old city as the horses clopped along. Having grown up riding alongside their father in a machine shop's delivery wagon, the Sharpe women felt right at home perched atop an omnibus.

"We noticed that some people stared," Mary wrote to her fiancé, "but we are so used to that we did not think much of it, but a Dutchman told Papa it wasn't a customary thing for ladies." The stares did not put off Mary, herself a good wagon driver; she liked to drive the horse-drawn roller and mower over the lawn sprouted from Central Park Seed at Nayatt. "It was great fun," Mary wrote of her trip atop the omnibus, "and I intend to do it again."

In Brussels, Mary and her 19-year-old sister, Amey, were "ignominiously turned out" of a small town's chapel after they clambered over an altar rail so Amey could sketch "a bas relief of somebody over his tomb." On a hot train bound for Frankfort "in our efforts to secure a breeze" they accidentally broke a window. "Poppa paid for it.... It served to provoke a good laugh." In London the four young women again alighted atop an omnibus, prompting their baby brother Henry to pipe up, "Ladies don't ride on top, do they?"

Their father, Lucian, replied: "It looks as if they did."

Mary, at 23 more than a decade older than her baby brothers, showed no qualms in acting as a surrogate mother to the boys. She nixed Henry's morbid desire to sneak a peek at a catacombs and morgue, and in Amsterdam praised the boys for holding their own in an "earnest" discussion with an Englishman about the relative merits of the United States vs. England. "Henry's memory for details, and his habits of careful observation, are as active as ever," Mary noted. And, "The man asked what writers we had, and Lucian suggested that James Russell Lowell wrote 'pretty good poetry,' and [John Greenleaf] Whittier."

Mary also enjoyed drinking good beer with her father, and in Manchester, England she and her mother reluctantly joined him in touring the machine shop of his friend and fellow toolmaker, Alfred Muir. During his first European tour, Lucian worried that English factories were not sophisticated enough to employ Brown & Sharpe's tools. Now British machinists instantly recognized his company whenever they heard the Sharpe name. "Papa ... is so manifestly happy in visiting different machine-shops and meeting people who know him by name and use his tools," Mary wrote.

In Glasgow, Lucian even wrangled a tour of the Singer Sewing Machine works, despite that company's deserved reputation for secrecy. The truth was, Singer had little to fear from Willcox & Gibbs. By the 1880s, Singer's dominance of the sewing machine market was so complete that many smaller companies quit the sewing machine business and converted to making other goods such as guns, and the spokes and frames for high-wheel bicycles, first imported from France in 1878.[14]

The Willcox & Gibbs machines that Brown & Sharpe made through the American System were arguably better than any model produced by Singer, which employed "Old World" techniques of hand-fitting the pieces to each machine. If a Willcox & Gibbs part broke, the company could ship a replacement part that would fit the machine, as all parts were built to meet a model specification; when a Singer part broke—and their needles, gears, and shuttles often did—replacements had to be hand-filed or honed with a whetstone in order to fit.

Singer's boasts of building the best machines were hyperbole, but the firm did beat all of its competition in marketing. Edward Clark, a young New York lawyer when he bought a one-third share of Singer, invented installment purchasing where a customer could put money down on a machine and pay off the rest through monthly installments. Singer marketed the machine by hiring women to act as live mannequins, demonstrating its sewing machines in storefronts; from the company's beginnings in the mid–1850s, Clark acknowledged that "a large part of our success we attribute to our numerous advertisements and publications."[15] Marketing trumped mechanics: At the time of the Centennial Exposition in 1876, Singer sold 272,000 machines a year—more than Willcox & Gibbs had sold in its entire 17 years of production.

While Singer continued to grow market share, sales of the Willcox & Gibbs machine peaked in 1871 and never really recovered from the Panic. Rather than view Singer as a competitor in the sewing machine market, Lucian groomed the behemoth as a potential customer in the machine tools field. As Singer moved to adapt principles of the American System in the 1880s, Lucian sold Brown & Sharpe screw machines to the larger sewing machine company.

The Sharpes shipped for home on October 13, and after a rough passage of one week arrived in New York. The new bridge spanning the East River had opened while the Sharpes were abroad. As the steel ship *Servia* nosed up the river toward Manhattan, horses, trains, and pedestrians crossed above it on a steel truss 85 feet wide, suspended from the iconic neo-gothic towers of the Brooklyn Bridge by nearly 7 million pounds of steel cable.

8. Providence, Paris, Chicago

> To the sewing machine work, in particular, we are indebted, as the requirements of this industry have had an important part in stimulating the invention and development of the milling and grinding machines, the cutters that can be sharpened without change of form, and the gauges and exact measuring instruments, which have established our reputation and materially modified and improved machine shop practice throughout the world.—*Brown & Sharpe circular distributed at the 1893 Columbian Exposition*

Slowly, steadily, a new neighborhood sprang up around the factory on Promenade Street. What had been a sloping field behind the Brown & Sharpe plant was, by the mid–1880s, gridded with streets and lined with privately built houses for the company workers. Richmond Viall, the factory superintendent, had moved his family over to Smith's Hill in the 1870s and he now had his eye on a parcel atop the hill that would give him a good perch for viewing the factory at the base of the hill. Viall supervised nearly 700 workers, and he made it his business to know their business—in the plant, and in their homes.

"The most interesting character of all was Mister Viall," recalled longtime worker Henry Corp, for a company history in 1917. Viall, the farmer's son fond of quoting Shakespeare and the Bible, ran the shop in what Corp called a "dictatorial" manner. He was a stickler for punctuality, and he practiced it. Though he had office work to tend to, Viall always made sure to be walking the shop floors when the whistle blew at 7 a.m. to begin work, six days a week, and when the 6 p.m. whistle blew to end it. At quitting time on Saturdays, workers drew the window shades in all of the factory's 40-pane windows to the top of the lower sash to give the plant a tidy appearance to end one work week and begin the next, a practice Brown & Sharpe employed for decades.

For Viall, "Order and neatness were almost a religion," recalled Zechariah Chafee, Mary Sharpe's husband. "He had no use for a shop with dirty windows, dirty walls and ceilings, and dirty equipment. Tools must be cleaned when the men were through with them." Every day at noon, workers gathered on the third floor of the shop for a religious meeting led by a thickly-bearded preacher named Joseph D. Brown, no relation to the company's founder. Brown "wore his hair very long, and had a full beard," recalled Henry Corp, a Brown & Sharpe machinist in the 1870s and 1880s. "He used to hold religious meetings in a corner under the patronage of Henry M. Leland, who belonged to the same church," the Second Adventist Church on Hammond Street.

Brown often railed against Catholicism, angering the increasing number of Catholics coming to work at Brown & Sharpe. Corp recalled:

> One of them, an oiler, had to go around and oil the shafting at noontime. His place to oil the shafting was just where the meetings were held, at the southwest corner of the third floor. ... This oiler, an Irishman, objected very strongly to these meetings being held, and used to drop monkey wrenches on the floor, and hinder in various ways the meetings quite a little. ... He was told that he could not hinder these meetings which had the sanction of Mr. Viall.

Viall himself worshipped in the Congregationalist Church; he did not insist that workers join his church, but "did most strenuously believe that a man should have some kind of religion," Corp wrote. Henry Leland, who ran the sewing machine department on the third floor, always attended the noon meetings, and urged his workers to join him. "It was not true as some have said that you were obliged to attend religious services," recalled Corp. "But it was true that if one had 'Agnostic' views, it was impolitic to air them, at that time."

Even after the 6 p.m. whistle, Viall "insisted that the home life of his men was a part of his concern," Corp recalled. If he heard of drunkenness or domestic problems in his neighborhood of workers, he dropped in after hours to pay a troubled family a visit. "It is possible that this intimate interest was obnoxious to some," Corp wrote.

> His enemies said he had no business prying into the home affairs of his men. But this criticism never came from the families into which he came. ... No one but a man like Mister Viall could do this without offense. In his case the interference was successful, not so much from any kindly or gentle manner (he never assumed that pose) as from the commonsense view that his suggestions were such that everyone would approve.

Though he insisted on punctuality, cleanliness, and religion, Viall applied a flexible approach to enforcing shop rules. For example, Oscar Beale, the ingenious machinist from Maine, was an agnostic, and a bit of a daydreamer. He kept an office in the drafting room where many days he idly sat, striking up conversations with passersby but doing very little in the way of actual work. His white

One of the true geniuses at Brown & Sharpe, the Maine-born Oscar Beale joined the company as a bench-hand in 1869, and quickly worked his way up to chief inspector. He invented his first gear-cutting machine in his attic in 1878, and patented his last one, an automatic bevel gear generating machine, in 1901. In between he developed a measuring machine that Brown & Sharpe used as a master for most its gauges, wrote a handbook for apprentices, and developed a line of "blue books" to explain concepts of gearing and grinding to budding machinists.

apron was as clean on Saturday as it had been on Monday. But Beale did his best thinking at home, off hours. Some mornings he arrived at the shop bursting to discuss some idea that had come to him in the night. Viall appreciated Beale's genius, and was willing to overlook his agnosticism, and his apparent idleness around the shop.

Shortly after Viall took the job of superintendent, Beale revealed that he had used his own attic machine shop to build an "odontograph machine" that precisely inscribed the outlines of double-curve epicycloidal gear teeth. Brown & Sharpe bought the machine from him and used it for decades to make some of the gears sold in its catalog. Joseph Brown championed the use of involute gear teeth, which make stronger gears with less friction between teeth, when marketing his formed tooth cutter. But tradition still favored use of the epicycloidal gear teeth into the late 1800s, so Brown & Sharpe continued to make them.[1]

Beale taught evening lessons in gearing at his house; as he taught he continued to learn, eventually inventing gear cutting machines that helped meet the increasingly complex demands of an evolving automobile market. In 1884, a student who was moving to Ohio asked Beale if he could mail the lessons in gearing to him. Beale bound

Measuring machine developed by Oscar Beale in the early 1890s to act as a master gauge for multiple Brown & Sharpe gauges and products. By turning the wheel at left an operator could narrow the gap between the two rounded points to the right of center, which acted as a pair of calipers. A Vernier scale is mounted to the front of the machine, beneath a vertical eye piece which acts as a microscope allowing an operator to measure to the millionths of an inch. The machine is now in the collection of the American Precision Museum in Windsor, Vermont.

the lessons in a book that he first called *Beale's Blue Book on Gear Wheels*. Viall and Lucian Sharpe liked it well enough to buy the copyright, and the firm reissued it with an addendum by Beale as *A Treatise on Gearing*. From this, the company began a line of Blue Books with titles such *Construction and Use of Grinding Machines, Formulas in Gearing*, and Beale's *A Handbook for Apprenticed Machinists*, printed by J.W. Pratt in New York in a quality befitting the era's Arts & Crafts Movement, with leafy imagery lining the inside covers and gold leaf lettering stamped into the blue binding.

Besides writing the books and working on gear cutters, Beale paid close study to metrology, the science of measurement. He imported a set of Whitworth plugs and rings, measuring devices developed by the English machinist Joseph Whitworth to standardize screw threads. Beale suspected that the plugs contained discrepancies, so he set up a little shop on the second floor to test his hunch. Working with Henry Corp, he "discovered that Whitworth was not infallible," Corp recalled,

> our ordinary micrometers showed that these much revered standards were far from accurate and the desire to improve on them made an interesting game for us that seemed more like fun than work. ... He always came in with some fresh ideas each morning that he had thought up in the night, and we would try out each new scheme with much enthusiasm.

In working with Corp, Beale confided that his career was "hampered" because he was agnostic. "This was probably true, to some extent," Corp concluded. But Viall had the good sense to leave Beale alone during the day to foster his creative genius at night.

To commemorate the 250th anniversary of Roger Williams' banishment from Boston to what would become Providence, a local historian published a book: *The Providence Plantations for Two Hundred and Fifty Years: The People and Their Neighbors, Their Pursuits and Progress, 1636–1886*. In his book, Welcome Arnold Greene sought not only to chronicle the state's past, but also to provide a snapshot of its present. In describing the Brown & Sharpe plant, Greene wrote of an early 1880s addition that boosted the plant's square footage to 83,000 feet, more than two acres of manufacturing space:

> The factory is commodious, well ventilated, admirably lighted from all sides, and arranged especially for the work executed upon its various floors. ... Attached to the main building is the boiler-house, containing a battery of six boilers, of fifty horse-power each, also a case-hardening room with three furnaces. The first and second floors are devoted to the manufacture of heavy machinery, and the third floor to the manufacture of hair clippers and the Wilcox & Gibbs sewing machines, and to the finer classes of light machinery. Each floor of the building is a complete workshop, and is connected by telephone with the office, the other floors, the case-hardening room, and the foundry. The company use 586 machine tools, nine wood-working machines, nineteen polishing wheels, five smith shop hammers, seven foundry rattlers, and nine grindstones in the works.[2]

Greene described the foundry, the "pattern shop" where workers carved molding sands to cast molten metal, the smith's shop, and a factory library with several thousand carefully selected volumes; he concluded: "In these various buildings, under most skillful direction, and in the most thorough system, the manufacture of the various products of this company is conducted."

The various products now numbered about 240 different items. Brown & Sharpe published a new catalog in 1887; the index began with "American Standard Wire Gauge," the notched disc developed by Lucian for measuring wire in 1858, and ran for five pages to "Yarn Testers," used in textile mills to determine the breaking points of various yarns. In between were "Grindstone Truing Device, Hair Clippers, Height Gauges, Hole Gauges, Horse Clippers, Index Head for Milling Machines." Many of the 240 products, such as the micrometer, came in dozens of variations, bringing the product line to nearly 1,000 different pieces.

Stifled by Singer's market domination, sales of Willcox & Gibbs machines hit a plateau; but sales of Brown & Sharpe machine tools exploded. In 1886, the company's sales of machine tools—milling machines, screw machines, gear cutters, and grinders—more than doubled from the previous year, exceeding sewing machine sales for the first time. The growth of industries producing more sophisticated products such as cash registers, calculators, bicycles, and electrical equipment ignited the boom. In 1879, the year before Brown & Sharpe built its own foundry, the firm grossed $284,414. Seven years later the company pulled in more than $700,000, expanding its product line to include staples for other foundries such as furnaces, kettles, and foundry rattlers. Brown & Sharpe introduced its own rattler—a steel drum rotated to clean molding sands and to burnish castings—after trying existing rattlers at its foundry and finding them "so unsatisfactory that we have designed and constructed new ones." The company was so pleased with the self-flushing, trough-like urinals it made for its factory, that it offered its own line of "Sanitary Closets for Shops and Factories."

As more and more products tumbled out of the factory, more money rolled in. By the time of the 1889 Paris Exposition, sewing machines had become the poor stepchild of Brown & Sharpe's product line. They sold as well as ever, but their gross revenue trailed machine tool sales as the company's gross revenues crossed the million-dollar mark for the first time. The company consistently paid dividends, and Lucian used some of his money to diversify, buying land on Prospect Street near Brown University, shares in silver mines, and in 1885, he bought significant stock in the new Providence Journal Company, chartered to buy the oldest and best known newspaper of the dozen that then published in Providence.[3]

A few days shy of his 18th birthday, Lucian Sharpe, Jr., stood on the top platform of the newly built Eiffel Tower, fashioning paper "darts" (there being no such thing as air planes) to lob onto the crowded fairgrounds of the 1889 Paris Exposition. Only the boys, Lucian Jr. and 16-year-old Henry, accompanied their father on this European trip, as he groomed them to take the reins of the family business. Lucian's older girls might ride atop omnibuses and drive the horse-drawn mower at Nayatt, but running a machine shop in the nineteenth-century was strictly a man's business.

Lucian's trip to Europe with the boys was more businesslike than the 1883 trip had been. He hauled the boys to a stop that would not be on most tour itineraries: the Manufacture d'armes de Saint-Étienne, a government armory that had just begun making the Lebel Model 1886 rifle, a transformative event in the history of violence. The Lebel rifle fired a new kind of copper jacketed bullet, loaded into a metal casing that contained smokeless gunpowder. The copper jacket insulated a leaden core, making it possible to fire rounds at high velocities without the bullets shedding lead and fouling

Henry Sharpe, left, and his older brother Lucian in their teens. The boys accompanied their father to the 1889 Paris Exposition, as their father cultivated them to take the reins at Brown & Sharpe after his passing (courtesy Peggy Sharpe).

the barrel's rifled grooves. The French held the chemical composition of the smokeless powder as top secret. The propellant packed three times the combustible power of black powder and burned with barely a trace of residue, making it possible for the moving parts of newly automatic rifles to fire in rapid succession without fouling. Between 1886 and 1888, French government arsenals bought nearly 200 Brown & Sharpe machine tools as they geared up for production of these new rifles. Within 30 years they would churn out 2.8 million of them, many for use in the trenches of World War I.

At Saint-Étienne, Lucian and the boys boarded a steam tramway that carried them up to the door of the works. In his journal of the trip young Henry wrote:

> We first called on Col. Robert, the chief man of the establishment, to whom Papa gave his card. After waiting a few minutes, a man came and conducted us through the works. They employ from seven to eight thousand workmen, turning out 1400 [Lebel] rifles daily. The [Lebel] is a new magazine rifle, carrying ten shells at once, having a small caliber and firing about two miles. The powder, which is used produces little or no smoke, and [the] way it is made is kept secret by the government.

Though the powder recipe was closely guarded, the Colonel freely showed Lucian and his sons the rifle-making process. "I think the most interesting thing was making the forgings," wrote Henry, who saw drop forging for the first time. "They would place a rectangular piece of hot iron in a form, and drop a huge weight upon it, thus pressing it into the shape of the form."

Henry saw multiple Brown & Sharpe machines turning out gun parts, while his father showed his hosts how they could benefit from buying more: "Papa pointed out several machines where they lost a great deal of [motion]; for instance they had a screw machine that made 400 screws a day, while they make one at the shop, I think it is, which makes 1800." Brown & Sharpe's new catalogue advertised a light screw machine for sewing machine makers, a medium weight machine for "fire-arms &c," and a heavy machine to turn "all kinds of screws and studs" from bar iron.

At the Paris Exposition the boys spent time in the *Galeries de Machines,* or Machinery Hall, an iron-ribbed building that stretched for more than a hundred yards. "It is a very large and imposing structure," Henry observed. "The exhibits are countless— everything that one can conceive of, so many that while you see a great deal you really comprehend very little." The morning after Henry comprehended "very little" about the machines, Lucian again marched the boys into Machinery Hall, where Brown & Sharpe exhibited. "We wandered slowly about looking over machines that Papa called our attention to," Henry wrote, "but they were perfect Greek, as they say, to me." He liked a machine, "or rather a process" for making chocolate drops, but for the most part, Machinery Hall held little interest for a teen being groomed in the family business. Henry might have been happier had his grandfather stayed in the livery business. He was mad about horses, and expressed far more enthusiasm for the performing teams he saw in the Hippodrome to the tedium of machines chuffing away in Machinery Hall.

The trip was not all armories and machines, and the two Sharpe boys found plenty of fun in ascending the Eiffel Tower, the fairground's main attraction: a lattice of iron stretching more than a thousand feet into the sky. Its creator, Gustave Eiffel, hailed it as "the tallest edifice ever erected my man." The artists of Paris, decidedly less impressed than Eiffel, derided it as a "giddy, ridiculous tower dominating Paris like a giant black smokestack."

"We had no trouble in reaching the second platform where we arrived at 10.20 [p.m.]," wrote Henry, who scaled the tower with his brother and two other teens, "but had to wait there about an hour and a half. ... When half-way from the second to the third platform, you have to change cars. They don't allow you to go way up to the very top, but only 20 or 30 ft. below it."

"While up there we made one or two paper darts, like what we make in school, about a foot or a foot and a half long, putting coppers in the one end to weight them." The weight fell out of the first dart before it sailed any distance, but the second one flew beyond the roof of the tower's wide, arched base. "Then we threw down visiting cards; they would flutter about for a long time, taking considerable time to reach the ground."

From Paris, the Sharpes took a train into the French Alps to clamber across the glaciers at Chamonix. Along the way Lucian Jr. hired a man to fire a cannon used to trigger controlled avalanches in winter. "It made a big report," Henry wrote, "which after a pause reverberated back and forth among the mountains." After passing some mountain goats, the party reached the glacier's edge.

> We had not gone but a few rods when we had to pass one or two pretty dangerous crevasses; the guides helped us past these.... Towards the farther side we walked over considerable gravel which over[laid] the ice; this made better travelling. Where it was rather steep and slippery on the glacier the guides had cut out stairs beforehand. I think Papa felt somewhat relieved when we had gone way across; he says that this is his fifth time of crossing and the worst one at that.... I have forgotten all about last time, except that we had no difficulty. The guide says that it is getting worse every year.

In early 1890, Henry Leland told Lucian Sharpe about his dream to move westward and start a new business. After 18 years at Brown & Sharpe, two-thirds of them as head of the sewing machine department, Leland felt ready to strike out on his own. Unlike Joseph Brown and Oscar Beale, Leland held no passion for machinery design. His interest lay in production systems, how workers put a product together. He had brought piecework to the sewing machine floor, and had organized a storage system for screws and small parts so workers could churn them out in large batches rather than turn a set for each machine. He also introduced "operations sheets" into sewing machine production, which set in writing all the parts, tools, gauges, et cetera required to build each of the many models that Willcox & Gibbs sold.[4] At a glance, factory planners, procurement managers, and machinists could track the production of every job.

In the mid–1880s Brown & Sharpe began sending its foremen on business trips as salesmen and repair experts, the idea being to impress customers by sending them the best men. On one such trip Leland met a man in Michigan, Robert C. Faulconer, who had made a fortune in the lumber business. Faulconer, impressed with Leland's manufacturing know-how, wanted to take him on as partner in a new business out in Michigan, making trimmers to cut the wood frames that iron foundries used to mold sands into patterns.

Leland also mentioned to Lucian that he would like to take Charles Norton out west with him, Brown & Sharpe's expert on grinding machines. Norton had never liked the original design of the Universal Grinding Machine. Joseph Brown had died before seeing the first one, and the job of bringing it to fruition had fallen on a machinist named E.H. Parks. Norton found Parks' design too light, and had recently overhauled the Universal Grinding Machine to make it heavier, more rigid, and more stable. He also developed a new spindle for grinding the hard-to-reach inside surfaces of products such as cylinders in the Westinghouse airbrake. Through his work at Brown & Sharpe, Norton had set aside enough savings to invest with Faulconer, but Leland was a little short on his stake. He wondered: Could Lucian loan him some money?

Lucian agreed to the loan, with a strict payback schedule, and late in 1890 the firm of Leland, Faulconer, and Norton launched in Detroit. The company hit a snag when Leland could not find a Detroit foundry to make castings to his liking, forcing him to bear the expense of shipping castings by rail from the Builders Iron Foundry in Providence, owned by Lucian's son-in-law.[5] Norton became discouraged and returned to Brown & Sharpe, but Leland doubled down: when he could not make his payments to Lucian he asked his old boss, Richmond Viall, to pay Lucian and assume the loan. Viall agreed, underwriting a partner in a firm that would quickly morph from making wood grinders to bicycle gears to marine engines into the Cadillac Motor Company.

While Leland and Norton plotted their move to Detroit, Lucian's thoughts also turned to the west. He never missed a chance to promote Brown & Sharpe at a world's exposition, and the United States would host the next one in 1893. The question was: Which city would hold it? In the spring of 1890 a bidding war broke out between four cities that wanted to stage the fair, New York, Washington, St. Louis, and Chicago. With $15 million in pledges from the likes of J.P. Morgan, Cornelius Vanderbilt, and William Waldorf Astor, New York gained an early edge. But Chicago needed this World's Fair. A few years before a protest by tens-of-thousands of the city's workers, marching for an 8-hour day, had ended in a murderous riot when someone threw a bomb into the crowd gathered at Haymarket Square. The Haymarket Riot, which killed seven policemen and about as many civilians, exposed the volatile class conflict of the Gilded Age. The first World's Fair, London's Crystal Palace Exhibition of 1851, had successfully galvanized the British public in support of global expansionism and against political radicalism. The United States, and Chicago's elite in particular, hoped to use "the medium of the world's fair to provide the cultural cement for their badly fragmented societies."[6] The meat packers Philip Armour and Augustus Swift, along with Cyrus McCormick of McCormick reaper fame and department store magnate Marshall Field, pooled their formidable resources to put Chicago into the mix. In a frenetic, 24-hour period in April of 1890, Lyman Gage of The First National Bank of Chicago raised $5 million, securing "The Columbian Exposition" for the booming, stockyard city.

"As soon as we decided to go to Chicago, which was about a year before the opening, Mr. Zechariah Chafee was very much impressed with the fact that we should put our catalog into shape," recalled William Viall. "…The plan was laid down that the catalog should be put into shape, and I was given the job." The existing catalogue was a hodge-podge of fonts and various type sizes. Whenever the company brought out a new product it added a new page without regard to standardizing style, resulting in a catalogue that "had little to commend it as a good piece of typography." William Viall set out to rectify that, ordering new engravings of tools and machines, and setting completely new pages of like fonts and sizes.

One of the fresh woodcuts showed Brown & Sharpe's newest machine tool: an automatic screw machine. Throughout the 1880s, Brown & Sharpe had built automatic screw machines that relatively unskilled workers could operate after their initial, complex setup. But the firm only began selling them to other companies in 1890, making it a late-comer in an increasingly important market for easy-to-use screw machines. The Hartford Machine Screw Company, founded by Christopher Spencer shortly after he invented a revolutionary automatic screw machine, had been manufacturing the machine for nearly two decades. Brown & Sharpe had some catching up to do, and the Columbian Exposition would be a good place to do it. William Viall, son of the factory superintendent, and Henry Sharpe, a Brown University student, rode a train to Chicago on a scouting trip. They arrived late at night, checked into a small hotel, then took a few drinks before wandering to see "the lay of the ground." Viall wrote: "Later we learned we had done about as unwise a thing as we could have done, for the neighborhood was a pretty rough one."

Samuel Darling wanted out. He was now, in 1892, 77 years old, ready to step back from the demands of daily work. Lucian Sharpe was happy to oblige Darling's request to leave the company, for as skilled as Darling was he was sometimes as much trouble as he was worth.

Darling waged a public campaign against vaccination. William Viall recalled, "He hammered the Legislature with literature and valuable books, and worked in season and out of season butting his head against a stone wall." He sometimes went off on tangents, inventing useful products—a pencil sharpener of the type that became ubiquitous on schoolroom walls, built of interlocking, sharpened steel cylinders, and an improved inkstand purchased by the U.S. Army—but he held these patents independently of Darling, Brown & Sharpe. Earlier in their partnership, Darling had agreed to pay Lucian $500 in compensation for time spent on his own business.

According to the reminiscences of William Viall, Darling had also committed a business blunder which must have been galling to Lucian. In 1877 a deaf farmer and inventor, Leroy S. Starrett, asked Darling whether he would be interested in manufacturing his new invention: a combination square, which Starrett billed as an improvement on the try square. With a try square—two thin flat surfaces joined to form a right angle—a machinist could mark and measure straight lines and 90-degree angles; but a combination square came with multiple heads that allowed the two surfaces to swivel, giving the tool versatility as a protractor, or to determine whether a surface was perfectly flat, or find the depth of a hole in a surface, and multiple other uses. Darling thought that the attaching heads would wobble ever so slightly, introducing inaccuracies into the measurements, he was not alone in this belief. The tool builder that made Starrett's first model told him, "I would not give a damn for it."[7] Darling, too, "would not think of making" the combination square. "It simply was not square, and that is all there was about it."[8]

After Darling turned down his offer to make the combination square, Starrett decided to produce it himself. He borrowed money from a bank and hired a shop in New Britain, Connecticut, to make a batch of 5,000; after initial resistance from tool dealers he sold enough to open his own shop in Athol, Massachusetts, then branched out into other types of tools. By the time of Darling's retirement, Starrett Tool had become Lucian's nemesis, presenting a real threat to Darling, Brown & Sharpe's share of the precision tools market.

In December 1892, Lucian bought Darling's interest in their partnership. He now owned all of Darling, Brown, and Sharpe, and everything except for the three-sixteenths share of Brown & Sharpe that Joseph Brown had left for his wife, Jane, and daughter, Lyra. He had elevated his son-in-law, Zechariah Chafee, to corporate secretary, but still held the stock in his own name. Lucian also owned more shares than anyone else in Willcox & Gibbs, a healthy company that had past its peak but was still steadily selling quality sewing machines churned out by Brown & Sharpe. Lucian, once an 18-year-old apprentice and the son of a bankrupt livery owner, had weathered two panics; a civil war; the death of the company's co-founder, and an attempted hostile takeover to build a global business that made him a very wealthy man. But even as he bought out Samuel Darling in 1892 the seeds of the next panic were sprouting in the December darkness, seeds sewn by a fellow Rhode Islander—the powerful senator Nelson Aldrich.

9. Spoke Wheels, Turning

Nelson Aldrich confided to a friend that he planned to give up his seat in the United States Senate. For a man with eight children and grand ambitions, the job just did not pay enough. So in 1890 he told his friend, a Rhode Island utilities magnate named Marsden Perry, that he planned to retire from the Senate when his term expired in two years.[1]

The possibility of losing Aldrich in the Senate represented a real threat to Theodore Havermeyer and John Searles, Jr., of the American Sugar Refining Company, better known as The Sugar Trust. Aldrich hailed from the smallest state but with a seat on the Senate Finance Committee he wielded enormous power, particularly when it came to influencing votes on setting tariffs. The Sugar Trust held a wish list of changes in the tariff schedule, including elimination of tariffs on imported raw sugar, and Aldrich had proven to be a reliable ally of the big sugar producers.

As a wholesale grocer who had dealt in bulk sales of sugar, Aldrich was well-versed in the arcane rules governing the importation of sugar, raw and its multiple grades of refined. He spoke authoritatively on the subject, and his fellow senators gave his words a lot of weight. In 1883 and again in 1887 he almost single-handedly managed to win approval of proposals backed by Havermeyer and Searles. With another tariff act coming due in 1890, the Sugar Trust wanted to ensure that their best ally, who now chaired the Finance Committee, remained in the Senate.

Aldrich again came through for the Sugar Trust in 1890, shepherding through the McKinley Tariff that entirely eliminated raw sugar from tariffs while slightly increasing the cost of importing refined sugars, policies that hurt importers of fine sugars so they could not compete with the Trust's big, domestic refineries. The Senate appeased Louisiana sugar growers, who favored tariffs on raw sugar grown overseas, by offering them government subsidies worth millions of dollars for their homegrown crops. By completely removing tariffs on raw sugar, lawmakers stripped the federal government of $50 million a year, 15 percent of its annual revenue, a problem the Senate compounded by paying the domestic growers millions in subsidies.

Naturally people began to wonder: Why was Aldrich so obviously doing the Sugar Trust's business? The answer lay in the papers of Nelson Aldrich. In a contract dated February 2, 1893, Searles and his associates in the Sugar Trust agreed to give Aldrich between $5.5 million and $7 million in cash, money that he used to buy up the four horse drawn rail lines that carried people around metropolitan Providence.[2] In a single day Nelson Aldrich, wholesale grocer and U.S. Senator, became a multi-millionaire by cashing in on the power of his office. By giving Aldrich millions the Sugar Trust got what it wanted: a reliable, powerful friend in the Senate. Aldrich would not relinquish his Senate seat for another 21 years.

With the Sugar Trust's money, Aldrich and his associates bought the four Rhode Island street car lines for less than $3 million, giving them a monopoly on metropolitan Providence's public transit in a time when almost everyone relied upon it. They incorporated in New Jersey, where incorporation terms were most favorable, as The United Traction and Electric Company. And they brought in Marsden Perry, Aldrich's friend, as a partner. As owner of both the Narragansett Electric and the Rhode Island Electric Lighting companies, Perry held a monopoly on electricity service in Rhode Island, a perfect partner for switching the trolleys from horse drawn to electrical power.

The United Traction and Electric Company built vast improvements into the public transit system, laying down 218 miles of electrified steel rail that stitched together Providence, Pawtucket, Cranston, North Providence, and Warwick, carrying by the end of the decade 53 million passengers a year.[3] The UTE's monopoly proved vast, efficient, and immensely profitable. Its stock more than tripled in seven years, making Aldrich a multi-millionaire without investing a cent of his own money. The problem was that in building a public transit monopoly with graft obtained from the Sugar Trust in exchange for a tariff waiver, Aldrich and his cronies depleted the finances of the United States, precipitating the Panic of 1893.[4]

For a few years the sharp drop in federal revenue precipitated by the loss of sugar tariffs passed largely unnoticed, propped up by high wheat prices, a commodity America grew in abundance. Farmers, banks, and railroads did a booming business while an early 1890s famine wracked southern Russia. But in 1893, Russia produced a banner wheat crop, glutting the international market and depressing wheat prices; American farmers could not pay mortgages.

J. Henry Drury, one of Brown & Sharpe's traveling salesman, arrived in Chicago in March of 1893, in order to install the company's exhibition on the grounds of the Columbia Exposition, the decade's World's Fair. In 17 years with Brown & Sharpe, Drury had worked his way up from polishing screw heads to bookkeeping to establishing a network of dealers to sell the firm's small tools in upper New York State. He was, in the words of William Viall, "a typical salesman in that he was always happiest when he was on the road."

While setting up for the World's Fair, Drury ran into George Beale, brother of the mechanical genius Oscar Beale. Both Beale's grew up in Maine, and for a decade they had worked together at Brown & Sharpe before George moved west, to Chicago. While Oscar was studious and serious-minded, George was "anything but studious and always ready for a fight or a frolic."[5] Drury noticed that people who stopped by the Brown & Sharpe booth liked talking to George Beale—he possessed a dry, Maine wit, and through his 10 years as a Brown & Sharpe worker he was well-versed in the company's products. So Drury hired him back as part of Brown & Sharpe's Chicago exhibition team.

Chicago's Columbia Exhibition opened in May, but the fair did not really get cranking until its star attraction, the first Ferris Wheel, finally began turning in mid–June. The exposition's directors wanted a skyline feature that would rival the Paris Exposition's Eiffel Tower, and they felt they had it when they commissioned a great, turning wheel from a young engineer named George Washington Gale Ferris. But Ferris's wheel turned out to be an ungainly thing to build.

The axle required to turn the whole contraption weighed nearly 90,000 pounds; forged at the Bethlehem Iron Works, it arrived on the fairgrounds in mid–March when the ground was still frozen a yard deep. Workers smashed through the frost to lay 700 feet of pipes connecting the axle to the big steam engines that would power it. When the turnstiles opened on May 1, pieces of the wheel lay scattered about the Midway Plaisance, a park on the south side of Chicago set aside to host the fair's amusements. (The Midway as synonymous with amusement park became a new term in the American lexicon.) Workers connected spokes fastening the axle to the great, iron wheel, then day after day hung the 36 passenger cars on it. Each car stretched 24 feet wide, stood 10 feet tall, and held 60 passengers.

Ominously, a week after the World's Fair opened one of Chicago's largest banks failed. Besides stripping federal coffers of sugar tariffs, the Congress of 1890 had also approved a bill requiring the treasury to buy silver from western mining interests at inflated prices. As the U.S. Mint collected over-priced silver, investors cashing in government bonds and other securities demanded payment in gold rather than silver, skewering the ratio between the two metals held by the treasury. International investors, aware of the depletion of America's gold supplies and its dwindling revenue stream became leery of the nation's ability to guarantee its currency; they tightened credit extended to American businesses. Throughout the month of May, stock prices plummeted.

With his big wheel finally assembled on June 21, Ferris blew a golden whistle. The Iowa State Band struck up "America," and the Ferris Wheel began its slow turn. The wheel itself stood 264 feet at its apex; at dusk, thousands of Edison electric lights blinked from its frame, and the Columbian Exposition finally showcased a feature of spinning, blinking, iron and steel to rival Eiffel's tower.[6] More than 70,000 exhibitors set up displays at the Columbian Exposition, and for the first time a World's Fair included national pavilions, with 46 nations participating. "The Germans made a considerable exhibit in Chicago," recalled William Viall, who scrutinized the exhibits from the German metal working firms to see how they stood up to Brown & Sharpe's. A firm called Krupp showed off its work with a "Gun Pavilion" bristling with artillery, including a shiny cannon 46 feet long.

Brown & Sharpe hoped to promote its new automatic screw machines and its latest line of milling machines, and had ample opportunity to do so: 26 million people, drawn in part by the amusements along the Midway, toured the fairgrounds, a good crowd; but as the summer wore on Drury and Beale found it hard to make a sale. In fact, during the six months of the fair Brown & Sharpe sold just one machine tool, a small lathe bought by the Burroughs Adding Machine Co. of St. Louis. William Viall recalled of the Summer of 1893:

> It then looked as though the whole business fabric of the country had gone completely to smash. … The wretched business conditions carried on all of the summer of 1893 and the following winter, bread lines were formed throughout the country and here in the city [Providence] men were set to work shoveling gravel in order to obtain a pittance.

Brown & Sharpe, which began the year with nearly 1,000 employees, cut its workforce to 617. "It was a winter of many tragedies," Viall wrote. "High-grade toolmakers and artisans were offering their services at helpers' pay, if only they could get something to do."[7]

Unlike the previous panics, the Panic of 1893 was largely confined to the United States; though acute it was relatively short-lived, particularly for Brown & Sharpe which had done a good job cultivating overseas markets. Even during the height of the Panic, when sales of machine tools and other goods dropped, sales of the reliable Willcox & Gibbs sewing machine increased, snatched up by foreign knit-goods factories to stitch underwear, hosiery, and lace curtains. Contacts made at the World's Fair, particularly among a growing Chicago firm called Western Electric and the German metal firms, helped lift Brown & Sharpe out of the doldrums of 1893. In the latter half of the decade, Germany became Brown & Sharpe's third largest foreign consumer of machine tools, behind France and the United Kingdom. The Czar's Russian arsenals also bought more than 300 machines.

American companies that packaged food further drove sales of machine tools. Through the late 1880s and 1890s, U.S. exports of manufactured food far exceeded foreign sales of manufactured goods.[8] In the mid–1890s A lawyer named Henry D. Perky contracted with Brown & Sharpe for customized machines that could shape his concoction of wheat—boiled, baked, and formed into a biscuit—that he marketed as a breakfast food called "Shredded Wheat."

"In 1895, business revived to such an extent that it became a boom," Viall recalled decades later. "The effects of German visitation were apparent, and the bicycle had come in a similar manner to what we, of this generation, know of the automobile."

As America came out of its self-inflicted panic, its factories also began snatching up machine tools. Companies making telephones and other electrical equipment, such as Western Electric and the new General Electric of upstate New York bought the most machine tools; rival machine tool firms also bought Brown & Sharpe machines to help them build their own tools, and the "Bicycle Craze" of the mid–1890s spurred the sale of at least 267 machines while literally paving the way for the rise of the automobile.[9]

Whatever the United Traction and Electric Company wanted, it got. When the monopolistic UTE asked the Providence City Council's Committee on Railroads for permission to install raised streetcar tracks in lieu of more expensive recessed track, the Committee acceded, a decision that sparked a massive protest by the city's bicyclists. On a May day in 1897 a swarm of bicyclists, 40,000 according to a newspaper report, rode through the streets of Providence to denounce the council's decision.[10] The protestors rode at the apex of the Bicycle Craze, a mid–1890s phenomena that revolutionized transportation and industry. The bicycle proved to be a "transitional technology" that laid the technological groundwork for automobile manufacturing in the early twentieth century.[11]

Brown & Sharpe draftsmen in the mid–1890s. With the "Bicycle Craze," bicycle manufacturers ordered at least 267 machines from Brown & Sharpe, quickly bringing the company out of the doldrums caused by the Panic of 1893.

The bicycle's roots in America stretched to the Centennial Exposition of 1876, where a Boston venture capitalist laid eyes on a British "ordinary," a bicycle with a front wheel nearly five feet high. The capitalist, Albert Pope, steamed to England where he toured the countryside on one of these high-wheel bicycles; he enjoyed it so much he shipped eight of them back home. He then set about trying to manufacture them in America, hiring the Weed Sewing Machine Company of Hartford to cast and assemble the parts. Weed's factory reluctantly accepted the challenge because it needed the business, owing to Singer's dominance in the sewing machine field. Weed tackled the bicycle job in the summer of 1878, producing by year's end 50 bicycles for Pope's Columbia Bicycle Company. Through trial and error, workers grew more adroit at building the bicycles so that by 1881 they churned out more than 1,000 a month to meet a demand almost entirely created by Pope.

Pope proved a master marketer. He hired a young Maxfield Parrish to illustrate eye-catching bicycle posters, published magazines called *Bicycle World* and *Wheelman*, organized popular trade shows, and formed a group called The League of American Wheelman. Still, the high-wheeled bicycle was a tough sell. Expensive, fast, and risky to ride, the high-wheel had a limited market as a play thing for rich young men.

In 1887, England exported another kind of bicycle: this one featured two wheels of equal size connected by a chain drive, the "safety bicycle." Anyone could ride these bicycles, and for awhile it seemed as if almost everyone did. Pope's American patents for bicycles expired in 1886; once new manufacturers no longer had to pay him $10 per bike, New England metal workers began using some or all of their capacity to make safety bicycles: firearms makers, the Ames shovel manufacturer, sewing machine companies, and Waltham watch makers all began bicycle production.

Bicycles presented complex technological challenges. They required cogged sprockets connected by riveted chains, with axles and ball bearings and spoke wheels that had to be ground smooth so as not to puncture the new, pneumatic tires that provided a cushioned ride. To produce bicycles, shops that had been tooled up for other kinds of metal work needed to add a "number of tools made by the New England machine tool companies such as Pratt & Whitney and Brown & Sharpe."[12] Joseph Brown's invention of the universal grinding machines made it possible to produce ball bearings in commercial quantities for use in bicycle wheels; his universal miller was of inestimable use in "hobbing" or cutting sprocket teeth. Screw machines, hand-fed and automatic, evolved beyond their initial use as makers of screws, proving useful for shaping metal bar stock into wheel hubs.

At the peak of the Bicycle Craze in the mid–1890s, manufacturers made 1.2 million bikes annually, which explains how metropolitan Providence was able to turn out tens of thousands of riders for a protest.[13] With financial backing from Pope, the League of American Wheelman (LAW) grew from a cadre of elite young men into an influential lobbying group similar to the later American Automobile Association. The LAW successfully pushed New York City to allow bikes on streets and in Central Park, and launched a Good Roads campaign that resulted in widespread legislation to pave and extend roadways, creating America's first highway network.

In the final three years of the decade the Bicycle Craze came skidding to a stop, a victim of its own success. Bicycles freed people from the rigid time tables of railroads and streetcars—you could ride when and where you wanted, and right up to the door. Hiram Percy Maxim, hired by Pope to develop an electrically powered, four-wheeled "quadricycle" wrote that the bicycle "directed men's minds to the possibilities of independent, long-distance travel over the ordinary highway. ... Then it came about that the bicycle could not satisfy the demand which it had created. A mechanically propelled vehicle was wanted instead of a foot-propelled one...."[14]

Inventors tried different methods of propelling their vehicles: Pope bought the Weed Sewing Machine works in Hartford, and in 1895 converted it to making electric cars; that same year a German engineer named Gottlieb Daimler improved an internal combustion engine to drive a modified stagecoach; and the Maine twins Francis and Freelan Stanley employed steam engines to propel a car, driving one of their hissing contraptions to the summit of Mount Washington in 1899.

Susan Hale, a summer resident of Matunuck, Rhode Island, tried a *horseless carriage* (her emphasis) in 1898, and wrote from New York: "they are quite common here now, and no dearer than a cab." She felt awed sitting in the car "looking out into space with no dasher, nor reins, nor tail, nor any other part of a horse, in front ... rubber tires, noiseless springs, the thing glides along avoiding teams and everything. It's glorious."[15]

The most efficient, or most popular, way of powering mechanically driven cars still needed to be worked out, but for the first time in human history people could roll over roads without the sweaty, flatulent flanks of horses twitching in their faces. It was "glorious." The bicycle could not compete with this; more quickly than it had burst on the scene the Bicycle Craze faded, leaving the bicycle as more of a novelty and a child's plaything than as a serious source of transportation. But in its brief life the Craze had spawned a new technology and a greater demand for machine tools.

Charles Norton felt humiliated. He had asked for a meeting of his bosses and colleagues at Brown & Sharpe to explain what he thought was a good idea—a larger, heavier grinding machine—and they shot it down as wildly impractical. "I cannot tell my feeling when I realized I was being ridiculed," Norton recalled. "That conference broke up very soon."

Norton may have wished he had never come back to Brown & Sharpe after leaving to help his friend, Henry Leland, start a metals working firm in Detroit. Now in the waning months of 1897, Leland and Faulconer was doing well making bicycle gears and a one-cylinder marine engine for boats plying the Great Lakes, while he was struggling with entrenched bosses who refused to see the worth of his new idea for grinding machines.

At the time he called for the meeting, Norton was already on the hot seat with the factory superintendent, Richmond Viall. Norton had talked Viall into letting him make a set of six heavier, plain grinding machines on the promise that they would weigh no more than 1,900 pounds, the weight of the firm's largest Universal Grinder. When the new machines topped 2,000 pounds Viall scolded him for unnecessarily making a machine heavier and more expensive without giving the customer anything extra to justify the expense.

The crux of the issue was Norton believed grinding machines could serve as cutting machines that could lop off metal and increase production by saving time, while Viall viewed grinding machines only as a useful last-step for attaining precision. Norton felt that a heavier machine with a bigger grinding wheel could more than make up for its added expense by taking bigger bites out of steel so a machine operator could finish an entire cut without having to pass the piece onto a lathe. By reducing one step, a bigger grinding wheel could save time and money. Norton's idea ran contrary to the Brown & Sharpe canon, which held that a large grinding wheel could cause the piece being ground to "chatter," or vibrate, compromising precision. It was set in black-and-white in the firm's Blue Book, *Construction and Use of Grinding Machines:* "A wide cutting surface will cause the work to chatter more or less and, therefore produce an irregular surface."

A month or two after being laughed out of the meeting, Norton asked for another conference; this time he suggested that if you built a grinding machine heavy enough, you could not only make it more productive as a cutter, but more precise. "We could make an actual micrometer out of the machine itself," Norton said, "and when the operator moved the index .00025 [twenty-five-thousands of an inch] it actually reduced the piece .00025." His colleagues told him his idea was "ridiculous."

Norton skipped a day of work and took a train up to Worcester, where he found a new job with the Norton Emery Wheel Company founded some 30 years earlier by a Franklin Norton, no relation. The company was spinning off a new business, the Norton Grinding Company, to make grinding machines, and they wanted his expertise. When he told Viall of his new job, Viall retorted, "If the Norton Emery Wheel Company are going to make grinding machines, we will give them away." The Norton Grinding Company did proceed to make grinding machines, big, heavy machines that spun grinding wheels far larger and thicker than the wheels on Brown & Sharpe's machines. A centrifugal pump drew coolant from a large tank to cool the steel and prevent heat distortion. Car companies found them perfect for grinding crankshafts. A Norton grinder could grind in 15 minutes work that had previously taken 5 hours; Henry Ford eventually placed an initial order for 35 Norton grinders, ushering in the era of true mass production.[16]

By the time of Norton's departure in the mid–1890s, Brown & Sharpe was no longer the collection of smaller, almost artisanal shops it had been when the new factory opened in 1872. Oscar Beale had carte blanche to create whenever the mood struck him; but most of the ingenious old timers were gone: Joseph Brown, the mind behind the universal milling machine and universal grinder, was dead; Samuel Darling, the eccentric inventor who could not help inventing, had retired, and the talented crew he had brought with him from Maine had been dragged out of their warren in the No. 1 Building and forced to mix with the rest of the small tools builders in the new Building No. 3.

Norton ran afoul of a new conservatism in the shop that encouraged productivity over creativity: machinists got paid by the piece, putting a premium on speed; the operations sheets pioneered by Leland laid out all of the tools, fixtures, gages, and steps necessary for a job, requiring machinists to work more by rote than by feel. Since moving to Promenade Street, Brown & Sharpe had added two buildings and a foundry covering three city blocks. The enormous size and complexity of the factory required systematic change and scientific management to impose order on the old, chaotic way of doing things.[17] As soon as he returned from the Chicago World's Fair, William Viall had noticed the change. Though the Panic of 1893 had still held the nation in its grip, foreign sales of sewing machines kept that department busy, and Viall's father placed him in charge of it. "My job was to see that the material moved in sequence and kept moving," Viall recalled. "It was probably the first time that the shop had ever attempted systematically to follow up on a job in the manner that is so well known now."

With Darling out of the factory, Lucian Sharpe, Jr., now in his 20s, decided to increase the efficiency of the small tools department by moving all of the graduating machines, the micrometer workers, and gauge makers onto the same floor. This required breaking up Darling's quirky old shop on the first floor of the No. 1 building, and moving the men and machines into the new, No. 3 building rising on the hill behind it. The change was more than some of the old-timers could take. In August 1895, Calvin Weld told William Viall, "Now we have stayed with you until you are moved, but we should like to get through." He retired, and his older son, George, left with him. Charles Weld stayed on the job, running the large wet grinder into the mid–1930s. Had the Welds left in the 1870s, they would have been irreplaceable; no one could run their balky old grinding machines the way they could. But with the invention of the universal grinder, much of the skill for grinding light pieces had transferred from the mechanic to the machine.

All aspects of the factory were fair game for the "new factory system," including the age-old apprenticeship program. As soon as Lucian Sharpe, Sr., became a partner in Joseph Brown's shop, he began keeping detailed records of apprentices hired by the firm. A leather-bound book of blank pages, neatly lined in columns of red pencil, held the names, hometowns, and service terms of every teenage boy who signed on as an apprentice between 1858 and 1902, more than 300 names in all.

For most of the nineteenth-century, the apprenticeship agreement was nearly identical to the one that Wilkes Sharpe signed to indenture his son to Joseph Brown. For a fee, Brown & Sharpe would take a boy into its shop and teach him the trade. The length of the apprenticeship fluctuated from three to four years, but the basic parameters were the same: apprentices would learn to use the tools of their trades, but were on their own to develop others skills such as drafting or algebra that might help them advance.

While Lucian Sharpe had the chance to work directly under Joseph Brown in a small, artisanal shop, apprentices joining the firm after the Civil War were supervised by shop foremen. The foreman of each floor and department—sewing machines, milling machines, the foundry—shaped the overall apprenticeship experience, often with vastly different results for the apprentices and for the company that hoped to train them as good workers. In the mid–1890s, Richmond Viall decided that the apprenticeship program needed standardization across departments. As his son, William, recalled: "He became convinced that the system, as we were then carrying it on, was subject to too much chance, that boys were not systematically moved about...." Boys in the program received too much training in one area such as grinding machines, and not enough in overall shop practice. That worked well when grinding required the deft touch of the "Silk Hat Machinist," but a person of modest mechanical ability could now learn most any machine. Factory managers no longer needed specialists; they sought workers with the versatility to work across all departments. Smaller companies drifted away from apprenticeship programs, leaving new workers to learn the machinist's trade through schools. Viall did not want to scrap the apprenticeship program, he wanted to overhaul it so that all apprentices reported to a single instructor who could ensure that the boys learned how to work on multiple machines. Foremen viewed Viall's plan for a supervisor of apprentices as a threat to their authority. "At first blush it did not take at all well," William Viall recalled. But Richmond Viall persisted, winning over the foremen one by one to adopt a company-wide apprentice system.

The new system codified some past practices, and changed others. Applicants for admission to the apprenticeship program still had to be white males 16 to 18 years old; now they needed at least a grammar school education. The company preferred to interview applicants in person, a visit that included measuring height, weight, hearing ability, and sight. If an applicant could not visit, he had to submit a photograph with "particulars" of his health—sight, hearing, and weight. There was no written rule against hiring blacks as apprentices, it just was not done. Brown & Sharpe did not employ a black worker until it hired Jesse Bradley as a foundry worker in the 1920s. The International Association of Machinists did not admit blacks until forced to do so by the Roosevelt administration in the 1940s.

A "boy" accepted into the program worked an eight-week probationary period; if he passed, his parents or guardian paid Brown & Sharpe $50 for four years of training in the machinist's "art and trade." Apprentices worked regular shop hours, 60-hour weeks, earning 6 cents an hour the first year, increasing each year to a high of 14 cents an hour. At the end of the apprenticeship, each received a lump sum of $150, a bonus that protected the company's training investment by encouraging the apprentice to complete his term.

The company could fire apprentices for "improper conduct within or without the shops." Apprentices could not smoke or chew tobacco at work or even at home, lest Richmond Viall find out about it. Viall, according to his son, kept "in close touch with the apprentice boys, [and] knew most of their life history...."

Shop rules included no "long coats or other substitutes for the jumper and overalls," no "sneaks or tennis shoes," and "to leave the floor permission must be obtained from the Foreman." Requests for leaves of more than a day required parental consent. Oscar Beale wrote *A Handbook for Apprenticed Machinists* that warned the boys: "If a workman makes a mistake when running a machine the machine never excuses him." He wrote of one workman who loosely clamped a piece of steel; when it broke loose, he lost his thumb. Another man in the Promenade Street factory had "bent over to look into the place where he had just been planing; his leg pushed the operating lever, the planer started, and the tool plowed deep through his skull."[18]

Besides instructing the boys in the practical matters of safety, calculating gear speeds, and precisely subdividing screw threads, Beale's handbook advised them on the finer points of shop culture: "The use of a monkey-wrench for a hammer indicates poor taste; and to jam a piece of finished work in a vice or under a set-screw proves that a man lacks mechanical ability.... Any man to whom a bad job is not a lasting mortification, shows himself deficient in self-respect." Conceding that the machinist's trade was difficult, dirty, and dangerous, Beale impressed upon the boys the importance of their chosen career: "the comfort, convenience, and safety of every person depends more or less upon well-made machinery."[19]

As part of the apprenticeship overhaul, Brown & Sharpe encouraged the boys to form a self-governing Apprenticeship Association that met bi-weekly for 90 minutes on company time. During their meetings the teen apprentices held lively debates, listened to guests lecture on the latest technology, and learned the hard work of running a democratic organization with elected officers, recorded minutes, and rules of order. The Association kept its minutes in a leather-covered book of blank pages bound with string; over the years its pages darkened with machine oil smeared on by palms and finger tips of boys fresh from the shop writing and reading accounts of their meetings:

"Mr. C. R. Burt spoke of the accident of Geo. Hanna about the losing of his time," an early entry read. "Motion was made and seconded that the investergating [sic.] committee see Mr. Viall about it. Carried. An amendment was [made] to same to see about getting his pay carried."[20]

Professor Herman Carey Bumpus of Brown University spoke to the boys about newly discovered X-Rays. "In order to illustrate his subject the Professor brought with him an electric machine and other apparatus for producing X-Rays," the secretary wrote. "The machine, however, failed to work. The atmosphere was so damp that it absorbed all the electricity the machine could produce, and consequently left none for the X-Rays. To make up for the failure of the experiment a number of X-Ray photographs were passed around the room."

Professor Estes of the Rhode Island School of Design spoke on a less sexy subject—Anti-Friction Bearing, "and the lecture was very interesting," the secretary wrote. "It was a subject with which most of the apprentices will have to deal with sooner or later and strict attention was paid to the speaker with the exception of two or three."

Meetings could be quite lively, and keeping the peace at a meeting of a few dozen teenage mechanics sometimes proved to be a challenge. "A suggestion to appoint a sergeant of arms to keep order in the back of the room was laid over until the next meeting," read an early minutes book entry; though the boys rejected that suggestion, they may have regretted it in July of 1899, when Oscar Beale spoke about worm gears, hob cutters "and there [sic] relation to one another and also some experiments and their results." The meeting adjourned "After some very ungentlemanly conduct by some members...." Two apprentices washed out of the program the following day, and Viall suspended the Apprentice Association for awhile, until the boys petitioned for "the restoration of our meetings. In return for same we agree to carry on such meetings, in a business like manner."

As long as the boys behaved, Viall gave them wide latitude in the subjects they chose to debate. They tackled the question: Is the boy who learns his trade with Brown & Sharpe better qualified and more capable of filling positions of responsibility than one who learns his trade in any other similar establishment in the United States? That debate "proved to be a success there being a large attendance [and] the speakers remarks called forth applause frequently." At the end of the day, Lucian Yeomans and Charles Osgood provided the better argument that Brown & Sharpe boys were not necessarily better than those trained anywhere else. In January 1896 the boys debated: "Do labor organizations benefit the working classes or not?" The boys, working a dangerous job, without insurance for, at the most, 14 cents an hour, decided that labor organizations did indeed benefit the working class.

10. Death and Succession

The lecture room inside the Brown & Sharpe factory looked like an arms depot. Spread across tables were rifles, the Springfield and the Krag-Jorgensen models, revolvers, and belts of bullets for the newly invented "machine gun." An array of artillery shells stood at the front of the room, packed with explosives and weighing from 1 to 800 pounds.

The Spanish-American War was on, and the boys of the Brown & Sharpe Apprenticeship Association had asked George Owen, Jr., a former colleague who was now a government ordnance inspector, to lecture about the latest in weaponry. The oil-smeared minutes of that September 23, 1898, meeting reported: "That the subject of his lecture was a popular one, was evident from the fact that besides a full attendance of the members, there were present a number of Visitors as well of reporters from the Providence Daily papers."

Owen demonstrated weapons showing the latest in metallurgy—nickel-plated bullets which could spiral through a rifled barrel and "not tumble over"; the new style "knife bayonet" of the Krag-Jorgensen, forged from a much lighter metal than bayonets for the Civil War–vintage Springfield; artillery shells that held primer, smokeless powder, and projectile all in one casing for quicker, cleaner firing; armor-piercing shrapnel with "a good deal of killing power" cut from "very carefully treated" steel.[1]

Among the visitors in the gallery that day stood Lucian Sharpe, Jr., who, at the end of the meeting, "requested that the members of the association be allowed to examine the different fire arms & projectiles" by hefting the guns, cartridges, and bombs arrayed about the room. As the older of Lucian Sharpe's two sons, Lucian Jr. was the obvious choice to take over the company's reins from his father. But the elder Lucian began to worry that mental illness might be affecting Lucian Jr.'s judgment. For example, one day at lunch hour young Lucian appeared on the factory roof with a box of 100 expensive, Mandarin oranges at his side. As workers milled about on lunch break he lobbed oranges into the throng of mostly men, who violently wrestled each other for the taste of good fruit.

Lucian Sr., expressed doubts about his eldest son to Richmond Viall, the longtime factory superintendent. Lucian had placed both of his boys under Viall's supervision for "good shop training," a kind of abbreviated apprenticeship. In *his* time on the floor, Lucian's youngest son, Henry, had expressed to Viall serious doubts about following his father into the metal-working business. Henry "had times when he felt that he was not fitted for the business...," recalled Viall's son, William, who worked with both of the Sharpe boys. At those times Richmond Viall stepped in with encouragement that "kept him at the task," helping Henry cultivate a "liking for the business."[2]

With Lucian seeing psychiatrists and Henry overcoming his initial reluctance to working at Brown & Sharpe, Lucian confided in Richmond Viall that he had attached to his will a memorandum "intended to seriously curtail" Lucian Jr.'s participation in company business.[3] But Viall stepped up to support Lucian Jr. as a competent successor; Viall's support, along with Louisa Dexter Sharpe's maternal "entreaties and hopes" for her elder son, convinced Lucian to withdraw the codicil from the envelope holding his will.

Writing a will was very much on Lucian Sharpe's mind in the late 1890s. He had long suffered from "Bright's Disease" or nephritis, a chronic inflammation of the kidneys. The symptoms of Bright's Disease were often painful: high blood pressure, swelling in the feet or face, vomiting, difficulty urinating. Doctors generally prescribed bed rest, fasting, and laxatives, but there was no successful treatment for chronic Bright's Disease. To alleviate his symptoms Lucian employed his time honored cure for most

Henry Sharpe, left, with his brother, Lucian Jr., and an errand boy name Patty McNiff, outside the factory in the late 1890s. Lucian Sharpe, Sr., placed both of his sons under Richmond Viall's supervision for "good shop training," an apprenticeship to help them learn the business from the shop floor up.

any ailment: an overseas sojourn, this time to the famous, curative salt springs at Bad Nauheim, Germany.

Lucian and his wife, Louisa, sailed for Europe on May 9, leaving the factory in care of Viall and his own two sons just as a crisis unfolded in Providence: the molders' strike of 1899. Molders occupied the top tier of workers in the foundry—they made more money than the blacksmiths, pattern makers, and chippers. They performed the hot, physical work of pouring molten metal into molds to make castings that machinists perfected through cutting and grinding to greater precision. Much of a metal company's work began with the molders.

In the spring of 1899, most of Providence's 400 molders decided to flex their muscle. From their union hall in Market Square they sent out representatives from the Molders' Union of North America to negotiate with most of the city's 15 foundries. The molders sought a minimum wage of $2.75 per day for their 10-hour days; abolition of piece work; and recognition of the union. The key issue was the minimum wage. With Brown & Sharpe there was no need to negotiate: the firm already paid the desired minimum. When the strike began on May 1, Brown & Sharpe was one of only four companies where workers reported for duty with the union's blessing.

At the other 11 metal working shops, foundry ovens largely remained idle. Skeleton crews at some foundries made "small heats" to make their most important pieces. The Providence Journal observed on May 2: "The clink, clank of the hammers sounded right merrily in the bright sunshine, as workmen were cleaning up castings, but there was not the general air of a beehive that often pervades a busy shop in some of the big buildings."

Everyone hoped for a short strike. If the stock of molded parts ran out, as many as 6,000 machinists and other workers would have nothing to do, potentially crippling the city's metal working force at a boom time for the industry. But both sides had a very different view of the strike: for the union, it was mostly about a minimum wage. From the foundry owners' perspective, the strike was about the power to manage the businesses. If they surrendered the right to set prices for piece work, if they gave up the right to hire non-union workers, they would no longer be in control. And so the strike dragged on throughout the summer, quietly at first, then protests grew a bit more public: in September about 400 molders marched through the streets flanked by two billy-goats bearing signs reading "I.M.U. 41, Hard Thing to Buck."

The strike lasted into the following year, when Brown & Sharpe's molders joined it. They walked out to protest the hiring of "a molder who had made himself notorious as a scab" at strikes in other cities. "No self-respecting molder cared to associate with such a fellow and union and non-union molders alike protested."[4] A union official spoke to an unnamed Brown & Sharpe superintendent who "assured a very fair position ... with a view to an amicable adjustment" on the scab issue. But Brown & Sharpe maintained piece work, and continued to set wages without input from the few union cardholders in its shop.

Lucian Sharpe, Jr., joined his ailing father abroad in that restless summer of 1899, traveling first to England. He lodged in Manchester's Grand Hotel, which boasted of telephones and electric lights "supplied from an outside station" so guests would not be troubled by the chuff of an onsite steam engine. From his room in Manchester, Lucian wrote to his father over in Germany, trying to convince him to expedite construction of a new Number 4 building at Brown & Sharpe: "Now as to the new building foundation, I feel like making a 'try' to have it up by next Nov. that is closed in. In these days of quick erection, it can be done."

Henry Sharpe and Richmond Viall were not enthusiastic about trying to erect a new factory building in five months. They planned to have a new building up by June 1900, but that was not fast enough for Lucian, who argued that in the current boom times, quicker was better: "First of all, is the monetary side. It might be built for $100,000, building alone [and] remember we are practically sold up till Jan 1900 now, and were we to receive no more orders, would have to work 5 mos. more to 'get up stock.' So there is bound to be a lot of money flowing in."

If they built earlier they could make machine tools there throughout 1900, "ready for delivery the following January 1901," necessary, he wrote "for delivery to meet the increased demand.... Our four big lines (millers, grinders, gear cutters, screw machines) in all of which we excel beyond a question now in design involve increased output. Then there is that cutter business, the sewing lockstitch machine [and] the hair clippers, in which I expect to see a gratifying increase of sales."

Lucian Jr. advised his father: "One short cable" from the patriarch to Providence "will do the business."

A couple of weeks after breathlessly writing about the advantages of expedited construction, Lucian Jr. eased up his attempt to swing his father to his view. From the Royal Victoria Station Hotel in Sheffield he wrote his father:

> There is no need of you're [sic] writing more. It is all understood now at home. The best you can do for all of us is to give all attention to getting well [and] strong [and] back to your old vigour It can be done. We will do our best to help you. I can not think of anything now, with which we shall further bother you, and we shall work on the lines you have laid down for us.
>
> Work to get well, father, and we will be so thankful.

Lucian Sr. stayed at Bad Nauheim through the summer's heat, occasionally receiving updates from the factory via mail. In mid–September, Richard Viall wrote him from the National Hotel on Block Island, to tell him that "the boys"—Henry and Lucian—apparently had things well in hand. "They are doing splendidly and our faith in them daily grows stronger," Viall wrote. "To me the future is all hopeful and full of promise for the boys."

In October, Lucian Jr. linked up with his father in Germany's famous spa town, where the two of them sat side-by-side on a couch in the elder Lucian's rooms. Here they talked quietly about how to conduct the business after Lucian's impending death. The elder Lucian said:

> I believe you will get on alright in the business, you two boys; guess you're started rightly; don't believe you'd do otherwise than work hard now. You have a trust to perform to your mother, sisters, and the estate. Stay on and do it. You remember Nelson's watchword: "England expects everyone to do his duty."

Lucian softly told his son that he did not have the strength to draft a formal plan of succession for the management of Brown & Sharpe. "The burden is upon you two boys," he said.

"Well," said the younger Lucian, "there would be no cause for feeling on the part of the girls, would there?"

His father replied, "No, it would be all talked over."

"Is that your wish?"

"It is my suggestion."

The awesome responsibility of inheriting a global business that employed thousands overwhelmed Lucian Jr. consuming his thoughts. Before he went to bed that night he fired off a letter to his brother, Henry: "I must write. I cannot do otherwise. I must unburden another anxiety."

Then he laid out the conversation he had just had with his father, the elder Lucian "sitting on the sofa and I by his side talked quietly to me.... The burden is upon you two boys." Lucian wrote Henry that he felt as if overseeing Brown & Sharpe with Henry were "the best thing I can perform to the world & a sacred duty laid down upon us by we know not what mind."

"Heavens what should I do without you!"

The next morning, Lucian revisited the topic of succession in a second letter to his brother:

> Dear Henry
>
> I am sorry I cannot wait for your response. Father spoke very plainly to me last night & I must write it down, as it occurred I shall write freely & frankly & little know how to express myself; but believe it my duty to do the best I can.

He then repeated the conversation of the night before, word for word: "You have a trust to perform to your mother, sisters, and the estate. Stay on and do it. You remember Nelson's watchword: England expects everyone to do his duty."

Lucian carried both letters on the train from Bad Nauheim to Berlin. But before posting them the next morning, he felt the impulse to write his brother yet another letter, scrawling across the top: "I have rewritten this letter to make more clear." Lucian had spoken to his father the previous morning, and he had "reiterated" his suggestion that the two boys take control of the factory.

Lucian Sharpe, Sr., who rose from Joseph Brown's apprentice to his business partner in five years. Lucian opted to build the Willcox & Gibbs sewing machine using the American System of interchangeable parts, one of the first applications of that system outside of an armory. By taking care of the company's business needs, he helped create an environment for Joseph Brown's inventive genius to flourish. On his deathbed he asked his son: "Are my pecuniary debts paid?"

My how I long to know my dear Henry, if you agree with me heart & soul & will act with me; whether you believe me urged by the right motives believe me straight way thru'! When I contemplate what evils may creep in that will destroy or wreck the mighty [power] for good in the community, that concern is exacting. I confess I am trembling! ... We are here on earth to bear each other's burdens, & you & I, my brother, will have our share.

About a week after his heart-to-heart conversation with his elder son, Lucian Sharpe, Sr., set sail for home aboard the S.S. *Saale* of the North German Lloyd line.[5] As the rocky coast of Newfoundland rose on the horizon, Lucian turned to his namesake and said, "Lucian are my pecuniary debts paid? Because you should know." Those were his last words. This millionaire who, as a boy, had endured the indignities of his father's bankruptcy, worried to the last about keeping on the positive side of the ledger. On October 17, 1899, he died at sea.

Six days later, hundreds of workmen filed past Lucian Sharpe's coffin inside the First Congregational Unitarian Church, where he had worshipped. After the workmen paid homage the church opened for public view, followed by a private service. The boards of directors for the Providence Journal Company, Rhode Island Hospital Trust, Providence Institution for Savings, and the Providence Gas Company attended en masse, honoring a colleague who served on boards with all of them.

The *Providence Journal* reported, "The black broad cloth casket rested in the half light of the stain glass windows." Roses of red and yellow, chrysanthemums, and ferns framed the black casket. Since Lucian died at sea the Rev. Augustus M. Lord read the 107th psalm: "They that go down to sea in ships, that do business in great waters; These see the works of the LORD, and his wonders in the deep." After the service a carriage carrying Lucian's body made the trek along Hope Street out to the garden cemetery at Swan Point.

The heirs of Lucian Sharpe wasted little time in calling a meeting of the stockholders to carry on the company's business, which had never been better. In 1899, the firm sold more than a million dollars worth of machine tools, more than a 500 percent increase since the darkest days of the latest Panic. Machine tools accounted for nearly half of the firm's sales with hundreds of other items—hair clippers, sewing machines, calipers of all kinds, rules, and foundry parts earning the rest. In Lucian's final year the company's few stockholders shared dividends of $99,550, adjusting for inflation in the consumer price index through 2016, about $80 million.

A group of the stockholders—Jane Brown and her daughter, Lyra Nickerson, and Louisa Dexter Sharpe—met with Louisa's sons in mid–November to elect new officers. They chose Lucian Jr. as treasurer, and Henry as secretary. Tellingly, they left the office of president vacant. Richmond Viall acted as a de facto president, running the day-to-day operations while the Sharpe boys, now in their late 20s, continued to grow into the business.

Lucian inherited his father's desk, and as treasurer held a slightly more powerful position than Henry. The two brothers shared office space on the second floor of the old No. 1 Building, a factory of brick and steel built for production with offices almost an afterthought. The two brothers uneasily worked together in one small room in the northeast corner of the building with their desks back to back, facing each other. Each hoped to ascend to the presidency, a decision that was entirely out of their hands. A small group of powerful women would make that choice, Jane Brown, Lyra Nickerson, Lousia Dexter Sharpe, and the four Sharpe sisters led by the indomitable Mary Chafee, a surrogate mother to her younger siblings and a woman who was used to getting her way.

The year after Lucian's death the company idled along, coasting on its inertia. He had left the firm in great shape with record sales, plans for the No. 4 building well underway, and a new 800 horsepower steam engine powering belts looping to more than a thousand machines. The steam engineer, Robert Freer, called the big engine "the Sewing Machine" because it ran as reliably and as quietly as a Willcox & Gibbs.[6] Brown & Sharpe designers continued to patent key inventions: Oscar Beale created a new kind of gear cutter that finished both sides of a gear at the same time, halving production time, and a team added hydraulic controls to increase precision of the heavy grinders that the firm had at first resisted.[7] The problem was, Brown & Sharpe's sales remained flat just as the machine tool industry in general was emerging as an important, independent sector of the economy.[8] Machine tool makers no longer built just a few hundred machines for New England armories; they now made the machines that made just about everything else: cars, cans, cereals, trains, telephones, phonographs, electric lights.

Brown & Sharpe's sales fell in every sector in 1900, and dividends paid shrank by more than 25 percent. And Lucian, Jr. the company treasurer, was not a well man. Lucian complained of a bizarre array of symptoms: he could not taste anything. His hearing was bad. His feet felt swollen, "mere lumps of yellow clay" on which the toe nails would not grow. He had trouble looking people in the eye; when he did, their faces disappeared into a blur. "To talk was often impossible, and to look a person in the face was absolutely so," he wrote of his condition.

At times "blood rushed to my head in quantities" precipitating nosebleeds, "and at those times thinking was impossible, and at such times action was also impossible. At other times the blood surged in some other direction, when to keep quiet was absolutely out of the question."[9]

The Sharpe siblings had a problem. Their brother could not look people in the eye; sometimes he could not speak and at other times he could not keep appropriately quiet. His symptoms seemed crazy—he claimed his hair and fingernails would not grow, that he could not taste his food, that his body could not follow his brain, that he could not hear them speak. And this obviously unwell man was the treasurer of the company on which their fortunes rested, on which more than a thousand Rhode Islanders depended for their jobs.

By Lucian's own account of his troubles his family sent him to see "physicians for the insane." And the more he went to such physicians, the worse his symptoms grew. Pressured by his family he sought "the advice of many physicians, but always not only without benefit, but with a marked increase of difficulties accentuated by their methods."

In December 1901 Lucian lost perhaps his strongest advocate, his mother, Louisa, who reportedly had convinced his father to let him take a leadership role in the company. She died two days before Christmas. After another year of flat sales, Lucian's siblings agreed that he had to resign his position as treasurer, a post he felt was his "sacred duty." The job of telling him fell upon the indomitable Mary Chafee. In early spring 1902, she invited Lucian to her new home at 5 Cooke Street, up on the city's East Side. Here she squired him into her second-floor bedroom and told him, as he recollected it, "I am the oldest sister, I speak for all the family."

Mary told Lucian that he needed to resign, to go away, not only to restore his health but to protect Henry's. Being ill he "wore on Henry," causing stress that might sicken him too. For Lucian this was maddening. If he, an eldest son, surrendered his rightful place in a family business he would essentially be affirming the rumors of his mental instability. He would lose face in his community. Lucian felt at his core that he was stable, that "my brain was the most normal and efficient part of my body." He could see "distinctly how wrong were all the diagnoses of all those physicians and of my entire family." Yet in his condition he felt "helpless to help myself."

"Tis a good deal you ask me, Mary," Lucian recalled saying. According to Lucian, his sister won him over with a solemn promise: When he regained his health, he could resume as company treasurer. "I give you my word," he said she said.

And so, on April 10, 1902, Lucian acceded to his family's wishes. He wrote two letters of resignation, a terse, one paragraph note:

Gentlemen:
In accordance with your expressed wish I hereby resign my position as Treasurer of the Brown & Sharpe Mfg. Co.
Very truly yours,
Lucian Sharpe.

His longer resignation letter was a little more revealing. He wrote that he submitted his resignation *cheerfully*, while acknowledging that he did so "at great personal cost. ... I have been ill and my physician does assure me that my next step, which is to sojourn & is at my suggestion & my choice, to which he heartily agrees, will render my health as 'impregnable as it used to be.'"

Board members accepted his shorter, one-page resignation letter on May 14; they named him a vice-president, a titular role that gave Lucian no authority but gave him significant shares of company stock and allowed him to draw a hefty, five-figure salary. Workers moved his father's desk out of the office, giving 29-year-old Henry his own space, though the president's position officially remained vacant with Richmond Viall assuming some of those duties in the interregnum.

Richmond Viall liked to pass the winter in Florida, and while there in early 1902 he came down with a terrible illness. When his strength allowed he returned to Providence, where a team of movers had recently jacked his house from its foundation atop Smith Hill and set it down at the corner of Holden and Jewett Streets, making room on the hill for Rhode Island's gleaming, new marble State House.

Richmond's son, William, also a company executive, thought that his father looked frail. The elder Viall, who would soon turn 68, had always made it a point to be in the shop when workers punched in at 7 a.m.; now he was coming in later and leaving earlier. He reluctantly eased back on overseeing day-to-day operations, becoming in his son's words "an advisor and counselor which could not have always been easy."

10. Death and Succession

Front office clerks, stenographers, and errand boys on the front steps of the plant in 1904, two years after Henry Sharpe, the youngest of the six Sharpe siblings, took on responsibility for running the large manufacturing plant on Promenade Street.

As Viall's duties decreased, Henry's increased. Sales and profits also increased, tremendously. Despite the American Federation of Labor's suggested boycott of Brown & Sharpe's non-union products, in 1905 sales of all goods topped $3 million for the first time. Stockholders split $208,150 in dividends, another record, and in 1906 the board elected Henry Sharpe as president. Henry celebrated the promotion by buying a new house, a duplex on the East Side, as the siblings prepared to donate their childhood home on Angell Street to Brown University.

From afar, Lucian Jr. seemed happy. He retained his share of the company's dividends, and in 1903 he paid for Brown University to install in Sayles Hall a huge pipe organ that boasted more than 3,000 pipes, the largest of which stood 32 feet. He also paid to install the carillon bells that hourly chimed over the campus from the University's Carrie Tower.

The crisis of succession seemed to be averted, but Henry faced another emergency in a title inherited from his father: director of the Providence Journal Company. Lucian Sr. had purchased a "significant" share of the Providence Journal Company's initial stock offering at $100 per share, making him one of the three largest shareholders. When the Providence Journal Company began in 1885, its long-established newspaper faced competition from 35 other newspapers in Rhode Island. Lucian's shrewd investments in linotype machines and in the Hoe printing press, manufactured by a machine tools customer of Brown & Sharpe, gave the *Providence Journal* a technological edge over competitors; its insistence on remaining politically independent (if predictably Republican) played well with readers in an age when many newspapers were the publications of political parties. By the time Henry took over his father's seat on the Journal's board of directors in the early 1900s, shares of the privately held Providence Journal Company had appreciated by 15 times. One of his fellow board members, Stephen Metcalf, was the brother of his brother-in-law, Jesse Metcalf (who had married Louisa Sharpe), heirs to the Wanskuck Mills textile fortune.

As a prosperous newspaper with a good reputation, the *Providence Journal* caught the eye of a formidable cabal of local politicians that included Senator Nelson Aldrich and wannabe senator Samuel Colt, son of the famous pistol maker. Colt owned the United States Rubber Company's Providence mills, the world's largest rubber plant, just upriver from Brown & Sharpe. The Aldrich-Colt partnership likely had the silent financial backing of Marsden Perry, the utilities magnate.

When some members of the Providence Journal Company's founding families sold out in 1904, the syndicate that included Aldrich and Perry bought their shares, giving the two politicians seats on the board of directors. Clearly they intended to use the *Providence Journal* to advance the political careers of themselves and their favored candidates. Henry Sharpe, Stephen Metcalf, and William Hoppin saw this both as a threat to the newspaper's journalistic integrity and to its profitability.

In a scene reminiscent of a young Lucian Sharpe, Sr., hustling stockholder votes to stave off a hostile takeover of Willcox & Gibbs, Henry Sharpe, Metcalf, and Hoppin devised a plan to thwart the syndicate's plans for the *Providence Journal*. In December 1905 the three men formed a trust in which they deposited their sizable shares of Providence Journal Company stock. Then they buttonholed smaller stockholders, trying to persuade them to add their shares to the trust to form a voting bloc that could stymie the efforts of Colt and Aldrich to turn the paper into just another publication of political agitprop.

The Colt-Aldrich syndicate could do the math—they knew that with the considerable shares held by the Sharpe and Metcalf families, its chance of obtaining a voting majority was thin. So they hatched a secret plan that came to fruition on a Saturday night in February. After putting the *Providence Journal's* evening edition to bed, the top editors, reporters, foreman, and compositor, popped champagne corks in the Journal's newsroom to toast their new, high-paying employer: *The Evening and Sunday Telegram*, owned by Samuel Colt. After draining their champagne they strolled down Westminster Street and into the offices of a new rival to the *Providence Journal*, leaving the *Journal* scrambling for a new team of reporters and editors.

Driven by a new managing editor, John Rathom, a big, cigar smoking Australian with a talent for self-promotion, the Journal quickly recovered, successfully fighting off the *Telegram's* challenge. On April 11, 1907, the Providence Journal published a rare, page 1 announcement saying that Stephen O. Metcalf, Henry D. Sharpe, and William A. Hoppin had indeed established a clear majority of stock placed in a trust so the newspaper would remain forever independent from political influence.[10] With this takeover averted, Henry again turned his attention back to his duties as president of Brown & Sharpe, where crisis flared again.

The crack of a pistol shot sounded inside the payroll department of Brown & Sharpe, followed by at least two more shots. Women, who had recently joined the company's workforce as payroll clerks and stenographers, started to scream.

The *Providence Journal* of June 1, 1907, reported: "The shooting occurred at 5:30 o'clock yesterday afternoon and in the presence of a large number of clerks, many of them girls, caused a panic, during which several of the young women were thrown into hysterics and some fainted." The gunman fired shots into the torsos of two men, who were hustled over to Eddy Street for surgery at Rhode Island Hospital.

The gunman, a 21-year-old clerk in the timekeeper's office named Fitzroy Willard, told police that the workplace shooting had its roots in sexual harassment. One of the men whom Willard shot, a clerk named Amos N. Gorham, had been temporarily suspended for, in the words of the *Journal*, "interfering with some of the young women employed in the timekeeper's department."

Walter H. Willard, the gunman's brother, managed the cost department, and employed 18 "girls" on his staff. He had reported Gorham to company bosses for harassing some of the women. After Fitzroy Willard came to work at the shop, Gorham determined that the Willards were brothers, and began annoying the younger Fitzroy by making "sneering references" to a third Willard brother named Charles, who suffered from a deformity in his spine.

"Gorham has been calling me a sneak and a member of a family of sneaks and humpbacks," the younger Willard told police. "To-day I started to talk to him as he had been talking to me. I returned his sneers and told him that I would give him as good as he had given me. I told him that he was a sneak...."

A half-hour before quitting time, Fitzroy Willard slipped into the coatroom, pulled a gun from his jacket pocket, and shot Gorham. A clerk named Thomas Edmundson tried to disarm Willard. He too, was shot. The elder Willard calmed his brother, locked him up in the timekeeping office, and held him there until the police came. At the *Journal's* press time it looked as though Edmundson would survive the shooting; a surgeon described Gorham as "a very sick man."

Brown & Sharpe weathered the temporary disruption of a workplace shooting; the big steam engine dubbed "the Sewing Machine," reliably chugged along, turning the belts on more than 1,000 machines that employed more than 1,800 workers. Brown & Sharpe now made machine tools powered by attached electric motors, freeing factories from the clutter and noise of leather belts looping and slapping overhead. But the company's own plant ran smoothly on the old-fashioned system of a big steam engine turning multiple shafts. One day a bolt loosened in the piston head which, left unchecked, would blow the engine. But even above the racket of sump pumps and an electric generator buzzing in the engine room, Robert Freer detected an unusual clicking in his steam engine. Within a couple of hours he had it chuffing away again.

Three years into Henry's presidency, the factory complex sprawled across 25 hilly acres clicked along as smoothly as the workings of a pocket watch sold by his father a half-century before, taking in sheet, bar, and wire steel, and turning out hundreds of products sold around the world. Henry Sharpe made plenty of money—more than $32,000 in Brown & Sharpe dividends in 1909, the equivalent of about $800,000 today, plus an annual salary in the range of $15,000 (the equivalent of an additional $380,000), and additional dividends through family holdings in the Providence Journal, Hospital Trust, and in the gas company. Henry easily earned more in a single year than a good machinist, who pulled down less than $2,000 a year (the current equivalent of about $50,000,) could make in a 25-year career.

And yet, Henry's family life was wanting. At 36 years old he had no partner, no children, no domestic life. From later letters it is clear that a forced estrangement from his brother, Lucian, caused Henry and his sisters real pain. At times it seemed as though the factory did not exist to serve the family, as much as the family existed to serve the factory.

PART TWO: CENTRALIZATION

11. "Clean the Damned Place Out!"

The wild rhododendron bushes bloomed along the freshwater ponds of southern Rhode Island, and a full moon made for fine nighttime bathing in early July 1909. One of Henry Sharpe's nephews, Zechariah Chafee, enjoyed this summer nocturne while on summer break from the Builder's Iron Foundry, his father's busy metal-working firm. Zech, now 24, did not like metal work, and referred to the foundry as "the family graveyard."

The moonlight and flowers put thoughts of romance into Zech and his host, Christopher Greene. They organized a "Philanthropic League" with matrimony as its object, and the marriage of Christopher's sister a particular interest. "Whom do you suppose we selected for the lucky man?" Zech wrote to a sibling. "Uncle Henry Sharpe!"[1]

By 1909, three of the six Sharpe siblings had married. Louisa, the middle child, now 40, had recently acquired two step children by marrying the widower Jesse Metcalf, heir to the Wanskuck Mill fortune and, with Henry, a stockholder in the Providence Journal Company. At the time of his marriage to Louisa he was up-and-coming on the political scene, a Providence City Councilor who later served in the U.S. Senate.

Amey, the second-oldest daughter and Lucian's favorite sister, was the only Sharpe woman to settle out of Providence—she married William Peters, a busy Boston architect who designed many of the brick houses in the Back Bay, as well as the Weld Mansion. They lived in Cambridge. Lucian Jr. who still earned handsome dividends as a titular vice-president of Brown & Sharpe, rented a mansion near Amey and her three children. Other than occasional holidays, the Peters children rarely saw their Uncle Henry down in Providence.

Mary Chafee, Zech's mother and the stern, family matriarch, may have been better suited for managing a factory than for mothering a family. Her eldest daughter, Elisabeth, developed Scrofula at age 8, a type of tuberculosis that lodges in neck glands. Mary found her sick daughter "nervous and irritable." Rather than nurturing her, she exiled the girl to live with an older nanny in Massachusetts, offering the excuse that "the necessities of younger children make it impossible for me to give her the attention she needs...."

As Elisabeth, noted in 1902: "I keep discovering the disadvantages of being a girl," Mary Chafee also felt the disadvantages of being a woman at a time when women lacked the legal standing to vote and the cultural standing to run factories, so she turned her talents to managing a growing family. After four years, Mary took Elisabeth back into the Chafee family, which had grown to include a sister, Mary, two brothers, Zechariah and Henry, and a constantly rotating staff of servants in the Cooke Street house: Emma, "the loyal black laundress"; the Irish maids Annie and Annie; "Miss Dugane" the housekeeper; Miss Graves, governess to the Chafee boys; Josephine the nurse. After being raised by governesses and maids, Elisabeth observed: "The wealthy do no not love their children."[2]

While her brother, Zech, was trying to marry off Uncle Henry Sharpe that summer, 21-year-old Elisabeth was struggling to find a family life of her own. To this end she enrolled in The Garland School of Matrimonial Bliss, a Boston school which boasted of having a "group of girls [from] many of the best families in the city." In a 1910 brochure the school's founders explained that the industrial era had so radically changed the domestic sphere that women now required reeducation in the art of running their households:

> The old New England home was the center of many small industries by which the daily needs of the family were supplied, and in which each member of the family, young and old, had a share. The daughters of the house helped their mothers in the baking and the brewing, the spinning and the weaving and the care of the younger children. By their helping they learned something, at least, of the art of homemaking. The problems of education were simple, because the means of education were limited. Boys and girls alike learned trades and entered professions by this system of apprenticeship.
>
> In the present day the home has a relatively small share in the production of that which it consumes. Its material needs are largely supplied by factory, market and shop. And it is often with some difficulty that the younger members of the family are busied with the daily tasks which are essential to the development of home feeling.

The School of Matrimonial Bliss displayed a Progressive Era passion for scientific method, technology, and education to solve new challenges in and out of the home. To help the modern woman meet "new economic and educational problems," the School supplanted classes in cooking, serving, sewing, and shopping with "some of the recent advances in science, especially in the chemistry of daily life; [and] the development of the work done in cleaning the world from dirt and disease."

Many Progressive Era reformers obsessed over cleanliness, at least partly as a response to squalid conditions in industrial cities bursting with successive waves of immigrants finding work in factories. Besides changing domestic relations, industry had remolded larger social relationships as well. In the 30 years between 1880 and 1910 the population of Providence—driven largely by immigration from Ireland, French Canada, Portugal, Scandinavia, Cape Verde, and Italy—had increased by more than 250 percent. The city now stood on the cusp of a quarter-million people, 70 percent of whom descended from foreign-born parents.

Newcomers sought out their own kind, settling into densely packed ethnic enclaves: Italians crowded into triple deckers on Federal Hill and in Silver Lake; Irish and French Canadian immigrants lived and worshipped in separate parishes near Olneyville and on Smith Hill; Cape Verdeans lived on Fox Point; African Americans, by no means newcomers, also settled together in the seventh ward, between Cranston

Street and Elmwood Avenue. The city that Henry Sharpe navigated was a vastly different place from the largely homogenous Providence where his father had walked the single main street observing the first telegraph wires, nail factory, and gas lights. Providence ranked as the United States' 20th largest city, and held the nation's second-highest income when averaged on a per capita basis. At one point Providence boasted of being home to "the Five Industrial Wonders of the World": the Corliss Steam Engine Company; Nicholson File; American Screw Company; Gorham Silver; and Brown & Sharpe, all of them being among the largest businesses of their kind, manufacturing steam engines, files, screws, fancy silver, and machine tools. Men with capital had raised factories that built all manner of things, from automobiles at the Burnside Locomotive works to Brown & Sharpe's machines that shredded and packaged wheat. With money and through political influence industrial capitalists had woven networks of public roadways and private rail lines, telephone, telegraph, and electric wires, creating profound changes to both the domestic sphere and to the political, economic, and social fabric of the city, the nation, and the globe.

The year 1909 proved to be a busy one for Brown & Sharpe. John Parker, an engineer charged with redesigning milling machines, had found a way to regulate the speed of milling cutters so they spun at a constant rate regardless of any reasonable load put on them, or the thickness or hardness of the steel they were cutting.[3] Small electric motors powering the cutters made possible the constant speed drive milling machine, the latest in the company's machine tools line. The company's workforce steadily increased from 2,710 in January, to more than 4,000 by year's end.

Early that October, Henry Sharpe received a phone call from his brother, Lucian, who announced that he was back in town and prepared to resume his rightful position as treasurer of Brown &Sharpe. After taking this call Henry, in his own words, "did not know what to do, and what not to do." His brother had resigned as treasurer seven years previously to seek treatment for his mental health. What made him think he was sane enough to tackle those responsibilities now?

Lucian had checked into a guest room in the Hope Club on Benevolent Street, a brick, four-story Victorian building raised in 1886 as a clubhouse where the upper class could gather "for the purposes of social and literary culture." From here he began dropping into the factory, which resurrected the problem of office space. During his seven years away the firm had built new offices, but no one had thought to make one for Lucian. And nobody could find his desk, the one previously used by his late father.

To Henry, Lucian seemed insolent beyond belief. Richmond Viall, now in his 70s and a trusted counselor who still liked to come around the shop, reported that Lucian would not acknowledge him. Henry advised Viall "not to make any advances," to wait until Lucian said hello to him. But "following a generous impulse," Viall spoke first, offering a hello to his erstwhile student in the shops. Lucian rebuffed his old mentor, telling him not to call him Lucian but "Mister Sharpe." Viall felt crushed. Henry spied Richmond shortly after the exchange; "the brightness of his eyes, and the flush of his cheeks, told me of his deep feelings." But he spoke kindly of Lucian, and Henry did not take up the snub with his brother. Instead he told Viall's son, William, to steer clear of Lucian, in order to spare him a similar rebuff.

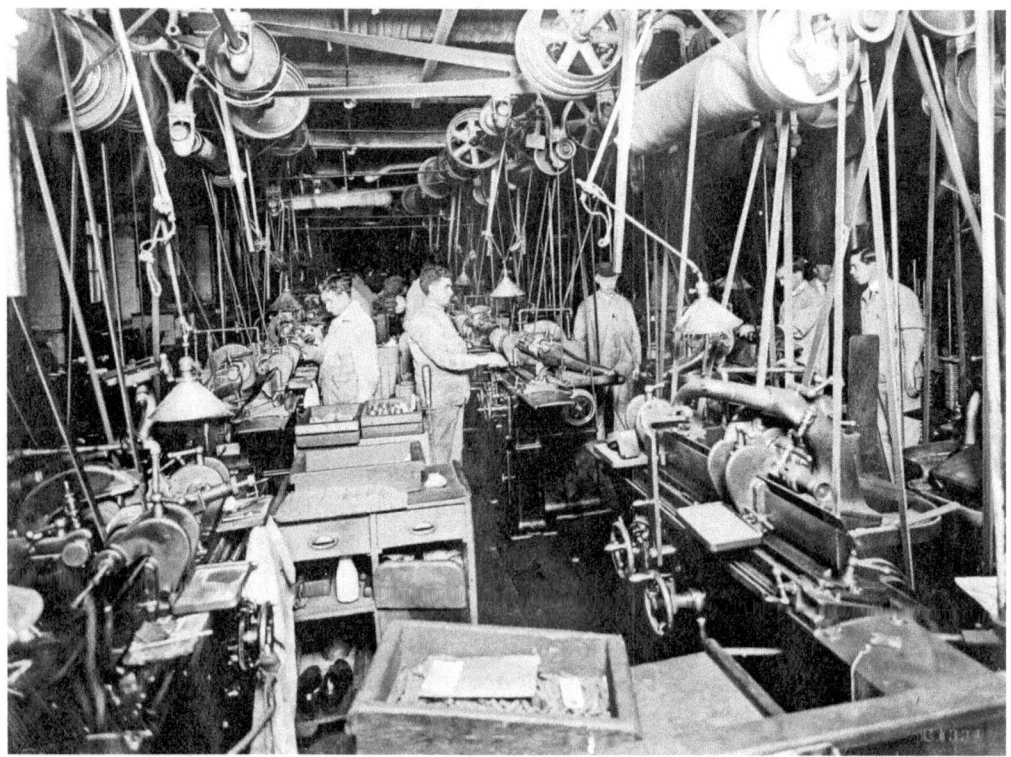

11. "Clean the Damned Place Out!"

Lucian, of course, viewed events inside the factory in a different light than his brother. He had recently begun seeing a doctor, Morris Longstreth, who had really helped him. Longstreth, a Philadelphia physician who also owned a home near Lucian on tony Brattle Street in Cambridge, suspected that Lucian's problems stemmed from "impacted blood circulation." He put Lucian on a different regimen than previous "physicians for the insane" had prescribed. Lucian's symptoms—the deafness, the blind spots, his loss of taste, began to subside. Lucian only returned to Providence after he truly felt that his health had come back.

At the time of his resignation in 1902, Lucian had almost certainly been suffering from lithium poisoning. Psychiatrists began using lithium to treat mania long before John Cade's 1949 studies on its effectiveness led to its widespread use in the latter half of the twentieth-century. A London internist began treating "brain gout" with lithium in 1847, and in 1871 a professor William Hammond at Bellevue Hospital Medical College in New York, became the first physician to prescribe lithium for mania. By the late 1800s psychiatrists were well aware of lithium's effectiveness as a "prophylaxis of depression."[4]

A problem with lithium is that it has a tight "therapeutic window"; the dose at which it is effective is awfully close to the dosage in which it is poisonous. Chronic lithium toxicity causes "optic neuritis" resulting in blurred vision; edema, or swelling of the feet and ankles; loss of taste, tinnitus (a loud ringing in the ears), abnormal heart rhythms, slurred speech, tremors, and brittle fingernails. The symptoms that Lucian complained about in a letter to his family—loss of taste, swollen feet, blurred faces, uncontrollable limbs, bad hearing, difficulty speaking, perfectly mirrored the symptoms of lithium toxicity at a time when psychiatrists were well aware of the drug's effectiveness in treating mania. And in his letter he noted a "marked increase of difficulties accentuated" by his psychiatric treatment.

The Sharpe siblings had no way of knowing what ailed their brother. He had needed psychiatric care for episodes such as the orange tossing incident, and once he had begun treatment he spoke of a bizarre array of problems—brittle nails, hair that would not grow—that forced them to remove him as a company officer. As their letters make clear, they did not love him any less, they just could not trust him with the family business.

When Lucian had been at his sickest, he had found it "cruel" that no one believed his problems had been physical. Now that he felt clear-headed, ready to resume his role, he expected a warm welcome, "an expression of salutation that I had regained my

Opposite, top: **In the early 1900s, Brown & Sharpe began making machine tools, such as this vertical spindle milling machine, with attached electric motors that freed factories from the clutter and noise of leather belts looping overhead. Motor-driven machine tools allowed factory designers to place machines without worrying about attaching them to a central power shaft. Ironically, though Brown & Sharpe made motor-driven machines, it continued to use many belt-driven machine tools into the early 1960s.** *Bottom:* **Workers in the grinding machine department circa 1910, when Brown & Sharpe employed more than 4,000 workers operating well over 1,000 machines powered by belts attached to a central power shaft. By grabbing one of the rope loops like the one hanging in the left of the picture, an operator could attach a belt to an overhead shaft to engage the machine, or detach it to cut power.**

health, and been relieved from the physical condition in which I had so long, and had so terribly suffered."

No such salutation was forthcoming. He had been sent to psychiatrists for a reason, and no one was sure that his claim of good health was genuine. Lucian's siblings did not welcome him back into the factory with open arms so he bulled his way back in, insisting on office space and a desk. His siblings gave him these, but refused to grant him what he most wanted, at least until he proved himself capable: the treasurer's job.

On December 12, 1909, Henry Sharpe's 37th birthday, Lucian broke out some Hope Club stationary and wrote a 27-page letter to his brother blasting him and their sisters for their treatment of him, and demanding that they allow him to fulfill his sacred duty by restoring him as treasurer. Clearly Mary Chafee wielded the power in the family as he cited her promise to him "on behalf of the family" that he would resume his position when he recovered his health.

"You have usurped my position," he wrote accusingly to Henry. "I have scanned closely your manner since my return and have found it [betaking] a keen exceeding delight in possession." He claimed that "the two Vialls, father and son, have been imprudence personified," refusing to speak to him, and eavesdropping on his conversations. "They have become the proprietors of the factory, pulling the wires needed by them to make you hop in whatever direction agreeable to them.

"...You have put me in the position of being neither fish, flesh nor fowl. I am neither in the Factory, nor out. I am 'the outlawed member of the Sharpe family.'..."

The Vialls had indeed been refusing to speak to Lucian—at Henry's request. He had advised them to avoid his brother, in order to avoid a repeat of Lucian's snapping at the elder Viall. With tensions high, people may have been eavesdropping on Lucian; or his belief that they were may have been a symptom of the mental illness for which he had been prescribed lithium in the first place. Whether caused by a clear-headed belief that his family had not been supportive when he had been physically ill, or by a recrudescence of mental illness, Lucian barged back into the factory with an attitude.

Three days after Lucian wrote his acerbic letter, Henry wrote a response that their sister, Ellen, copied in cursive for the family's files:

> My dear Lucian
> Your long and feeling letter was read by me with a great deal of sorrow for it showed me better than before, how deeply you have suffered during past years. In view of your absolute inappreciation of my attitude towards you, whatever it may really be, and your extreme views of events and facts, it would seem to be impossible to convince you that what has been, and what is, is for the interest of all the stockholders of the Company.

Henry said that he had offered Lucian a job "to give counsel when it was sought," and if his judgment proved sound "authority would come as it was deserved." But Lucian insisted on resuming his role as company treasurer. Henry made it clear that that was not going to happen:

> You are certainly very outspoken and frank with me in your letter, and I shall try to be so with you. I do sincerely hope, in fact I beg of you—that you will give up your demand for the treasurership of the Company. This aspiration has been a lure which has embittered your prolonged absence from the business; it will be your ruin if you don't check it at once. No one in the Family

has the slightest idea you should be placed in that position, or [that] you would be comfortable in that position if you were returned to it.

Henry's letter made it clear that the Sharpe family still loved Lucian as a brother and felt great sorrow about his suffering; they offered him what they felt was a rational entry back into the business; and if he refused the offer, they and the other stockholders would *finally do what is best for the business.*

Henry and Ellen signed off: "With the kindest of feeling, and the hope that clouds which now seem so dark will blow away."[5]

The darkness looming around Lucian did not blow away. Two years later, at a Thanksgiving Dinner in 1911, the Sharpe siblings felt the pang of Lucian's absence—and tried to rectify it. Together they worked on drafting letters, one to their estranged brother and the other to his doctor, first writing them by hand before typing final copies in an attempt to bring Lucian back into the family.

In a letter to Lucian both carefully worded and heartfelt, the siblings wrote:

> At our Thanksgiving gathering we had an opportunity of talking over the fact of your separation from us, which unhappily seems to continue as time passes. Our desire is to have you resume those relations of family life and affection in which we were brought up and which ought to enjoy as the years roll by.
>
> We can assure you there is nothing on our part—nor has there ever been—which will prevent such a renewal if you are prepared to have it brought it about. Between us—brothers and sisters—can there not be a breadth of dealing and strength of feeling strong enough to accomplish this? ... We, who have chosen to think of ourselves as the Sharpe children are children no longer but are in the prime of life with all the responsibilities that life brings. The real children of the family—the grandchildren, are fast growing up. ... This year the old house on Angell Street is to be abandoned, which certainly marks an epoch in our family life. As we gather together from time to time, we are ever impressed how these changes of time only tend to tear us apart unless we strive to live into each other's lives by frequent intercourse, personal contact, and above all, by a generous interest in each other. This last element we especially wish to call to your mind now, Lucian. Cannot we get together in some way so as to revive our cordial family feeling, which used to be so real and binding? We make this appeal to you in all sincerity and genuineness.

To ease Lucian back into the family they would give him a job overseeing the stock list of machinery and researching the company's insurance policies, meaningful work, they assured him, "which requires judgment and foresight." Brown & Sharpe was considering buying insurance to cover workers injured on the job, a progressive step for 1911, and wanted him to become familiar with "the scheme of Accident Insurance."

In signing off the siblings asked Lucian "to respect truly, and without hesitation, our wishes as to the management of our own property. Our affection for you as our brother leads us to hope that you will again make our family one and undivided." Lucian could again unite the family, but only if he agreed to a lesser role in the family business.

The Sharpes also tactfully told Dr. Morris Longstreth to stop meddling in the business. While praising Longstreth for making Lucian "a more healthy man," they warned him to cease encouraging his "belief that he would be reinstated in his old responsible position in the business...."

"Lucian's character," they wrote, "repeatedly observed by us, has a strong tendency to lean on others and to neglect to rely upon his own judgment and efforts. Therefore, we think continuing to encourage this belief and hope of his only assists the prolongation of an attitude which is already a tragedy in his own life and our family life as well."

During the spring of 1913, business fell off at Brown & Sharpe; finished machinery piled up day after day in the works, and Henry Sharpe felt it necessary to begin laying off men. Layoffs continued into the next year—the company began 1914 with 4,132 workers, and in the face of the worst recession since the Panic of 1893, continued letting them go. The slack business conditions, the estrangement with his brother, and the recent death of his mentor, Richmond Viall, wore him out.[6] On June 11 he chaired a stockholders' meeting, before taking an extended leave.

A couple of weeks after Henry took his leave, newspapers reported the assassination of Archduke Franz Ferdinand of Austria, heir-presumptive to the Austro-Hungarian throne. Austria-Hungary, a union of Austria and the Apostolic Kingdom of Hungary, was a large and powerful nation. In Europe, only Russia held more land, and the empire trailed only Russia and Germany in population. A group of six men hoping to break off Austria-Hungary's south–Slav provinces to form a united Yugoslavia hatched a plot to kill the archduke on his scheduled visit to Serbia, and though their plans went all awry—the first two assassins failed to throw their bombs at the archduke's motorcade, and the third missed his target—one eventually shot the archduke and his wife when their motorcade took a wrong turn.

In response, Austria-Hungary sent Serbia a long list of demands, including the right of its agents to conduct investigations inside Serbia. The Austro-Hungarians gave Serbia three days to comply. When the smaller nation protested the tightness of the deadline, Austria-Hungary declared war on July 28. Russia, still smarting from the Austro-Hungarian annexation of Bosnia in 1908, then mobilized against Austria-Hungary and its ally, Germany. Within days the conflagration of war spread with astonishing swiftness. All of the elements on display at Chicago's Columbian Exposition of 1893—the nascent nationalism evident in each country's separate pavilions, the militarism exhibited in Krupp's park of artillery—coalesced into a nightmarish scene of total warfare.

By August 4, Germany and the Austrio-Hungarians had invaded Belgium and were fighting France, England, and Russia. U.S. President Woodrow Wilson urged Americans to be "neutral in thought as well as in action." Military leaders on both sides, confident in their troops and advanced weaponry, believed the war would be short, a belief put to rest when the allies halted the German advance into France at the bloody Battle of the Marne. With the warring factions literally entrenched on the Western Front, the fighting turned into a war of attrition. Constant bombardments required constant replacement of shells, guns, tanks, ships, all of which put great demand on the makers of machine tools. By August, the long, slow business decline of the previous year was a thing of the past; American business stood at the threshold of an unprecedented boom.

An early consequence of the war in Europe was high prices, as merchants anticipated that wartime demand for goods would limit supply. In late August the combination of inflationary prices in staple goods mixed with high unemployment from the year-long economic crisis, proved to be a volatile mix. On crowded

Federal Hill the Italians, incensed by price-gouging grocers, smashed display windows along Atwells Avenue and helped themselves to the goods. The causes of the so-called Macaroni Riots that roiled the Hill for days were, of course, far greater than the price of pasta. They were rooted in the long-term disenfranchisement of the average Rhode Islander.[7]

Seven of ten people were first or second-generation Americans, and most of these newcomers lacked legal standing to vote. Of course no women could vote, native or newcomer; and both the state and the city maintained laws that disenfranchised the urban poor. Every Rhode Island town sent the same number of representatives to the State House as Providence did, allowing long-established rural interests to dominate the concerns of more populous urban newcomers. And Providence held onto a law that required foreign-born residents to own real estate before they could cast a ballot, a restriction that prevented two-thirds of the city's adult men who qualified to vote in state elections from voting in city elections. Despite repeated attempts to change conditions through the ballot box, most city and state residents lacked political standing to make change through political means. The resultant Macaroni Riots were symptomatic of social unrest that would soon flare in a variety of ways.

Another immediate consequence of the war was an uptick in orders to American factories. Right away Brown & Sharpe received a "mass of orders" for machine tools from Germany. Workers rushed to fill the order. A freighter scheduled to depart Providence for European ports was held late so stevedores could load it with machine tools bound for Germany; the order never made it beyond Copenhagen, where British agents seized most of the machinery.

In December, William Viall sailed for Denmark where he met with a German sales agent named Mr. Kretschmer, to discuss the seized shipment. The agent, Viall recalled, "was very much incensed to think that the British confiscated goods that went through their lines." Kretschmer "begged and pleaded with me to urge the American manufacturers to issue an ultimatum to England that either they would let machinery go through to Germany or we would not sell them any machinery." Viall, acting on Henry Sharpe's orders, told Kretschmer that Brown & Sharpe was done doing business with German firms until the end of the war.

Factories in England, France, and Russia swamped Brown & Sharpe with orders to equip their armories and engine makers. In American ports goods piled up on the wharves as dock workers struggled to lade ships with food, arms, and machinery bound for allied cities. In June of 1915, Viall wrote to a Chicago sales agent: "We are still busy here at the works, everything going, and orders continuing to come in to a degree that we can not possibly accept them." The company actually had to turn away work for want of workers to produce it.

Such high demands for goods resulted in an equal demand for workers. Agents in France asked Brown & Sharpe to send five screw machine operators to instruct machinists there, offering to pay $50 per week plus expenses—a 34 percent boost in the $33 weekly wage that Brown & Sharpe paid its average skilled machinists.[8]

Connecticut companies began sending recruiters to pilfer Providence workers, offering enticements to machinists, molders, and toolmakers to relocate to New Haven and Bridgeport, a gambit that incensed Henry Sharpe. One August morning, Henry spied a classified ad placed in his *Providence Journal* by the Winchester Repeating Arms Company, seeking machinists for its New Haven plant. He dashed off a letter to New Haven that afternoon, chiding the Winchester Arms Company with a vague threat to hold up its orders with Brown & Sharpe:

> There are just so many men to go around here in New England, and attempts to 'pull' men from one city to another if practiced by all concerns, would have obvious results.
>
> Generally speaking, this concern employs a large proportion of the mechanical help in this community, and incidentally, you have on order, from this Company, one hundred and thirty (130) Machines, more or less. We presume you would not like to have your deliveries deferred, and certainly we should not like to look forward to having you suffer any deprivation.

In Bridgeport, a city booming with war-related work, wildcat strikers demanded, and received, eight-hour days that summer. But Brown & Sharpe would not be moved into accepting the eight-hour day. In writing to a Newark firm in late August, William Viall explained: "we are working fifty-five hours per week, ten hours daily, excepting Saturday, when we close at twelve."

> We have not under consideration the reduction of working hours. The air seems to be full of it in some respects and in some places, but we believe it is a wave that is going over, influenced largely by the labor unions, who are attempting to take advantage of the situation. ... Of course we do not know what may happen, but we think we shall be able to carry on business as it has been carried on in the past.

Henry Sharpe held little tolerance for unions. When the American Museum Society sent him a fundraising request on stationary bearing the union label, he shot back:

> In our opinion the use of the Union label is not only un–American but it is useless for the purpose of the Museum. Inasmuch manufacturers are expected to support the Museum in the good efforts which it proposes, we think it only fair that the same manufacturers should be free to make known their position on this un–American practice.

With strikes roiling Bridgeport in that summer of 1915, Henry prepped for strikers even as he predicted that Providence manufacturers would succeed in thwarting them. In August, Brown & Sharpe hired the W.J. Burns Agency to post guards around the plant, ostensibly to prevent German sabotage. The firm also fired, in William Viall's words, "quite a number of men who have pronounced Union proclivities."

The firing of union sympathizers backfired. As workers filed into the Promenade Street plant on Monday, September 20, a committee of local union leaders joined them in hopes of making their way to the front office to meet with executives. Henry was out of town that morning; when William Viall sent word that he would not meet with the labor leaders, they sent back a letter that read, in part:

> The employees of the Brown & Sharpe Company having lost confidence in their ability to receive fair treatment by the Company, owing to certain recent discharges of men who have served the Company long and faithfully, therefore hand these demands to the Metal Trades Council to present to the Company with power to call a strike if they are refused....

The list of demands included an eight-hour day at the same weekly wage; non-discrimination against union members; and shop committees to negotiate grievances with management. Viall refused to send an answer to the demands. At lunchtime, as men milled outside the factory, O.L. Preble of the American Federation of Labor stood on a public street that ran up Smith Hill between Brown & Sharpe's brick buildings. From here he made a "somewhat impassioned address," asking lunching workers what should be done. "If strike it is," Preble intoned, "then by God, I'm with you: get in there and clean the damned place out!" Workers briefly returned to their posts and, shortly after 1 p.m. more than 2,000 men made a break for the streets.

To William Viall, the walkout seemed "governed by hysteria.... The crowd marched about the works and then about the city until the police headed them off." In the evening the strikers massed about the works, calling to those who remained inside until horse-mounted members of the Providence Police rode into the strikers, sending them scattering.[9]

Henry Sharpe returned to Providence on Tuesday night, and on Wednesday morning he took control of the company's response to strikers. The morning newspapers handed him a perfect cudgel with which to beat the metal workers union: News that Austria-Hungary had just recalled its ambassador to the United States for serious misconduct. The ambassador, Konstantin Dumba, had written confidential letters to his foreign minister—then tried to smuggle them across the ocean by giving them to an American war correspondent traveling overseas. British agents in Falmouth arrested the correspondent, seized the letters, and published them, revealing Dumba's plots to disrupt American war production by paying "soap-box orators" and union organizers to stage strikes at steel and munitions plants in the mid-west.

In writing his public response to the strikers Henry seized on the Dumba affair. "All of the strikers in so far as the action goes, contribute to delay of the delivery of machines needed by industries serving the cause of the Allies and thereby assist a scheme as was proposed by Dumba himself." Henry responded to the union's six demands with nine conditions of his own, including: "That the Company is firm in its decision to continue its regular working schedule" of five 10-hour days with five-hour Saturdays; and "the Company will continue to operate its business without dealing with labor unions, shop committees, or walking delegates." Union acknowledgment, he wrote, would "mean the surrender of everything of value in industrial dealing and practice which the company has stood for in its long history."

In his handling of the press Henry was ingenious. He welcomed all reporters into the factory for daily press conferences, where he issued brief statements spiced with "any little items of interest." By inviting reporters into the factory and making it easy for them to collect the news, he won them over. He explained to other factory owners: "Every politician learns to use the press in his peculiar necessity, and it is our honest belief that a studied use by employers will produce good results." Though workers accused the *Providence Journal* of bias, Henry gave equal access to all of the city's newspapers. He observed: "The business of the Press is news, and the more easy we make it for them to get news, the better they will serve us."

A federal mediator met with Henry that afternoon, but Henry turned down his attempts to broker a deal. The next day the president of the International Association of Machinists arrived in town; he too was turned away.

Henry did a little investigative reporting of his own, and found that the International Association of Machinists was deeply in debt. This rattled some of the strikers, who had walked out in the belief that the Association would pay them something while they were on the street. When strike leaders tried to organize a tag day to raise funds for strikers, police shut them down. Henry was so pleased with police response to the strike that he wrote a $2,000 check to the Providence Police Association.

By week's end about half of Brown & Sharpe's workforce of 5,500 was on strike, while the other 2,700 or so workers tried, with limited success, to keep up production. The story about the Association's strained finances broke over the weekend, and on the following Monday a few hundred striking workers returned to the plant, where the company welcomed them without punishment.

To avoid exacerbating class tensions, "It's important to extend a hearty welcome to those who wish to return," Henry counseled. "We employers must be broad minded enough to know that the element of intimidation and class feeling is so strong on the part of a large portion of the striking force, that we must not give the impression that a man by going out on strike commits an unpardonable offense."

With the company's inflexible position, no money forthcoming from the union, and police disrupting their fundraising attempts, strikers found themselves in a bad spot. After two weeks out, so many workers returned that Brown & Sharpe announced the end of the strike. Still, about a thousand workers remained out at a time when the company was so busy with war-related work it had to turn away orders. On October 30 some of the remaining strikers met in a hall on Kinsley Avenue, just across the river from Brown & Sharpe, where they listened to arguments from a few men who advocated an official end of the strike.

George Finnell, a 25-year-old shop worker and strike leader, told fellow strikers that if they stayed out, Brown & Sharpe would have no trouble training new workers. "The Brown & Sharpe Manufacturing Company has been a school of education for a long time," Finnell said.

> They bring a man in there and they educate him in a short while and he is making a living wage. That is what's going to happen. I say this, boys, I believe we ought to have eight hours and get proper treatment at the hands of the manufacturers, but the easiest way is the best way. On election day put men into the legislature that will vote for a universal eight hours. That is the way you want to get it, without bringing any misery upon our homes and families, and bear in mind, the workingmen are a large majority—95 percent. You can do it.

Finnell then put the question to a vote: "All those in favor of calling it off will say 'aye.' Say it loud, AYE." In response came silence.[10]

Workingmen, disenfranchised in the political process, had little faith in the ballot box. Officially the strike continued, but in fact nothing had changed and the strike would change nothing. A year later, with the factory again chugging along at full production, Henry told a gathering of the National Metal Trades Association: "We do not believe that any bitterness has resulted from the strike, and are confident that a firm attitude on the part of an employer goes to retain the real respect of the employee."

12. When Peppermint Creams Meet Steel

In her grandfather's closet there squatted a huge iron dumbbell that no one ever used. The thing only appeared when maids turned out the closet for cleaning. And one May day in 1897 there this dumbbell sat, propped atop the staircase in the upper hall. Mary Elizabeth Evans, a curious 12-year-old girl, tentatively placed one foot upon it. To gauge its heft she gave the dumbbell a little nudge. This sent it rolling; the weight of its own inertia carried it toward the stairwell. Down it went. In the solemn quiet of the house of a dying man, the tumbling dumbbell banged—bumpity, bumpity, bumpity, bump!

Miss McDill, grandfather's private nurse, came running out of the sick room in her starched collar and cuffs, her dark hair parted beneath a nurse's cap. "Don't you know your grandfather is very, very ill?" she scolded.[1] He died that afternoon.

It being May the big house in downtown Syracuse was rife with sweet-scented lilacs. The next day a great many people came to call because grandfather, Henry Riegel, had been the longest-serving judge of the Onandaga County Court. Judge Riegel had bought a lot of property. Besides the big house where he was laid out, he owned a gridded subdivision of 100 laborer's cottages that he had built on a plat of old farmland, small homes with outhouses in lieu of bathrooms. When the Syracuse limits crept out to subsume the subdivision, city fathers insisted that the judge install sewers, sidewalks, and streetlights, which placed the property under a heavy mortgage.

The mortgage came at a bad time. These small, cookie-cutter houses were meant for factory workers, people who labored in places like the Stickley furniture plant, a local machine works, and the New York Central roundhouse. The Panic of 1893 had drastically cut the factory workforce, and at the time of the judge's death in 1897 employment still had not rebounded, leaving the Riegels with a lot of vacant cottages.

Adding to the drain on the family's finances the judge's daughter, Fannie, lived in the big house with her four children including the eldest, Mary Elizabeth. Fannie was a free spirit who refused to wear corsets and donned split skirts in order to peddle a woman's bicycle. The judge and his wife had never much liked the children's father, William Evans, a music professor from Syracuse University. Though he complained of sickness, they diagnosed him as lazy. The tension between them had caused Evans to decamp for New York City, from where he had hoped to support his family through tutoring private music lessons. When that did not pan out he decided to join the Klondike Gold Rush of 1896, and then was heard from no more.[2]

With the death of her father, free-spirited Fannie was left with four children, lots of property, no husband, and no money. Immediately she sold all of the house's furniture except for the Steinway grand piano, which she brought with her to one of the little laborer's cottages on the city's outskirts. She took over two of these cottages, connected them with a passageway plumbed with a cold water bath, and moved in with her mother, four children, and Lizzie, the family's longtime Irish maid who insisted on coming for only the room and board.

Fannie's mother, Polly, was the type of person who, if she could not have the best dressmaker in town, was ashamed to be seen. As the family moved into their tiny cottages Polly paced, muttering, "We're all going to the poor house. We might just as well go now as later. We're all going to the poor house."

The family skimped by that fall—the boy, Henry, took a paper route; tenant families bartered for rent with fresh vegetables; Fannie knew a lot about mycology, and spiced up her dinners with hand-picked mushrooms. At Thanksgiving, Lizzie the maid exhausted her savings to buy a ham, which they cooked in cider and served with a fresh pumpkin pie. Come Christmas, the younger Evans children grew excited about Santa's yearly visit. At 13, Mary Elizabeth knew there was no such thing as Santa Claus, but her younger siblings did not. To head off their disappointment, Fannie sat her children down and explained: There is no Santa Claus. This year there would be no gifts.

The children figured that surely they would receive at least some stocking stuffers. At the big house the family used to hang stockings on the warm, flickering hearth. So on Christmas Eve they ignored their mother's dire prediction and draped their stockings from the dining room chairs. In the morning they spied the stockings, bulging with a ripe plumpness. Their spirits soared: there was a Santa after all! Expectant hands reached in for Christmas gifts, only to find the stockings filled with black lumps of coal wrapped in newspaper. Their mother told them they should feel grateful for even the warmth of that.

Polly's repeated imprecations of "We're all going to the poor house" were pretty much on target—the family was headed for homelessness, but not without a fight. Fannie observed that the nearest drugstore to her crumbling subdivision was two miles away, so she opened a little apothecary in one of the houses from whence she sold soap powders, syrups, salts, and talc. Lizzie baked bread and fried doughnuts, the latter being particularly popular among a gang of surveyors working the neighborhood. The problem was the neighborhood grocer, the family's largest tenant, took umbrage at the competition from bread and doughnuts and moved out, leaving a vacant storefront.

Fannie moved her maid and eldest daughter, Mary Elizabeth, into that space with their beds and clothes, and from there the two of them tried to make a go of it as grocers. In the grocery's kitchen, Mary Elizabeth got the idea to add to the store's product line by making candy. Her grandfather had always said that the children could eat all the candy they wished as long as it was not cheap, penny candy. He had installed a pull hook in the big house's kitchen so they could make their own taffy and candies of wholesome ingredients—butter, maple sugar, molasses, baker's chocolate and the like. At the age of 13, Mary Elizabeth knew a thing or two about making large batches of candy.

The local Sunday School teacher from the May Memorial Church liked the candy. To help out the struggling family with the little grocery store, she offered to sell monthly candy subscriptions to a group of her friends from the German Club—for $4 a month the subscriber received a weekly box of whatever kind of candy Mary Elizabeth had made. With a flourish, Mary Elizabeth signed each box: *Mary Elizabeth's Candy*. With Henry's paper route, sales from Mary Elizabeth's candy, bartered vegetables, and a little profit from the grocery store, the Evans family limped along for four years. Then on January 10, 1901, they received the following notice from a local bank:

> Dear Madame,
> We require possession without delay of the several houses on the Riegel tract now occupied by you, and will you please take notice that we hereby demand that you surrender and give to us possession of the premises ... or we will be compelled to take legal proceedings to dispossess you without further notice.[3]

The foreclosure notice gave the family ten days to find a new home, but they had no money to find one. With her mother on the edge of hysteria, 16-year-old Mary Elizabeth felt that finding housing was her responsibility. She remembered her grandfather's words—to "think big and not be afraid of the issue."

Mary Elizabeth resolved to "live life on a green light." Without asking her mother she called on a family friend, Patty Wright, who owned The Flora, a big apartment house in Syracuse. Patty had nothing available in The Flora, but she did have a vacant blockhouse with a brownstone front that she could let for $30 a month. The Evans' could not raise $30 a month, not yet, but they moved in anyway, turning the basement into a candy kitchen with a second-story display room. Back in their old downtown neighborhood, the family received more help from people who had known the judge. Whenever there was an overdraft or an overdue bill, Fanny dispatched Mary Elizabeth to see Miss Sewall, who would lift up her skirts, take out a lot of dollar bills, and cover the expense.

In 1903, the Stickley Brothers furniture factory agreed to build the family a mission style cabinet, gratis, that Mary Elizabeth installed in a busy, downtown office building. She filled the cabinet with candy and a little sign that read: *Open these doors, take what you will, leave costs of goods taken, make your change from my till.* Miss Jessup who ran the telephone switchboard kept an eye on the unstaffed candy booth, as did the half-dozen elevator operators shuffling office workers up and down the building; people bustling by bought candy and made the correct change without short-changing the till.

Newspapers caught wind of the story of a teenage entrepreneur supporting her struggling family through her trust and her candy booth. Papers in Syracuse and in Buffalo picked up on the story, as well as a national candy magazine. Within six months, Mary Elizabeth sold 16,000 boxes of candy. Mary Elizabeth's first hire was a 14-year-old black girl named Harriet who stirred 50-pound batches of boiling candy with a big wooden oar.[4] She eventually rose to the post of head candy maker. As the candy cooled the family gathered to cut it with scissors and to wrap it by hand, a process that intrigued young Henry Evans, a mechanically minded teen who had once built his own violin. He developed a machine that allowed Harriet and his sisters to drop the candy into a feed that caught and wrapped the candy for them.

As the family began to make money, Mary Elizabeth embraced another of Judge Riegel's aphorisms: Money is not a thing to possess—it is a tool to create things. So Mary Elizabeth plowed money back into her business in order to grow it. In summers she took to the rails, chasing vacationing office workers from the empty bank building in downtown Syracuse to the hotels where they stayed in New Hampshire, Maine, and Rhode Island. At Newport she spied an empty florist's shop across the street from the Casino, where the city's elite gathered to play tennis. She rented that for a candy showroom, and filled the extra space with prettily decorated tables for teas and cakes.

The tea proved even more profitable than the candy, so in 1911 she took the idea to New York, opening a Mary Elizabeth's Tea Room in an old brownstone facing Fifth Avenue. Within a year, business was so brisk that Mary Elizabeth took on a bigger property, paying $50,000 a year for two floors at the corner of 36th Street and Fifth Avenue. She moved the candy factory into the basement—Harriet the head candy-maker moved down from Syracuse, and they began pumping out candy, tea, cakes, and a full luncheon menu. In the busy season, Mary Elizabeth's served 1,000 luncheons per day, plus sales of cakes and candy. She was not yet 30 years old. One newspaper described her as "the pretty brunette, noticeable for her girlish, frank smile and expressive brown eyes" who earned twice as much as most bank presidents and "no less than the President of the United States."[5]

As a teenager in Syracuse, Mary Elizabeth Evans lifted her dispossessed family out of poverty by selling homemade candies. She expanded her first candy store in Newport, Rhode Island, into a tea room, then took the tea room idea to Manhattan, where she sold more than 1,000 lunches per day. Before she was 30 she was a self-made millionaire (courtesy Peggy Sharpe).

Mary Elizabeth loved horses. Even at her poorest she found access to them, riding with young men from New York's National Guard who boarded their horses near the laborers' cottages. She had even obtained a free carthorse named Ben-Hur that she had used to make grocery deliveries. Now that she had money, Mary Elizabeth decided to treat herself and her friend Rosie to a "wild west" adventure at the Eaton Family Dude Ranch near Sheridan, Wyoming.

The ranch required a three-week stay so guests could learn to ride with western saddles, cowboy style, before moving up to Montana for a ride through the new Glacier

National Park. A party from Rhode Island had also booked this trip: Henry Sharpe with two of his sisters and two of the Chafee children. At first Henry wanted his party's tents to be set up away from the main encampment for privacy; but after a few meetings with Rosie and Mary Elizabeth, he told the camp's owners that he would like to move closer to the rest of the group after all.

Though Henry was a dozen years older than the two women from Syracuse, Rosie felt an attraction to him. After a few days of flirting in vain she said to Mary Elizabeth: "Well I'm not making any headway at all. He hasn't looked at me. He seems to be more interested in you." Henry was a man whom everybody took seriously—he owned one of the largest machine tool companies in the world, was tall, square shouldered, with brown eyes, a full head of white hair, and he carried himself in a quiet way that commanded respect. Mary Elizabeth could not help but poke fun at him. She called him Harry after Harry Hotspur, a medieval English nobleman made famous in Shakespeare's *Henry IV*.

"But my name is Henry," he said.

"Well I don't care. I'm going to call you Harry."

Henry Sharpe enjoyed the spunk of this younger woman; she refused to take him as seriously as most people did, and that brought out his lighter side. When he returned to Providence he began going to New York to visit her at her Fifth Avenue apartment, beginning what he hoped would be a long-distance courtship. Her occasional presence in his life doubtless buoyed the trying days of his brother's estrangement, and the bust-boom cycle of 1913 to 1915 that first gave him too little work, then too much to handle.

From his small, third-floor office in the old section of the Brown & Sharpe factory, William Viall tried to soothe the angst of a war-harried customer in England. In June of 1915 an agent of the tool-making firm Buck & Hickman fretted about recent design changes in a screw machine that shaped wire steel. The firm was leery about buying Brown & Sharpe machines if the American company planned to keep tweaking them, because change produced a learning curve that slowed production. Viall assured a sales agent that "no changes of any kind, are contemplated at the present time. The line is in as fixed a condition as any of our lines."

The war years stymied innovation in machine tool design, but hyper-stimulated production. As the war ground on into the following summer of 1916, Brown & Sharpe faced a backlog of one year between orders and deliveries due to increased demand and the previous year's strike. A Chicago banker asked Henry Sharpe's opinion on the state of the tools business, prompting him to reply that in both the machine tools and small tools lines, demand was "excessive." The price for steels and other raw materials had jumped an average of 40 percent as the war devoured metals for ships, artillery, tanks, shells.

Henry attributed two-thirds of Brown & Sharpe's machine tools work to war-related business, and much of the remaining orders came from U.S. automobile plants—some of which continued making cars for civilian use during the war. War work often came through government contracts. Since government was spending tax money furnishing the Allies with weapons, pro-labor congressmen and senators felt that factory owners should share some of their windfall profits with workers. Through legislation they tried to obtain what workers often failed to win through strikes: eight-hour days and union representation.

Henry became incensed by Congressional efforts to dictate working conditions to the owners of factories who signed government contracts. He kept a sharp eye out for legislation such as the Tavenner Bill of 1916 which sought (unsuccessfully) to outlaw stop-watches and other time-measuring devices for efficiency studies in any government work. Brown & Sharpe was an early adapter of "scientific management"—Henry Leland introduced operations sheets and piece work into the plant long before Fredrick Taylor coined the phrase—and continually sought more efficient ways of manufacturing, as evidenced by William Viall's worn, annotated copy of *The Human Factor in Works Management,* published in 1912. Rhode Island held three seats in Congress, and Henry suspected that two of the state's representatives favored attempts to attach pro-labor riders to government contracts: the Irish-born George O-Saughnessy, and the liberal Republican Ambrose Kennedy. Henry told the Chicago banker that in his opinion "politics in this country are still a great menace to business. Congress, one after another, seem called upon to represent minorities instead of the real American feeling."

As the war in Europe raged on, the "real American feeling" was tough to discern. America was a polyglot of ethnicities riven by class and regional differences. Nowhere were these differences of class and ethnicity more obvious than in the industrial cities such as Providence, where a majority of people were first or second-generation Americans. Trying to marshal such a disparate people into unified action was often impossible. Just two years before America's entry into the war, 3,000 German and Irish immigrants paraded through downtown Providence demanding United States neutrality. Germans opposed American intervention out of loyalty for their homeland; the Irish opposed it because they feared their native countrymen could be impressed into fighting for England, to them an oppressor nation; and Catholics of all ethnicities were wary of Woodrow Wilson's recognition of Mexican president Venustiano Carranza, who was persecuting Catholic clergy.[6]

For multi-generational Americans such as Henry Sharpe, the nation needed to develop a unified American feeling in order to progress as one nation with common goals. In 1917, Brown & Sharpe employed 1,312 non-citizens, and the firm launched an in-plant campaign to "assist all employees who desired to become American citizens," by teaching English classes and civics. As a result of the push to Americanize workers, all but 201 applied for citizenship.[7] Henry also contributed to the North American Civic League for Immigrants, which sought to Americanize immigrants.

Henry chaired a housing committee created by the local Chamber of Commerce to study housing conditions in Providence, an issue that caught the Chamber's eye when triple-deckers began crowding the streets of ethnic neighborhoods such as Silver Lake and Federal Hill. The study's author found that

> Italians, like the Irish, may be expected to marry among themselves. ... With some of them it is esteemed a virtue to do as much trading as possible among themselves. They form little communities which consciously aim to be as self-sufficient as possible. As their numbers grow so does their ability to form these self-contained alien communities in the midst of an American community.[8]

Neighborhood enclaves with distinctive foods, music, and language made some of the long-established Americans nervous. The housing survey concluded that "so long as a foreign group lives as a group and maintains different, especially lower standards, it is not assimilated." New arrivals did not necessarily see their standards as lower, and many were loath to give up familiar rites in favor of assimilating; Polish and French-Canadian families often suspected attempts to Americanize them were attacks on their Catholic faith, and they zealously held onto their foods, festivals, and languages.

The 1916 housing survey found that living conditions were not good—a census counted 1,807 outhouses within city limits, and tenements that had indoor toilets often flushed into open cesspools. Worse than these, according to the study, were the "cellar water-closets" where "filth gathers in the dark corners behind the fixtures, [and] the constant emptying of slops saturates the woodwork and floor, so even when the cellar is dry the compartment is likely to be damp" and foul with human wastes.[9] Ethnic neighborhoods and substandard housing presented problems as America mobilized for war. "The future soldier must be born and bred in a wholesome dwelling," the 1916 report opined. The United States Marine Corps was finding a dearth of good men fit for enlistment. In Boston, Marine recruiters deemed only 107 of 1,110 applicants as fit enough join the Corps, though segregation and bigotry could have been limiting factors.

With German U-boats targeting American ships to stem the flow of munitions, civilian casualties aboard passenger ships such as the *Lusitania* mounted, tipping American sentiment toward entering the war. One year after the anti-intervention rally, Providence's civic leaders staged a Preparedness Parade with a slogan of: *"Americanism" means undivided loyalty to the US of A*. The parade drew 53,000 marchers.[10]

In April 1917, President Wilson addressed a joint session of Congress; famously intoning, "The world must be made safe for Democracy," he sought a declaration of war against Germany. Six senators and 50 house members opposed his request, but on April 6 the measure passed. America was at war.

There was no way that Mary Elizabeth Evans, a woman who lived on a green light, could miss out on witnessing history in the making. She was determined to sail to France. Her "little old Harry" was pressuring marriage, and though she was fond of him, she thought she really could not be in love with him or she would not want so badly to go to France. When she eventually did cross the Atlantic in a convoy in the spring of 1918, her "Harry" bought her the best life preserver on the market, in case her ship should be shot out from under her.

Worry about his girlfriend who wanted to go to France drove Henry Sharpe to distraction in the summer of 1917. Besides Mary Elizabeth's waffling on marriage, big brother Lucian was abroad, causing uneasiness by sending postcards of his visits to munitions plants in Italy and France, big customers of Brown & Sharpe's. Henry could only hope that his unstable brother would not embarrass the firm. Henry also received letters from lawyers seeking Lucian's whereabouts, because he was mixed up in an art-buying hoax. Lucian had purchased $400,000 worth of art and furnishings for his Cambridge house, decorative arts that his lawyer discovered to be nearly worthless; the lawyer got the money back, but was now suing Lucian for neglecting to pay him.

Lyra Nickerson—Joseph Brown's daughter, had died the previous August, leaving Henry scrambling to ensure that her significant share of stock remained in control of current stockholders. He succeeded, but the stock purchase required major capital from each stockholder to raise the $255,000 payable to the Nickerson estate, all of which was donated to the Providence Public Library and the Rhode Island School of Design.

Henry also served on the board of Rhode Island Hospital Trust, a bank that had just broken ground on an 11-story, U-shaped building of Italian High Renaissance style, fronted with white limestone. And of course the factory was busier than ever. Brown & Sharpe now employed more than 6,000 people. When America entered the war, workers left for the military, some 900 men. While Henry was trying to drum up workers to keep the plant running, the federal government was trying to get him to eliminate piece work on its contracts, and to cut the work day from 10 hours to 8. Henry could scarcely contain his outrage as he dashed off a letter to the federal munitions board:

> In anticipating co-operation in preparation for the great war this country has entered upon, the manufacturers of this country are generally patriotic, and wish to do everything needful in order to back the United States. In our opinion industries are a way in which to astonish the world by their effectiveness, but it will be most damaging if their efforts are to be lessened, and arrangements of years standing are to be jeopardized by injections of prescription of this sort.
>
> … We believe that the Government should place its orders for its requirements at prices which may be satisfactory to it, and that every manufacturer should be left to carry on his industrial arrangements as may be permitted by state laws, leaving arrangements for reimbursing of workmen untrammeled.

Besides chiding the federal government for meddling in business, Henry was also arguing with the State of Rhode Island about a similar issue: Workmen's Compensation Insurance. Brown & Sharpe had recently embraced an insurance plan for it workers—for an employee contribution of 5 cents a week, each worker would receive "blanket insurance" of up to $250. If the insurance pool ran a deficit, the company would pay the difference; if it posted a profit, the money would go to the Mutual Benefit Association for workers. Henry had no problem with contributing to a workman's insurance pool—he held a progressivist view of "factory welfare," providing a library for workers, sanitary locker rooms, and company-backed insurance pools; but he felt suspicious of state-mandated programs such as workers compensation. Henry agreed with the idea that an insurance company should pay the medical bills of a person injured on the job. But as long as the company was paying the bills, it should be able to choose the doctor. A proposed provision in state law allowed the injured party to hire a doctor at insurance company expense. Henry concluded that this would "encourage the existence of ambulance chasing doctors" and inevitably lead to fraud: "As is well known, there are a host of men called doctors who, under such a provision, are encouraged to exist through seeking the patronage of injured men."

The state also sought to pay compensation after a worker had missed just two weeks of work instead of after six missed weeks. Henry figured that would just encourage workers who were not seriously injured to fake an injury long enough to draw a free two weeks pay. "Malingering is the curse of compensation legislation the world over," he wrote, "and every encouragement will be given to this if the assembly proceeds to enact such a provision into our Rhode Island law."

All of this finally got to be too much for Henry—an excessively busy factory; too few workers to fill orders and a federal government mandating fewer working hours; state regulators meddling in the employer-employee compensation; negotiating a major stock buy-back; an ambitious bank building project; and a girlfriend who wanted to delay marriage until she got a first-hand view of the war. In April he wrote a harsh letter to the regional Internal Revenue Collector, demanding an explanation for the collections office's failure to protect the company's financial information, for him always a touchy subject. "It is a very annoying proceeding and, we are frank to say, very unbusinesslike," he scolded. After this uncharacteristic outburst, Henry took his doctor's advice and checked into the new Bethel Inn on the Maine side of the White Mountains for an extended rest. Brown & Sharpe was so busy that it did not have to chase down orders; the company was awash in profitable work, and in the hands of a good manager in William Viall.

Arctic air flowed into Rhode Island just before Christmas, and by mid–January 1918 the river and bay thickly froze all the way from Providence to Newport. Coast Guard ice breakers worked continuously to bust out a channel, but there was so much floating ice that the coal companies would not send their barges through to deliver coal to the city. Schools closed for want of heat; some of Providence's foundries resorted to burning cordwood in place of coal to stoke their fires. Besides wrestling with the fuel problem, William Viall puzzled over how to replace men marching off to war when orders continued to come in at an all-time high? The most obvious solution was women. In 1918 Brown & Sharpe began employing them in ever-greater numbers.

Until 1917, the firm limited women to clerical work. Now managers did not have that luxury; they began training women as riveters, assemblers, lathe workers, screw machine operators, gear cutters. The only place that remained off limits to women was the foundry. Managers felt some trepidation at employing women in these non-traditional roles, expressing worries about discipline, production, turnover, and safety.

In the early phases of the experiment, things went well enough for Viall to write to Henry up in Maine with a request to spend serious money on new locker rooms to expand the female workforce: "In going over the whole situation, they have made good and there is no reason why, as the necessity grows greater, we should not continue to add to their numbers. There is the problem of housing them, however, that is before us....

"The thing that presents itself to us as a solution is to put a floor on top of the #2 Building, and the sketches, which we are sending, show what could be done in this line." There were a couple of problems with adding another floor to the old No. 2 building. For one, the elder Lucian Sharpe had already added a third-floor to that wing to increase factory space; there was no assurance that the stone footing could handle the weight of another floor till workers piled 35 tons of iron over a column and determined that it could. The other problem was cost, about $4 per foot to include the heating and lighting apparatus, plus the cost of the toilets, lockers, sinks and chairs, which would make a total of $66,000, more than $1.1 million by early twenty-first-century standards.

"This is a very high figure for such a structure," Viall reported, "but as I have talked this matter over with Mr. DeWolf and Mr. Hatch, I do not see how we can very well avoid it, provided we are to extend the employment of women."

The company did "extend" its employment of women in 1918, ferrying groups of them via freight elevator to a locker room in the upper reaches of the factory, where they stuffed long hair into hair nets, changed into work clothes for 9-hour shifts, and showered off grease and machine oil before returning home, cutting a "neat and orderly appearance not only in the shop but also on the street coming to and leaving work."

All told, Brown & Sharpe hired 1,481 women during the war, all but a few hundred of them on the shop floor. All of them spoke English, but managers placed them in two categories: "English Speaking Races" defined as "American, English and Canadian, Irish, and Scots"; and "Non-English Speaking Races" defined as French and French Canadian, Swedes and Norwegians, Italians, Jews, and Russians." There were no Asian, Black, or Spanish-speaking hires. An assessment of the women's performance showed that "although women seemed to be more ready to lose time than men, they have on

Women machinists working the milling machines during World War I. Brown & Sharpe hired 1,481 women during the war, most of them as machinists and floor workers. A company report concluded that "women were doing better work and had learned more quickly than the men." Some foremen said that women outworked the men because the men did such a good job of training them.

the whole shown more stability" by staying on the job longer than men. More than half of the women, 57.7 percent, were still on the job after one year, whereas only 36.5 percent of men remained for a year.[11]

Supervisors found that men and women worked better and produced more when they worked together, rather than when they were segregated. Discipline proved not to be a problem. The factory hired "matrons" that supervised groups of "girls" in each department, to "prevent objectionable features creeping into the shop." As far as safety went: "the accident record shows that they have had four-fifths as many accidents, proportionally, as the men, and that, therefore, they have not proved to be an added hazard, as it was predicted that they would be."

The real test was production. Here a report by company historian Luther D. Burlingame concluded that "women were doing better work and had learned more quickly than the men." Foremen explained that they had given more attention to women than they had given to the average male employee, "the assumption being that a woman, having less mechanical background and intuition than a man, required more training and more specific instructions. This has been the reason advanced by some foremen in explaining" why the women out-performed the men. In other words, women out-worked men because the men did such a good job of training them.

In September of 1918 a representative from the new Lincoln Motor Company dropped into Brown & Sharpe, in order to discuss the building of "aeroplane" engines. Henry thought this was a great idea. After the visit, Henry wrote a letter to the Lincoln Company's founder, his father's one-time employee, Henry Leland, now a robust 67 years old.

"Personally I wish to state I am anxious that our Company shall be of every possible assistance to you in your efforts to assist the carrying out of the aeroplane programme, for patriotic reasons as well as considering the old friendship which you have always had for this company, and we shall continue to strive to assist you in every way we can."

Leland chartered the Lincoln company in 1917, after breaking away from his old firm, the Cadillac Motor Company. Before America's entry into the war, Leland had sold Cadillac for $4.6 million to Will Durant and his General Motors consortium that had been buying up car companies all over the mid–West. Leland stayed on as a manager until he and Durant had a serious ethical split over the looming war: Leland wanted to make airplane engines, not only for profit but because he thought Allied air power could hasten the war's end; Durant, a pacifist, did want to bloody his hands with war work.

Leland quit over the issue. He then negotiated a $10.6 million loan from the government to build and equip a plant to manufacture 12-cylinder "Liberty" engines to power airplanes. The complexity of building engines that were both light and powerful required expensive research and development. Leland negotiated a controversial "cost-plus" contract with the government that protected him from loss and rewarded him for success.[12]

The hawk-dove dispute that divided Leland and Durant also played out in Henry Sharpe's own family. His nephews, John and Zechariah Chafee came down on decidedly different sides of the issue. John first joined the war as an ambulance driver for the Red Cross, then came home to train as an artillery officer before returning to the Western Front. Zechariah, a pacifist and Harvard-trained lawyer, took a different route. Horrified by draconian federal actions and laws of the World War I-era, such as the Palmer Raids and the Espionage Act (which made it a crime to even speak against "the form of government of the United States,") he published a cogent article in the *Harvard Review* that set the cornerstone of his career as America's pre-eminent First Amendment scholar.

By the Fall of 1918, Henry closely knew four people in and around Paris—his nephew, John, the artillery officer; nephew-in-law, Jim Gamble, Elisabeth Chafee's husband working as a Red Cross doctor; Richmond Viall, son of William Viall, now piloting a Sopwith Camel over Europe with the Royal Canadian Air Force; and his future wife, Mary Elizabeth Evans, who ran a Red Cross kitchen in Paris while German guns shelled the city.

Mary Elizabeth found the night raids "most terrifying things, because you'd be sweetly asleep, and you'd be awfully tired, and you'd be suddenly awakened by this horrible, horrible noise." Sleep was hard to come by. Many nights she was called to the train station to meet cars full of displaced children in need of a hot drink and some comfort; she felt too exhausted to heed the sirens warning of incoming shells. "You get used to these things," she concluded. "You *have* to."

At Brown & Sharpe there was no ambivalence about the war. In late August of 1918 some 7,000 employees assembled outside the plant for the first of many Loyalty Mass Meetings. The Brown & Sharpe Band, led by J.J. Rooney, whipped up peppy, jingoistic tunes before Capt. W.A. Cameron of the Canadian Army gave a patriotic oration.

Along Promenade Street, shop carpenters built what looked like a 150-foot thermometer tipped onto its side, with the bulb painted to look like a globe. A 15-foot caricature of the Kaiser stood precariously atop the north pole, while a goateed Uncle Sam clutching a big stick snarled at him from the thermometer's tip. Whenever workers pledged to loan the government $25,000 through the Liberty Loans program, Uncle Sam plunged farther down the thermometer to "Drive the Kaiser off the Earth." On the October day that worker loans to the government topped $400,000, Uncle Sam and his big stick reached the Kaiser, who was knocked down into a throng of workers that tore the figure "to small bits, to be carried off as souvenirs."[13]

Brown & Sharpe workers eventually loaned $1.4 million to the government at 4 percent interest, receiving more than $65,000 in annual interest on their money. It was unusual for factory workers to loan money to the U.S. treasury, but the war had forged new economic relationships between Americans and their government, as well as between America and the world. In the fiscal year that ended in 1917, the federal government's budget totaled less than $800 million. During the two years America was at war, the government spent $15.5 billion each year, a 20-fold increase with the majority of it funded through the relatively new income tax. Excises and customs duties no longer provided the bulk of federal revenue—taxpayers did.[14] Outside of the 1860s, the World War I era was arguably the most transformative period in American history. Domestically, the federal government grew from a relatively small provider of post

To spur interest in buying war bonds, shop carpenters built a 150-foot thermometer tipped onto its side, with the bulb painted to look like a globe. A 15-foot caricature of the Kaiser stood atop the north pole, while a goateed Uncle Sam clutching a big stick snarled at him from the thermometer's tip. When pledges for the Liberty Loans program reached $400,000, Uncle Sam knocked the Kaiser "off the Earth" into a waiting thong of workers who tore the caricature to bits.

offices and customs collectors to a large, central power with its own Federal Bureau of Investigation.

When Europe first went to war, the United States was a debtor nation. America and its companies were still paying for 1890s improvements in industrial and transportation infrastructure. In 1914, the money stream gushed the other way, from Europe to the United States.

The war years also prodded changes in Brown & Sharpe's spending habits. Weeks before America entered the war in 1917, Henry Sharpe reminded stockholders "it has not been our general practice to contribute substantially of corporation funds for any charitable purpose whatsoever." However, he asked each shareholder to contribute a pro rata share to raise $5,000 for a Red Cross hospital for soldiers in Newport, explaining that "this is a movement entirely outside of the usual sort." A year later he asked stockholders for a corporate contribution of $25,000 to the Red Cross, citing the "patriotic character of the Fund" and "the fact that the vast majority of our business can be considered as War business." And in the October of 1918, Henry again pushed through another $25,000 donation from Brown & Sharpe to the United War Work Campaign, a consortium or seven war-related charities. In arguing for the expenditure Henry wrote to his company's board: "The large and profitable amount of business now enjoyed by the Company is practically all incidental to the conduct of the War: It may be considered as War business."

On November 11, the war business came to and end. Before sunup, word circulated that the Allied forces and Axis powers had signed an armistice. The Brown & Sharpe band met workers outside the shop, soon forming an impromptu parade. Automobiles loaded with employees circled the works, horns honking; soon the parade of marchers, band members, and cars flowed noisily away from the factory toward downtown where it was subsumed in a wild celebration. There would be no work done on Armistice Day.[15]

Henry's nephew, nephew-in-law, and Richmond Viall survived their time in France. Mary Elizabeth lived too, but unlike the others, she lingered overseas. The Red Cross asked her to turn the kitchen of the Hotel du Louvre into an officers mess hall with American food for officers on leave and waiting to be shipped back home. They wanted to have pumpkin pies and baked beans in lieu of French cuisine, and Mary Elizabeth complied, schooling a staff of 50 French cooks in the art of American cooking.

In May of 1919, Mary Elizabeth sailed home, this time freed of the worry that German U-boats might blast the ship out from under her. Henry, now 46 years old, was in California when her ship touched shore, and as soon as he heard she was home he booked passage on an east-bound train. On spying Mary Elizabeth in New York the first words out of his mouth were, "Now when are we going to get married?"

She replied, "I'm not sure we're ever going to get married, Harry."

But one day, returning to

Henry Sharpe, a man of contrasts who embraced progressive era ideas of "factory welfare" for workers, while resisting unions and government-mandated programs. He remained a bachelor well into his 40s, before proposing to self-made millionaire Mary Elizabeth Evans. "Mary," he said, "it would be so beautiful to walk down the path of life with a woman of imagination" (courtesy Peggy Sharpe).

New York from a trip to Boston, her train stopped in Providence long enough for Henry to board with an armful of magazines for her reading pleasure. "Mary," he said, "it would be so beautiful to walk down the path of life with a woman of imagination."

"I was so touched by this remark," Mary later said, "so out of character, that I knew *then* I would never meet a more gentle man."[16] That was the paradox of Henry Sharpe, Sr.: in casual conversation he came across as a gruff, cigar-chomping, taciturn Yankee. If a union or government official tried to negotiate change inside his factory, Henry pushed back. And yet he was gentle; he liked animals too much to buy his wife furs, doted on his dogs, and adored horses. He became an early adopter of company-paid pensions for salaried workers and workmen's insurance plans for laborers; he believed in a clean, sanitary workplace with a well-stocked library to make workers happy and productive; and he was a known soft-touch among both community fundraisers and for individuals down on their luck. Henry Sharpe gave generously—he was a charter member of the Community Chest organization that grew into the United Way—but casual conversation pained him, and he gruffly rebuffed anyone trying to tell him how to run his business.

Within six months of her return from Paris, candy-maker and restaurateur Mary Elizabeth Evans was sporting an engagement ring with a big diamond. Some wag in the Brown & Sharpe factory wrote a ditty for the occasion, sung to the tune of "Pack Up Your Troubles":

> When Henry plunges into married life it's good-bye Viall.
> Takes to himself a very charming wife and hookey's all the while,
> Oh, that will start 'em worrying
> When he sets the style,
> B. & S. machines all making sweets; oh, Viall, Viall, Viall!
>
> No more micrometers or fine machines; oh, Viall, Viall, Viall!
> Nine tons of candy, mostly nut pralines, wrapped in modern style,
> Good-bye good old factory,
> Won't it rouse your bile
> When peppermint creams get mixed with steel, oh, Viall, Viall, Viall.[17]

When Henry began introducing his fiancée around Providence, Mary Elizabeth cut a fine figure. His niece, Elisabeth Chafee Gamble, found her to be "solid gold, a real queen. She will bring the needed human touch into the Sharpe family."[18]

13. "If I Had Known This Was Coming"

The wedding was a hot and crowded affair, held in Mary Elizabeth's 5th Avenue apartment at 6 p.m. of a long June evening in 1920. "The bride and groom took altogether too much time preparing to go away," groused Jim Gamble, who got dragooned into attending the marriage of his wife, Elisabeth's cousin, Henry Sharpe. "It was after 9 o'clock when they left. For me this meant a very long interval of difficult conversation with a great many folks."

Prohibition had just become the law of the land, so Gamble could not even loosen up with a little liquor to get him through the long, hot evening. "There was something in the punch which made me drink hopefully of it," he reported, "but it gave me little encouragement." As for the bride, Mary Elizabeth, "I thought for my half minute acquaintance that she seemed a very honest and wholesome person. She was also handsome—even as a bride."[1] Prohibition was one of many radical changes sweeping through American culture in the post-war years. Though women in Providence had won the right to vote in presidential elections in 1917, women's suffrage for all offices was just two months away from becoming federal law.

So much had changed in the years book-ending the Great War. Prohibition and universal suffrage followed it. Pre-war, Congress had passed an income tax to lessen the federal government's reliance on duties and tariffs, and had signed a bill creating the Federal Reserve, a central bank comprised of member banks with the power to control the nation's gold supply, set interest rates, and bail out failing banks in times of crisis.[2] The Federal Reserve had a goal to create a stable, gold-backed currency that could rival the British pound's unquestioned acceptance as good money in international trade.

When Britain went to war in August of 1914, the Bank of England suspended all payments in gold until the end of the war. America's new Federal Reserve then began issuing a financial instrument to replace Britain's sterling bill of exchange: the banker's acceptance. A bill stamped "accepted" by a Federal Reserve bank was as good as gold on the international market. During the war years American bank acceptances, backed by gold and the regional banks of the Federal Reserve, quickly became the global standard. In August of 1914, Americans owed foreign lenders such as the Bank of England $3.7 billion. By 1920, foreigners buying U.S.–made goods and paying interest rates on

U.S.–backed acceptances owed Americans $12.6 billion, setting the foundation for a decade of prosperity. But for a couple of years before the Roaring Twenties got on track the global economy struggled to find its equilibrium. Government armories, no longer needing to push out war products, cancelled orders and glutted the market with surplus machine tools, putting a hit in what had become Brown & Sharpe's core business.

"The war has brought up conditions in business that were little dreamed of prior to August 1914," mused William Viall, Brown & Sharpe's corporate secretary. "How we are to solve the problems of this new world, I do not know," Viall wrote, but the company intended to seek solutions to these new global challenges. In the summer of 1920 it sent a representative on a tour of Japan, China, India, and Australia, with emphasis on using "the best possible judgment in regard to China."

After returning from his months-long honeymoon, Henry Sharpe got around to answering an inquiry from *American Machinist* magazine soliciting his outlook for the industry in 1921. He wrote: "During recent months we have suffered the cancellation of an exceedingly large amount of machine tools," leading to layoffs and shortened work weeks. Whenever a belt powering a machine snapped it was not replaced and that machine was simply taken offline, weeding out the forest of overhanging belts as the factory grew one machine quieter. "In our opinion, the machine tool industry is in for a period of dull times, the extent of which no man can predict. Some relief may be effected in a relatively short period and then again there may be a prolonged period of dullness, reaching into 1922, or beyond."

As Henry saw it, cyclical doldrums presented new challenges to large, technologically sophisticated businesses. In the old days a mill owner faced with a slump could just layoff unskilled workers until business got good again, then re-hire a new batch who could easily do the job. But Brown & Sharpe employed skilled machine operators working within a carefully choreographed system governed by operations sheets that carefully tracked the evolution of every product they made. If he let those workers go en masse, it could take years of training to replicate their skills and knowledge.

Henry explained to *American Machinist*:

> The more a business is developed ... the more is the organization, itself, of vital importance. Hence, during slack times such as are upon us, it is important that the skeleton of the employing force be kept together and utilized to it best until larger manufacturing operations can be performed. Doubtless all manufacturers of machine tools have before them the keeping of their organizations.

One key sector of the economy offered hope to machine tools builders: car manufacturers. The industry had been in a state of flux; though car companies continued to make models for civilian use throughout the war, many automobile plants converted to making war-related products such as tanks, airplane engines, and Eagle Boats to hunt submarines. Those plants would have to retool to resume making cars. "The automobile trade will be of a normal one in future time, and insofar as it is, the machine tool business will be more healthy," Henry predicted.

In the meantime, Henry diversified his investments. He bought from the Rhode Island Hospital Trust, which had just finished building its tall Italianate headquarters,

275 building lots on the old Elm Farm, a couple of miles north of the Brown & Sharpe plant. A stream ran through the acreage, flanked on either side by a wide median and parallel roadways called the Pleasant Valley Parkway. Brown & Sharpe created the Pleasant Valley Land Company to hold all of the company's real estate, and to develop bungalows along the Parkway for sale to the city's factory workers. Henry also invested $50,000 of Brown & Sharpe's money to buy 7 percent interest in the new Biltmore Hotel, to be built downtown adjacent to City Hall. This purchase, he wrote to the chairman of the campaign to build the hotel, was not a private investment as much as it was a public service: "Our belief is that the inauguration of a new hotel, such is contemplated, will be of the greatest value to the progress of greater Providence and that the large industries should subscribe liberally" to its erection.

On the floor of Brown & Sharpe, Henry dealt with hard times by reducing his workforce while keeping enough men—most of the women had been replaced by returning soldiers—to maintain a critical mass of skills. It was not easy. Between February 1920 and September of 1921, the company's workforce plunged from its all-time peak of 7,600 (as the company put returning soldiers back to work) to less than 1,000.

Henry saw a partial solution to hard times in the Winslow-Townsend Bill, a $360,000 million bailout for floundering railroad companies. In August of 1921 he wrote to his senator, LeBaron Colt, urging the bill's passage: "Considering the extreme depression in business, we consider the best avenue now lies through a rehabilitation of our railway systems....

"In ordinary times this company, among others, enjoys considerable patronage from railroads."

Unions opposed the railroad bailout bill, because they feared the rail companies would use that money to build new, non-union repair shops at the expense of existing shops where unions had established a foothold during the two war years that government had run the trains. In the end, Henry Sharpe and the railroad men won out over the unions as Congress passed the bill, giving railroads tax money to pay past bills, place new orders, and roll back union gains.

Business did get better in 1922, but Henry described conditions to a company stockholder as "still bad." The company continued paying stockholders, but through depletion of surplus, not with dividends from profits. "1923 has opened hopefully," Henry reported early in that year,

> but now there has set in an increase of wages and commodity prices which is discouraging.... Our foreign trade has been swept away by conditions abroad—when they will become better no one can tell, but probably not for some time. Generally speaking, the machine tool trade has been hard hit throughout the country, but is now more hopeful than before.

In his personal life, the year 1923 opened on a hopeful note for Henry Sharpe. At age 51, with his wife nearing 40, Henry and Mary Elizabeth had a baby. It was a boy: Henry Dexter Sharpe, Jr., a male heir who could fulfill Nelson's watchword about the expectations of duty.

The **1926** Ford Model T rolled off the assembly lines wearing coats of different colors: Highland Green, Gunmetal Blue, Phoenix Brown, Royal Maroon. These colors marked a sea change in the thinking of Henry Ford, who had famously

13. "If I Had Known This Was Coming"

In 1923, when he was 51 and she was 39, Henry and Mary Elizabeth Evans Sharpe became first-time parents with the delivery of a boy, Henry Sharpe, Jr. Mary Elizabeth developed a post-cesarean wound infection that kept her hospitalized for months (courtesy Peggy Sharpe).

declared in 1912, "Any customer can have a car painted any color he wants, as long as it is black."

For decades, Ford had built a "car for the masses," keeping prices down by employing a system of true mass production, which employed techniques of the American System to achieve interchangeability of parts—working off a model with custom jigs, fixtures, and gauges to hold metal stock firmly and consistently in place to achieve precise measurements. But Ford revolutionized the System by introducing the moving assembly line and the $5 work day, a doubling of wages that compensated workers for the boredom of being a cog in a moving machine.[3]

Throughout the war and beyond, Ford dominated the automobile market with his all-black, no frills Model T, cornering 50 percent of the automobile market by 1921. But by mid-decade other car companies, most notably General Motors' Chevrolet division, had begun employing the techniques of a more flexible mass production to build low-priced cars. They still cost a little more than Ford's, $400 versus $310 (the current equivalent of about $5400 versus $4,200) but they offered much-improved engineering in "ignition, carburetion, transmission, brake, and suspension systems," as well as stylizing that made the Model T look like the antique that it was.[4]

Ford's initial response was to give the masses color, balloon tires, and nickel-plated radiators. They did not buy it; and by 1927 Ford's market share had shrunk to less than 15 percent.[5] Ford's solution, in the words of one his supervisors, was to "get rid of all the Model-T sons of bitches" and replace them with an entirely new model, the Model-A. To do this, Ford bought 4,500 machine tools and scrapped thousands more, in what *Industrial Age* magazine called "an unparalleled example of scrapping machinery." Midwestern competitors of Brown & Sharpe got the lion's share of that business, but it escalated a machine tools race already begun among car companies bent on retooling the way Chevrolet had retooled to gain a head start on Ford.

General Motors stimulated buyer interest by tweaking car styles with every model year, and by establishing its own credit arm, the General Motors Acceptance Corporation. With the Federal Reserve setting low interest rates for its member banks, National City Bank of New York was only too happy to loan money to GM at a higher rate to create new credit customers. People deciding between a Ford or a slightly higher-priced but better-engineered Chevrolet learned that they could buy the Chevy with less money down than it took to pay cash for a Ford. For General Motors, easier credit made easier sales, not only of cars but of the washing machines and refrigerators made by its new subsidiaries, Maytag and Frigidaire.

Henry Ford initially resisted credit sales, complaining,

> I sometimes wonder if we have not lost our buying sense and fallen entirely under the spell of salesmanship. The American of a generation ago was a shrewd buyer. ... But nowadays the American people seem to listen and be sold; that is, they do not buy. They are sold; things are pushed on them. We have dotted lines for this, that and the other thing—all of them taking up income before it is earned.[6]

With approval of the railroad buyout bill, the auto industry's race to retool, easier consumer credit, and banner crop years for American farmers (makers of reapers, balers, and other agricultural equipment bought lots of machine tools), Henry Sharpe felt like the machine tools business was riding the crest of good times. In answer to a query from the National Trades Association he wrote in December of 1925:

> Outlook for 1926 we consider good, based upon the realization that business men of the country are getting more courage, doubtless following the bounteous crops with which we have been favored during the last two years. On the other hand, without doubt the condition in our industry more or less reflects the extraordinary prosperity of the motor industry [a development he had predicted years before].

Good times in the motor industry were symptomatic of American prosperity in general. With a stable American dollar and war-weakened European currencies, foreign investment flowed into American banks. American banks, barred from directly issuing bonds, spun off affiliates that allowed them to sell bonds to investors in order to raise $200 million to loan to Germany at 7 percent interest. American investors rushed to buy these seemingly safe bonds guaranteed by German cities, towns, and states, and then by entities in other war-wracked nations.[7]

In his 1925 prediction to the National Trades Association, Henry wrote with tongue-in-cheek: "In our judgment the greatest need in our line of business is more profits. Doubtless any man in America would say the same thing. The more profit he makes the farther off he is from his point of satisfaction. However," he wrote in a more serious vein, "it is true that since the war the share in gross profits coming to the owner of the plant is less, and the share of the wage-earner is more."

13. "If I Had Known This Was Coming"   141

Workers assemble precision tools in a "clean room," a climate-controlled room maintained at a constant temperature to prevent expansion and contraction of metals, subtle movements that could affect a tool's precision. By the mid–1920s, business was booming at Brown & Sharpe, driven by "the extraordinary prosperity of the motor industry," in the words of Henry Sharpe.

With factories ramping up production, workers were in a better position to negotiate concessions, such as pensions, arbitration to settle disputes, sick days, and holiday pay. Henry Sharpe was not a fan of negotiating with workers. To a fellow businessman inquiring about the firm's experience with pensions, Henry wrote that Brown & Sharpe had an "informal system" by which several employees of long service received an "honorarium."

"The plan works very well, but this thing must be kept in mind; that an old employee, generally speaking, wishes an opportunity to be kept busy, rather than to live at home with nothing to do." So Brown & Sharpe sometimes let an older employee "have a corner of the works where he can be kept busy, even though it may not be at his old job...." Brown & Sharpe also offered employees a chance to invest their own money in an insurance plan to provide income in old age; the company also managed a Relief Association that paid sick and death benefits.

Henry wrote to his business colleague,

> you may wish to know that Rhode Island Hospital Trust Co., of which the writer happens to be a director, three years ago adopted a formal pension plan, for which there have been regular

amounts set aside ... for the care of further charges. In this plan, sufficient experience has not been accumulated, except perhaps to see the possibility of future burdens as to which we cannot as yet estimate.

Worker demands for arbitration met with a similar skepticism from Henry Sharpe. "Arbitration," he wrote, "is generally a fraud in practice" that tended to prolong strikes by encouraging workers to hold out for an arbitrator rather than give in to company demands.

As far as sick days went, Brown & Sharpe had no formal sick day policy. If a "salaried man" got sick, the company generally paid him during his illness; if an hourly-wage worker took ill, the company might grant paid time off depending upon length of service to the company, the type of work he or she did, and whether the worker habitually missed work. Sometimes Brown & Sharpe paid generous pensions and carried workers through sicknesses, and other times it did not, depending upon the circumstances. As for holiday closures, Henry wrote one Christmas Eve, they brought "embarrassment" and hardship to the worker forced to "forego the wages he would earn on such days." Most people, he wrote, "would rather work on the majority of such occasions." Henry Sharpe placed utmost importance on the company freedom to make choices as he and fellow executives saw fit, without interference from unions or government agencies trying to dictate the way he ran his business.

The return of good times demanded more workers—as early as 1923, William Viall chided Pratt & Whitney for stealing Brown & Sharpe tool makers—and unions began to flex some muscle in pursuit of worker gains. Business pushed back. Henry organized a drive to raise $25,000 from Providence businesses to donate to a new headquarters for the national Chamber of Commerce in Washington. He donated $1,000 and prevailed upon his Metcalf cousins to raise another $1,000 through their Wanskuck Mill, and sought $250 from his Chafee cousins at the Builders Iron Foundry.

"Organized labor and organized agriculture have been so housed in Washington for several years," Henry wrote to Providence factory owners. "The ownership and occupancy of such headquarters cannot fail to bring an expression from the legislative and official Washington that business men throughout the country *are* organized and *are* prepared to resist radical influences."

The best way for Brown & Sharpe to find workers with the skills it needed was to train the workers themselves. By the 1920s, most firms had long ago dropped apprenticeship programs, leaving workers to learn trades through for-profit vocational schools, or through unions, which held an interest in controlling the supply of laborers. Brown & Sharpe believed that training new machinists, forge workers, and foundrymen was too important to farm out to vocational schools and unions. The company appointed an executive, J.E. Goss, to act as supervisor of apprentices overseeing recruitment of teenage boys for the program, a dormitory that housed some of them, the shop school where they learned mathematics and drafting, and the records compiled to track the progress of each one of the 200 boys in great detail.

In a speech at the National Metal Trades Convention in Chicago, Goss explained why Brown & Sharpe invested so much in apprentice training when most other firms had dropped it: "it is the *backbone of our* organization that we have in mind when we enter into correspondence with a farmer's boy in New Hampshire or a High School senior down on the Cape."

Brown & Sharpe largely recruited from the high schools and trade schools of small-town New England, though it often received applications from outside the region. The emphasis on "the farmer's boy in New Hampshire" or "the grandsons of seafaring men from Cape Cod" created a de facto, if unwritten, segregation that excluded African Americans.

An apprenticeship agreement signed by Charles Vevera of Newport, who finished his training in 1923, spelled out the terms of the program: applicants had to be between the ages of 16 and 18, "physically sound, of good moral character, and [had] to have received an education equivalent at least to that required for graduation from the public grammar schools of the City of Providence." Vevera and other applicants agreed to "perform their duties with punctuality, diligence and fidelity" and to refrain from chewing tobacco at work, and from smoking tobacco anywhere, "a custom generally regarded as deleterious to the health of young men."

The teens or their parents paid $50 to gain admission to the program; they worked 50 hour weeks that included 4 hours of classroom time in the shop school, supplementing lessons in math and drafting with "talks on gear cutting, grinding, lubrication, purchasing, routing, and the like." Apprentices earned 8 cents an hour their first year, then 10 cents, 12 cents, and 14 cents an hour in their fourth and final year. The company reserved the right to shorten hours or suspend the program during slack times.

"About a year ago we told the Cape Cod boys what we might do for those of them who had mechanical inclinations," Goss told the conventioneers in 1921.

> It was ten months before we received an inquiry, but it will not be long before boys from the Cape will be coming to us, and what better could the grandsons of seafaring men turn to than the building of machinery? We go after good material because we want to make something out of it that we could *not* make out of inferior stock.

The company opened a dormitory just east of the plant in 1920, the first of two, and Goss found company-run living quarters to be worth the expense. "In addition to attracting boys who are used to good, wholesome homes, it is functioning as a builder of apprentice morale. Many of the associations formed at the house are sure to develop an esprit de corps which will strengthen as time goes on."

One problem Goss encountered in recruiting was "the tendency of boys to seek the less manual kinds of work. ... For one thing, each year more boys are completing the secondary schools and colleges, and it is quite natural that they should expect a certain prestige, allowing them to evade shop training," Goss said. But there was no evading shop training for a Brown & Sharpe apprentice. "It seems to me that we might as well look for short cuts to *experience* as to short cuts for learning of trades, for is not experience a major factor in a sound and thorough knowledge of craftsmanship?"

In an article heaping praise on the Brown & Sharpe apprenticeship program, an anonymous writer for *Providence Magazine* observed:

> The [apprentice's] hands may be soiled, and his face grimed. What of it? At Brown & Sharpe's every employee has his individual locker in which, before going to his work, he places his street clothes. Quitting work, he takes a shower bath if he so elects or washes at a set basin which he alone

makes use of. He leaves for home perfectly refreshed and as clean as any clerk in town. In his jeans once a week he packs home an envelope that is fatter by considerable than that which the clerk receives. It is quite an object lesson to look over the long lines of automobiles that are parked near the big plant. They belong to the workmen many of whom also own the homes they occupy.

One tranquil day at the Sharpe summerhouse on Nayatt Point, a faint hum filled the air. Glancing skyward, 6-year-old Henry Jr. spied a yellow speck growing larger and larger, its sound growing louder, and louder, till its full-throated roar frightened young Henry. He was not used to seeing airplanes, and this one skimmed just above the rooftop then disappeared into blue sky over Narragansett Bay.

"That," said his father, "was your Uncle Lucian." That was the only glimpse Henry Jr. would ever catch of his estranged uncle.

Lucian had taken to flying with a passion. In 1921 he took one of the first long-distance flights in aviation history when he hired an unemployed RAF pilot named Alan Cobham to fly him on a joyride across Europe in a war-surplus plane. In three weeks they covered thousands of miles and 17 countries. Cobham liked to tell of Lucian pointing toward a smoking Mount Vesuvius and saying, "Cobham, let's fly over that crater."

"Do you really mean that?" Cobham said.

Lucian meant it, so off they roared in a fragile bi-plane to peer into the smoldering crater. "A whirling breath of poisonous gas" caught the plane, reported *Popular Mechanics* magazine, "and with terrific force shot it like a feather 1,000 feet into the air." Gas gagged the duo, forcing Cobham into amazing feats of skill to "keep himself and his passenger from frightful death." Years later Lucian claimed that Cobham, by then Sir Alan Cobham in recognition of his aviation exploits, exaggerated their flight over an active Vesuvius.

While Lucian was barnstorming across Europe on Brown & Sharpe dividends, Henry was steering the company through the short-lived Depression of 1921, then growing it through the Roaring Twenties till its factories spread over 32 acres up and across the face of Smith Hill. With foreign currency flowing into U.S. banks, per-capita income steadily rising, easy credit stimulating sales of cars, refrigerators, and washing machines; with harvesters reaping bumper crops, car companies engaged in a re-tooling race, and with the cars of his workers lining city blocks around his plant, the times looked good to Henry Sharpe and he decided to spend some money. The state's General Assembly became paralyzed by in-fighting in 1926, and could not even agree on passing a budget, leaving 300 foster parents without payment for the hundreds of orphaned children they had taken into their homes. So Henry Sharpe, the curt, cigar-chomping, business tycoon, stepped up and wrote a $15,000 check to ensure two months of payments for 300 foster families. He and one of his sisters also bought a historic 1600s stone-ender house across the city line in Johnston, and gave it to a local historical society.

With good times roaring along, Henry also decided that the time was ripe for building a big house for his small family. This was a dream house for Mary Elizabeth, a woman who liked to think big. While serving with the Red Cross in wartime Paris, she had fallen in love with all things French—architecture, art, furniture. She knew she wanted French furniture, and a palatial, Louis XVI–style house in which to display it.

For help in designing and decorating the house, Mary Elizabeth turned to her friend, Florence Koehler. Upon first seeing a house decorated by the woman she called "Mrs. Koehler," Mary Elizabeth insisted on meeting the decorator. Though Koehler swore that she was not a decorator but an artist, and that she was not taking on new customers, Mary Elizabeth prevailed upon her to help pick out art and furnishings for the first house that she and Henry bought after their marriage in 1920. One of the artworks Mrs. Koehler chose was a Grisaille painting, a gray monotone depicting the huntress Diana in her chariot, dressed in lacy robes with no clothes beneath them. It looked lovely against the sage green wall above the fireplace. But when Henry's sister, Louisa, saw it she was scandalized: "Oh Mary," she said, "we don't have naked women in our drawing rooms in Providence!"

Soon after she delivered her son in March 1923, Mary Elizabeth developed a postcesarean wound infection that kept her hospitalized for months; upon being discharged she took a ship to Europe to convalesce. She stayed with Mrs. Koehler in her Rome apartment. Koehler was a talented artist who worked in a variety of decorative media, including interior design, painting, china decorating, and jewelry. While living and working in various European countries, Koehler befriended many of the most influential artists, writers, and thinkers of her era, and they kept in touch with her, introducing Mary Elizabeth to Henri Matisse among others. Mary Elizabeth recalled of her time in Rome: "We had rooms on the side of the city where all the museums and so forth were. So we had a very lovely time there. Then my husband came over and picked me up, and we came home."

When it came time to decorate the big, Louis XVI–style manse with a sweeping view of downtown Providence from College Hill, there was no doubt whom the designer would be. Koehler sat in on the design plan, chose chimney pieces, French hardware for doorknobs and tap handles, and pieces of furniture. This was an authentically French mansion of three stories, with high vaulted ceilings in the foyer, a private drawing room for Mary Elizabeth and a nice library for Henry. The third floor was a warren of smaller rooms for the domestic staff, housing nine of the thirteen people serving there or at Nayatt as a butler, waitress, cook, maid, gardener, or nanny.

Workers finished building the house while Henry and Mary Elizabeth toured France, buying art and furniture for their new dream house in the fall of 1929. Times were good. In the year between the fall of 1928 and fall of 1929, the valuation of stocks on the New York Stock Exchange had doubled.[8] A key reason for the dramatic rise in stock prices was that banks had few other places to invest their money. The bond market had dried up in 1928, when overseas loans dropped by two-thirds, partly because there were few places left to loan money to. When the German finance minister publicly questioned whether some of his country's municipalities could ever pay back their loans the bond market trembled, causing investors to shy away from bonds and move into the stock market. Banks fueled the movement by loaning money to stockbrokers, who then lent it short term to stock speculators. The Federal Reserve tried

In 1929, Cadillac founder Henry Leland (center) came back to Brown & Sharpe for a reunion with some of the company's executives. From left to right are chief draftsman and company historian Luther Burlingame, Luther Burnham, Leland, Charles Richards, and Richard Wingo. Wingo had been in charge of the gear department before, like Leland, he moved to Detroit to open his own business, Superior Machine and Engineering Company. Men who trained at Brown & Sharpe in the 1800s formed many spinoff companies such as William Nicholson of Nicholson File, Leland of Cadillac, George Smith of Beaman & Smith Co., W.S. Davenport of Davenport Machine Company, and many others.

to deflate the bubble by raising interest rates, but banks circumvented this by issuing banker's acceptances guaranteeing the notes of their brokerage affiliates, which continued to lend money to brokers. In accepting as good the debts of their affiliates, banks were essentially writing themselves IOUs. As long as stocks stayed strong, everyone along the chain made a profit. But savvy investors began to suspect the high prices of stock in relation to assets and earnings; when they began pulling money out of the market, stock values dropped below the loans brokers had issued to buyers, forcing automatic sales which further depressed values sending the market into a death spiral. Henry and Mary Elizabeth were still in Paris when racks of newspapers such as the Continental Edition of the *London Daily Mail* headlined "Greatest Crash in Wall Street's History." Henry read the news of the Black Tuesday stock market crash, turned to his wife and said, "Oh, if I had known this was coming, I never would have built that house."

14. "Don't We Ever Play a Waltz?": The 1930s

As widow to the grandson of Samuel Slater, Mabel Hunt Slater was heiress of an estate worth $10 million in 1930; she was also an ingenious woman who held several patents, including a method for drawing clean drinking water directly from the melting ice in an ice box. Slater was a charitable person whose insistence on leaving her kitchen door open to feed the poor of Boston once led to the death of her butler, murdered by an unstable man who wandered into her open Back Bay mansion looking to kill her.

Traveling in the same social circles it is not surprising that Mable Slater and Lucian Sharpe, Jr., met; they fell in love and decided to marry in November of 1930. On the night before their wedding day, Slater's son and two daughters kidnapped her and hustled her off to the private sanitarium of Dr. J. J. Slocum at Beacon, N.Y., where they had her committed by court order. Slater's brother and two sisters stepped up on her behalf, challenging their nephew and nieces in court to overturn the commitment. In a hearing covered by a Boston newspaper their family's longtime doctor asserted that "Mrs. Slater is absolutely normal, mentally and physically."

For Lucian, the issue of his intended bride's sanity became moot. While she awaited marriage confined to Slocum's sanitarium, he took a sudden illness on a trip to Chablis. From his bed at an inn he called out to his housekeeper, "Mother I have caught cold." The housekeeper, a Mrs. E.V. Wortley, saw his forehead glistening with sweat though he complained of chills. She sent her husband to find a doctor, while she placed mustard plasters and a hot water bottle on his feet to draw the blood from his head. "He turned his dear head to one side and all was over," she wrote to a mutual friend. "Oh ... how handsome he is in this his last sleep." His siblings shipped Lucian's body back to Providence and buried him in the family plot with their parents in Swan Point Cemetery.

In a parting shot to his siblings, Lucian willed almost all of his $3 million estate to S.G. Archibald his expatriate lawyer in France, the current equivalent of nearly $45 million. The inheritance included a sizable share of Brown & Sharpe stock, creating a crisis for the family. Almost all of the company's stock was closely held by the six Sharpe siblings. A couple of times small shares of company stock had passed outside the family—in 1914 Jane Brown had made relatively small bequests to a houseman and to the

Rev. Marion Law, but the company had managed to buy back that stock. William Viall received 15 shares from Jane Brown as a company executive and family friend, but the Sharpe siblings held all of the rest of it, and Henry intended to keep it this way. In many ways Henry's situation in the early 1930s was eerily reminiscent of his father's predicament in the early 1870s: father and son both built big houses just before a financial panic; and both men faced challenges in securing stock critical to the success of Brown & Sharpe.

In his negotiations with the lawyer in France, Henry tried to persuade him that Lucian lacked the mental capacity to will his stock. In making this argument Henry held an ace: a long, rambling telegraph that Lucian had once cabled from Paris to the Brown & Sharpe factory in Providence. Lucian's manic, 4,220-word cable was full of looping language and non sequiturs, clearly the product of an unstable mind. In the spring of 1931, Henry struck a deal with the lawyer, who agreed to settle for $300,000 cash, the current equivalent of about $4.5 million, bringing Lucian's stock back into family hands. Henry's brother-in-law, Zechariah Chafee, then sent him the original copy of Lucian's cablegram with a note reading: "I presume this has no further interest but you may wish it for your files so I am returning it to you herewith. Congratulations are in order."

Henry likely was not in the mood for celebrating. His brother had died estranged, and by 1931 the value of his company's stock was plummeting with no bottom in sight. At first the stock market crash of 1929 did not rattle things too much—it was more of a symptom of an economy out of whack and could be viewed as a market correction.[1] For the first half of 1930 the company coasted on the inertia of orders already in the pipeline at the time of the crash; Henry nurtured hopes that business would resume as usual, though by December he conceded to a colleague in England, "On the contrary, business has been very quiet indeed and, as far as we are able to ascertain, we see no possibility of a change for many months to come."

In January 1931, Henry convened a meeting of stock holders and gave them the bad news: the quarter's dividends would be slashed by 75 percent. "The state of the business of this company is at a low ebb," he announced. "The year 1931 opens with a hope that things may become better as the months go on; as yet there is little to indicate what that improvement may be." Indeed there was no improvement at all. "The depression of the past year has been beyond any experience of any now with the company," Henry wrote in February. "Doubtless the experience of the middle '70s [the Panic of 1873] is the nearest analogy."

As the year progressed the big factory on Promenade Street hemorrhaged money, losing an average of $35,000 per week en route to posting a yearly loss of $1.8 million. On the day of the crash Brown & Sharpe employed 6,300 people working 50-hour weeks. By August 1931, two-thirds of the workforce had been let go through a steady series of layoffs. In deciding whom to keep, seniority was one factor, along with skill level and consideration of "home conditions" that gave preference to married men. Only 2,200 workers remained, and they earned only half of what they used to make, averaging 26-hour weeks at reduced pay. Stock holders, Henry reported on August 13, would receive no dividends.

With no social safety nets in place, Brown & Sharpe agreed to make several charitable contributions to help laid off workers, despite the company's own losses. Henry wrote to the Associated Charities of Pawtucket:

> We authorize you to give such aid as is may be considered proper in your judgment to former members of our working force, now living in your district, who have worked for us six months. Some of these parties you have already been helping, and we will stand back of such aid beginning as of January 1st last.
>
> In addition to the amount of this aid, you are authorized to charge us at reasonable expense for investigating and supervising such service.

Henry made similar payments for unemployment relief to the Family Welfare Society and the Unemployment Relief Committee of Providence. While Henry tried to steer Brown & Sharpe through what he quickly recognized as the worst depression in history, Mary Elizabeth Sharpe tried to keep her family's once-bustling tea room down in New York City from going broke. Two of her sisters and a brother-in-law had run the restaurant since 1920, but the depression was overwhelming their management skills. So early in 1930, Mary Elizabeth asked Henry for his thoughts about her spending weekdays in New York to rescue Mary Elizabeth's from insolvency.

"My sisters haven't got the courage to let anybody go," Mary Elizabeth told her husband, "and yet if they don't make some drastic changes, they're just going to lose everything they have in the business."

"Well," he said, "I think you're the only person who can go down there and do anything about it, have the courage to act."[2]

And so she went—every Monday morning she took the train from Providence to New York, leaving young Henry in care of his nurse, Ruth Erickson; and on Friday nights she took the train back home to be with her husband and son. At Mary Elizabeth's she revamped the menu, offering a different "peasant soup" daily, with a side of fried bread. A person could eat a tasty, satisfying lunch for 65 cents, half the price of the restaurant's average check before the depression.

Even with a decent lunch crowd, she later recalled,

> We had so much space that wasn't occupied. The place looked so lonely. Then I thought, why don't we divide up some of the space and serve dinners? I used some of the empty space and put in some card tables and ping-pong, a couple of ping-pong tables, and advertised for the first time in my life in the *New Yorker* magazine. I advertised that you could come and have dinner and stay for the evening and play bridge or ping-pong.

People responded. They came for dinner and stayed for the free night out, playing cards and ping-pong. Mary Elizabeth managed to keep most of her kitchen crew on by adding the dinner shift. She also renegotiated the restaurant's lease with a bank, knocking a few thousand dollars off the annual tab. With what she called "these odd little things here and there," the inexpensive luncheons, free dinner entertainment, a cheaper lease, Mary Elizabeth's managed to get by.

For more than two years Mary Elizabeth made her weekly commute to New York, and though she liked the challenge she hated leaving her husband and family. In March 1932 newsboys shouted about the kidnapping of Charles Lindbergh's baby, snatched in the night from the second floor of his family's New Jersey home. Police throughout the country worried about copycat kidnappings and for good reason—from Los Angeles to Cape Cod, from St. Paul to New Orleans, police reported kidnappings of socialites in exchange for ransom. In Providence, the police decided to escort a few of the wealthiest boys to their private schools. Every morning a polished police car purred into the circular drive in front of 84 Prospect Street to pick up 9-year-old Henry Sharpe, Jr., beneath the car port of his family's mansion; inside the car he met one of his Metcalf cousins and Danny Danforth, another boy from a wealthy East Side family. The boys loved it: "they thought it was just great," recalled Mary Elizabeth. "They felt very important." But the daily ritual with the police scared the daylights out of her. "This, of course, was terribly hard on me, thinking my son had to be guarded like that," she said, "and I wasn't even at home." And so, she gave up the restaurant, turning its management over to her middle sister, Martha, and her husband, whose children ran Mary Elizabeth's tearoom on Fifth Avenue into the 1970s.

Ensconced again in Providence, Mary Elizabeth took to decorating her house in earnest, buying artwork for its walls and overseeing the planting of trees and gardens. Though she was not all that fond of Matisse's work, at first, she took a shine to an up-and-coming artist, Henri Rousseau; to decorate her drawing room she bought a half-dozen of his paintings such as *The Football Players* and *L'Ombrelle Rose*, showing a woman holding a pink parasol in a jungle. *L'Ombrelle Rose* gave her a "tremendous feeling" because, she said, the woman reminded her of herself: "She is a person who is living immediately in the present but she really isn't in the present at all. She's in the world—the universe, you know?"

She spent as much as $2,000 for a single canvas. Henry balked at the price so she used her own Mary Elizabeth's Tea Room money to buy the paintings, the same way she used her own funds to buy mink because Henry, gruff as he could be, was an animal lover who found furs cruel. Before she wired funds to France for the Rousseaus, Henry insisted that she talk to the curator of the Rhode Island School of Design Museum; he told her, "Well, there's a great fad for those just now, but I think if you ever wanted to sell them you wouldn't get your money out of them."

"You see?" Henry said.

Mary Elizabeth bought them anyway; the Rousseaus brought her pleasure, and they proved to be an excellent investment.

The stock market collapse of 1929 was sudden and dramatic, but the steady losses in the bond market across the early 1930s played a greater role in sucking the lifeblood out of the global economy. Foreign nations began defaulting on their bonds at an alarming rate; by the summer of 1932, 20 percent of the $7.5 billion Americans loaned overseas was in default. Then the nascent Nazi party exacerbated the default problem by declaring that it would reexamine all loans made to Germany after the war and determine how much money, if any, it would pay back. Many Midwesterners had German heritage and held investments with a rebuilding Germany out of a sense of patriotic duty; when their bonds appeared at risk they returned to the banks that issued them to cash them in at a loss, only to find that bank asset sheets were full of bad loans. The Federal Reserve banks that pushed the foreign bonds onto smaller banks held vaults full of self-issued IOUs—banker's acceptances written to guarantee their own debts without sufficient assets to back up those guarantees.

Bank customers clamoring to recapture at least some of their money made runs on banks, forcing banks to close their doors. Just months into his first term the new president, Franklin Delano Roosevelt, declared a Bank Holiday to give bank examiners some breathing room in determining what went wrong and how to solve the problem.

The depression of 1930 and 1931 seemed mild in comparison to the desperation that followed in the next two years. By March 1933, 25 percent of the nation's workers were unemployed. In industrial Rhode Island, unemployment was even worse: 35,000 workers, some 32 percent of the state's total, had no job. In the Blackstone Valley, where textiles were king, the unemployment rate hit 50 percent.[3] People out of work did not buy cars, refrigerators, washing machines, or any of the metal goods made by Brown & Sharpe's customers. Drought in the southwest compounded the problem; winds swirled dry topsoil into dust storms, stripping fields, killing crops, and putting a hit on businesses that made agricultural machinery, another large sector of the machine-tools trade.

On Promenade Street, the 1 million-square-feet of Brown & Sharpe's mammoth manufacturing plant stood mostly quiet and nearly empty. A machinist named Gordon Ellis recalled: "Money was so tight in those days that we gauge makers used to carry an electric light bulb in our pocket. Each time we changed to another machine we would screw in the bulb and remove it when we left. You could not get a replacement without a major effort."[4] In June 1933 Henry pared the roster of hourly workers to just 420 of his most skilled workers. As he explained in a letter to a woman whose son had been laid off, cutting jobs was a painful process:

"I have taken occasion fully to review the matter of [your son's] lay-off," Henry wrote to Beatrice May Jenkinson, "and, frankly, letting any person go, especially one who has been here so many years, is not a pleasurable task.... May we look for happier days when every employer in this country will be able vastly to extend his list of employees."

In less than four years, nine-out-of ten Brown & Sharpe workers had lost their jobs. Yet in 1933 the plant still bled money losing more than $1.5 million for the year. That June, when employment at Brown & Sharpe fell to its nadir, Congress tried to stimulate the economy by passing the National Industrial Recovery Act. The new law gave workers the right to join unions without fear of retaliation by employers. It also gave businesses the right to set "fair codes of competition" for their industries so they could discuss matters such as price-setting without fear of violating the anti-trust provisions of the Sherman Law.

Given the dire emergency of the Great Depression, Henry decided that "good citizenship required that all individuals and all corporations cooperate with the Administration to the fullest extent by complying with the provisions of those laws." Despite the pro-union portions of the bill, which predictably stimulated union membership, he was willing to give the law a try. Brown & Sharpe agreed to a Code of Fair Competition for the machine-tool industry, and received a poster of a blue eagle clutching lightning bolts and a cogged wheel to show it was a member of the National Recovery Administration.

A Brown & Sharpe inspector with a portable cart holding the tools of his trade, including at right a large micrometer and gauge block. In 1933, the company laid off all but the most skilled workers as the factory hemorrhaged money during the Great Depression. One machinist recalled, "Money was so tight in those days that we gauge makers used to carry an electric light bulb in our pocket. Each time we changed to another machine we would screw in the bulb and remove it when we left".

By late August, Henry regretted the decision. The NRA became infamous for creating Byzantine regulations and issuing thousands of confusing administrative orders that presented real managerial headaches. To prove the absurdity of the federal government's attempts to regulate industry, Henry had workers inspect all of the many floors of Brown & Sharpe to make an inventory of "articles processed wholly or in chief value from Field Corn," as required by PT Form 33. He attached the inventory to the form along with a check to for $1.23 which he sent to the local IRS office as "a protest against the assessment of said tax."

Henry was an early critic of the National Industrial Recovery Act, but by 1934 he was just one of thousands of businessmen railing against the law. A New York City chicken processor charged that the act was unconstitutionally overbroad and in 1935 the Supreme Court agreed, striking down the law two weeks before its provisions would have expired. Roosevelt then came back with what he termed a "Second New Deal," legislation that created a Social Security Act to provide assistance for the aged and infirm; a Federal Housing Authority to stimulate home ownership; a National Labor Relations Board authorized through the Wagner Act to enforce employee rights to form and join unions, and obligated employers to bargain collectively with unions; and a Work Projects Administration to take over the work of the Public Works Administration that had built the Triborough Bridge, the Hoover Dam, and the Overseas Highway to Key West.

In pushing for his New Deal legislation, Roosevelt had the support of a British economist named John Maynard Keynes, who visited Congressmen, economics professors, and administrators to preach about the "multiplier" effects of government spending on public works. To get the economy working again, people needed to buy goods; to buy things, people needed government-supplied jobs to earn money. To appease southern Democrats, the Social Security Act would not pay benefits to agricultural and domestic workers—largely Latino field hands, black sharecroppers, and maids—and the FHA largely confined housing loans to white neighborhoods. But WPA projects and the Civilian Conservation Corps did put people back to work and paid them money to buy goods in an attempt to stimulate the shattered economy.

Social programs came at a price to the business owner, but they also provided benefits. The cost came in the form of increased bookwork to deal with regulations, not only on the federal level but on a state scale too. Rhode Island repealed property requirements as a mandatory qualification to vote in 1928, creating a bloc of new voters drawn from first and second generation Americans, many of them Catholics who supported Roosevelt. By 1935, Democrats controlled the state's House, Senate, and Governor's office for the first time; they used the power to pass New Deal–type legislation on the state level, such as unemployment insurance, and additional state holidays.

Henry Sharpe, a dyed-in-the-wool Republican, was none too happy about the state and national shift toward liberalism. He felt he had a sympathetic ear in Governor Theodore Francis Green, a pro-business Democrat. Green had recently called out the National Guard to quell a textile strike, and the Guard had fired on strikers in Central Falls, killing a 15-year-old boy who had turned out to watch. In a letter to the governor, Henry urged him to veto a 1935 bill to make Armistice Day a state holiday: "Most of the employed in the city of Providence would prefer to work on Armistice Day. The legislation in question would impose a real hardship on them. It's enactment, in our belief, is a species of petty tyranny on the part of the legislature."

When the House and Senate overrode the governor's veto of the holiday, Henry groused to Green:

> It seems plain that there will quickly follow similar Bills relating to Columbus Day, New Year's Day, St. John's Day, and even Pulaski Day. ... We cannot see why the Government of this State can not insist upon respect for the rights of the individual, whatever a mere majority may say.

In a memo titled "Burden of making tax and statistical reports" Henry complained that Brown & Sharpe annually spent 3,639 man-hours completing mandatory tax and regulatory reports to municipal, state, and federal agencies at a cost of more than $6,000, the current equivalent of about $108,000. That figure excluded time spent on maintaining records required for Social Security and Unemployment Compensation reports. These new social programs cost business money and manpower; but unemployment insurance also freed business owners from the burden of carrying the unemployed at slack times when their own fortunes were at an ebb, and Social Security lessened the cost of awarding pensions.

Like his father, Henry Sharpe, Sr., loved to exhibit his company's products at World's Fairs and trade shows. Though times were still tough in 1935, he rented a booth at the National Machine Tool Builders' Association show in Cleveland, spending thousands on freight, catalogs, and travel expenses to show off the company's wares. Automatic screw machines on display chewed through pricey stocks of steel and brass as they continuously churned out precision products throughout the convention. The expenditures paid off. In answering an October query about the state of the machine tools business, Henry reported "an upswing in business of late months."

> The exhibition was truly a wonderful event, certainly to anyone interested in the production or use of machine tools, being a portrayal of the hard work done by designers since the advent of the depression. Without doubt it will serve to stimulate added business, but we must remember that a great many of the designs shown at Cleveland had especially to do with automobile production. ... By and large the demand for automobile manufacture is the one bright spot as we look ahead.

By 1935 the darkest days of the Great Depression were in the past. "During the summer months our business has held up very well," Henry wrote, "and the Show seems to have stimulated orders. For the closing week of September our so-called productive hours were about 130,000, with employment of 3,100 hands."

As Brown & Sharpe called workers back to Promenade Street, it found that many of those it had trained were now working for more money down at the Newport Navy Yard, where the government was paying "excessive wages," in Henry's opinion, and keeping to shorter hours. "We have a Socialist and Labor administration which is in cahoots with the Labor Unions, and in this case the Union did its work sometime ago," Henry groused in a letter to a fellow machine tools manufacturer. Brown & Sharpe trained machinists and the Navy put them to work elsewhere at higher rates than the company wanted to pay. "In other words," Henry wrote, "the Government, which has no apprenticeship system, is sponging on the efforts of its industries, in order to equip itself to work in competition with private industry itself. It is a vicious circle."

Despite the difficulties of reconstituting the workforce, Brown & Sharpe's net income for 1935 topped $700,000, the current equivalent of $12.3 million. Not only were car companies again tooling up, but the firm had found "considerable business" from Britain and France. Henry felt cautiously optimistic about the coming year, though Italy's October invasion of Ethiopia concerned him: if France or England declared war on Italy, the welcome renewal of European exports could again be in jeopardy. Otherwise, Henry wrote in autumn: "If business can continue during the ensuing year, with no intrusion from these adverse elements of which I speak, we ought to show a profit every month. At the present time we are doing so, and we hope to continue."

Things did continue well, and the routine domestic rhythms of an upper-class household stole back into the French-style chateau at 84 Prospect Street. At exactly 7 a.m. each day, Henry's glass-encased alarm clock sounded in his second-floor bedroom.[5] Breakfast followed unerringly at 7:45; 'luncheon' was always at 1 p.m., when Barnes the chauffer arrived with Henry in one of his Pierce-Arrow automobiles. At exactly 5:30 p.m. the sound of Henry's door keys, echoing over the marble floor of the great front hall, signaled the evening's family time.

Heading straight into his library, Henry plumped comfortably into his great gray French armchair in the corner by the window. Davis, the elegant young English butler, appeared with a tray holding a steaming cup of his favorite French "Chocolat Prevost." Then, after a few pleasantries with his wife and son, the *Evening-Bulletin* covered his face while Henry caught up with the latest news.

Henry was a board member of the *Providence Journal* and *Evening-Bulletin*; despite the Depression the newspaper had just built a brick plant of Georgian design on Fountain Street, with a fully-equipped radio studio to keep abreast of new technology in delivering the news. At 6 p.m. the voice of John Edgren—"Broadcasting direct from the Editorial Rooms of the Providence Journal and Evening Bulletin"—boomed from the corner closet, where Mary Elizabeth kept her combination radio/record changer discretely out of sight. After Edgren recapped the news, Davis the butler appeared to announce, "It is time to go in for dinner."

Dinner at the Sharpe mansion was a quaint combination of intimate formality. Served by Davis and Stina, the maid, it was the time when the family discussed weekend plans and quizzed young Henry on "what happened in school today?" Occasionally the voice of Fulton Lewis, Jr., a radio commentator who offered 15 nightly minutes of right-wing, anti–Roosevelt commentary, would break through from a small radio that Henry brought with him to the dining-room table, much to Mary Elizabeth's chagrin. Young Henry suspected that his father, a good Republican, held an ulterior motive: a not-so-subliminal attempt to warp his political outlook, though in fact it had the opposite effect, instilling in him a wary mistrust of ultra-conservatives.

After dinner the three family members strolled back to the library for sporadic conversation, till Henry Jr. trundled off to do homework, Mary Elizabeth broke out a book and Henry unfolded a card table for a round of solitaire before the fire. Promptly at 10 p.m., Raymond Gram Swing barked from the corner closet with his roundup of the day's news. When Swing finally said, "Goodnight," so did Henry.

The generation gap between Henry, a successful man-of-the-world in his mid sixties, and his teenage son was sometimes a yawning one, as unforgettably illustrated one evening. As Henry Jr. later told it:

> My father, my mother and I were together one evening in a triangle across the library. My father, as usual invisible, curtained behind the full spread of the Evening Bulletin, which he held before him. Only the smoke of his cigar, which curled upward from behind, as from some gently dozing Mexican volcano, signaled his living presence. At the second corner of the triangle, across the room at the end of the couch sat my mother knitting. In the third angle, with head into the workings of my mother's intricate Capehart phonograph stood I, positioning a newly bought stack of Benny Goodman Trio jazz records into the machine. Finally, after the right sequence of buttons had been pressed, "The King of Swing" burst into the evening silence. I grinned with glee. At first there was no visible reaction, though I began to sense an aura of tension within the volcano behind the newspaper. Suddenly the volcano erupted and the newspaper dropped to his lap with a rustling crumple.
>
> "Mary," he announced, in a voice crawling with annoyance, "Don't we ever play a waltz?"

In the Fall of 1937, another profitable year for Brown & Sharpe, the time came for young Henry to enter boarding school at Phillips Exeter Academy in New Hampshire. Outside of trips with his father, he had never before been away from home.

"My father and I had driven all the way from Providence to Exeter together, when the full terror of this moment dawned upon me," Henry recalled.

> We sat alone on my cot in Dunbar Hall's stark and undecorated Room 10-A, at the head of the stairs on the second floor. We started to say goodbye, and suddenly the tears flooded down both our faces as we held each other in a farewell embrace. For a long time I couldn't stop crying. Finally, my father took me gently downstairs to make certain I was well acquainted with dear Mrs. Richardson, who served as Housemother to that building so filled that night with homesick souls. Were it not for her, I'm not sure my father would have had the strength to leave me. But at length he did, passing me on to the comforting reassurances of Mrs. Richardson.
>
> It was a moment that signaled the beginning of the greatest educational experience I was ever to enjoy. But it, and all that followed beyond it, would be founded upon a sense of the world that my father had already left inside me—without my having the slightest sense of its existence.

15. World War II: Defense Workers Wanted

A local rabbi, William G. Braude, from the Temple Beth-El on Broad Street sought jobs for Jewish people fleeing vicious persecution in Nazi Germany. In May 1939, Henry Sharpe answered Rabbi Braude's telephone inquiry with a letter, telling him that Brown & Sharpe would be willing to interview any mechanically minded man, regardless of religion. Exports to Europe were keeping the factory busy, and the company could use skilled help. "An experience with a limited number of prospects of this sort might be of value to you and to us," Henry wrote.

The rabbi wanted to know whether prospective workers would be able to count on stable employment if they resettled in Providence, and on this Henry was less sure:

> Foreign business now is very good, and is liable to continue during the summer months. It is impossible to foretell the eccentricities of people in Washington who frame our national laws, any change in which might alter the business prospects through curb of exports in process of completion.

Besides worrying about how America's political economy might affect his exports, Henry was then grappling with dramatically overhauling Brown & Sharpe's entire business structure. As his nieces and nephews—the Chafees, the Metcalfs, and the Peters' of Cambridge—began having children of their own, questions about the distribution of dividends and the best ways to transfer wealth to successive generations began to plague the company. Depression-era correspondence from Amy Peters revealed her growing unease at having most of her investments in Brown & Sharpe at a time when the company could not pay dividends. And Mary Chafee wanted to pass on shares of Brown & Sharpe's stock to her grandchildren, though as minors they could not vote in company affairs. Henry decided that the best solution for his siblings and their growing families was to take the company public in order to give them greater flexibility with their shares. So in June 1939, Brown & Sharpe agreed to sell 34,000 shares in the company, about 25 percent of the stock. In announcing the sale Henry told the *Providence Journal* that going public was "incidental to the wish of certain large stockholders of the company to dispose of a portion of their holdings in the business."

Brown & Sharpe was no longer a company that could hold its stockholder meetings in the living room of one of the six siblings, they way it used to do at the turn of the century. It now had a board of directors and hundreds of investors to whom the board owed a legal duty of fiduciary responsibility. Still, as Henry wrote to the firm's Paris sales agent weeks afterwards, the stock sale would not affect the company's day-to-day management: "The control of the company continues to rest with myself and sisters." Heirs who wanted to sell stock for the flexibility of cash could do it, but Henry and his sisters were still calling the shots.

As it turned out, Henry Sharpe had not needed to worry about the "eccentricities" of people in Washington leading to government-imposed curbs on exports. On September 1, 1939, German tanks, troops, and planes launched a blitzkrieg on Poland, strafing outdoor markets, shelling cities, killing civilians. Two days later Britain, France, Australia, and New Zealand declared war on Germany.

The war in Europe stimulated the machine tools business as if it were 1914 all over again. Brown & Sharpe's agent in France wrote, offering to pay premium wages to men who could operate the screw machines it was buying. A Boston newspaper reporter wanted Henry's opinion on bottlenecks slowing the production of warplane engines. Henry replied on Valentine's Day 1940: "This happening is but incidental to an overpressed wartime industry." That same day he told another French agent: "You will understand there is extreme pressure for making quicker deliveries for a very great many orders are on our books."

The Henry Sharpes, junior and senior, at home on 84 Prospect St., a house that Henry Jr. jokingly referred to as "the castle." Despite their 51-year age gap, father and son formed a tender bond (courtesy Peggy Sharpe).

World War II was on, though America was staying out of it. President Roosevelt had declared neutrality in the fall of 1939, and for once Henry agreed with the president. Noting the "obvious wish of 99–44/100% of the American people not to enter this war," Henry wrote: "I am in favor of assisting the War Department and other responsible people in the formulation of paper plans when we may go to war, if we do," but he still preferred to avoid a need to execute those plans.

In summing up 1940 for his company's stockholders, Henry Sharpe wrote that owing to the war in Europe orders for machine tools were flowing into the plant "on a scale undreamed of in the past." In that single year the company's net income jumped from $2.7 million to $4.9 million, nearly $83 million by current standards and a boost of 83 percent. Stockholder dividends reflected the results of this business bonanza; they soared by 75 percent to $2.9 million (nearly $49 million today), prompting one small investor from Maryland to send Henry a letter of thanks. Henry replied in January 1941, writing:

> The liberal earnings of our enterprise have been due to the excessive demands for our product. Since July 1st, 1939, when you made your investment, the threatened war, which a good many people denied would come, suddenly happened, since when our product has been in excessive demand.

By springtime Brown & Sharpe employed 8,125 people, an increase of 7,000 in a year-and-a-half. With other companies embarked on similar hiring binges there was little good help left to be found, yet companies needed more. The federal government was aware of the labor shortage, and passed legislation intended to jumpstart a federal apprenticeship program for machinists. An unnamed but earnest bureaucrat toured Rhode Island factories for six weeks in the early 1940s, trying to furnish metalworking companies with apprentices drawn from the rosters of high schools, the Work Projects Administration, and the Rhode Island School of Design. Henry viewed the attempt with equal part amusement and contempt. He figured the U.S. Federal Committee on Apprenticeship would never be able to do for industry what industry could better do for itself. If government ran internship programs "school men" would benefit by operating training schools, and unions would lobby to control the programs in order to influence wages and the supply of labor; but business would reap few trained workers as a result.

"The great enemy of industrial business at the present time is the legislature, whether state or national," Henry groused to a fellow manufacturer up in Woonsocket. He predicted the Committee on Apprenticeship would fail, and he was right. After a month of visiting factories and schools the bureaucrat assigned to building a federal internship program for machinists found four companies that would accept, in total 17 apprentices; he convinced only 5 young men to take IQ and mechanical aptitude tests to see if they could qualify for the program. At the time Brown & Sharpe enrolled 200 teens in its own apprenticeship program.

After America plunged into the war in December 1941, the demand for machine tools and machines of all types grew exponentially. In the worst three years of the Depression, 1931 through 1933, industries bought a total of $100 million worth of machine tools. Less than a decade later, in July of 1942, American machine tool companies posted a backlog of orders totaling $1.1 billion—11 times more business existed in that one month than they had seen in 3 years of the Depression. All of this new business created chaos throughout the steel-working industry. Every customer thought that its needs were the most important, forcing the government's War Production Board to assign priorities and dictate to factories which orders they should fill first.

With America's declaration of war in December 1941, the demand for machine tools grew exponentially. In July 1942, the backlog of orders to American machine tool companies crossed $1 billion. To meet demand Brown & Sharpe boosted its payroll to 12,100 workers, including these machinists who helped deliver this 5,000th No. 1 Universal Grinding Machine. Brown & Sharpe unveiled its first No. 1 Universal Grinder shortly after the death of Joseph Brown, its inventor, in 1876.

Setting priorities was not always clear-cut, as recounted in *A History of the Tool Division War Production Board*:

> Alloy steels were denied for anti-friction bearings, because Service members claimed that alloy steel was needed for tanks, only to find later that tank production was threatened because of a lack of anti-friction bearings. Shipyards were given all the [steel] plates they could use, but could not assemble a ship without cranes; yet the crane builder could not get delivery of plates.[1]

The federal government offered several incentives to help factories expand their production. For example, a company that built, bought, or leased a factory for defense work could protect 20 percent of its income from the excess profit tax so the building would eventually pay for itself. Brown & Sharpe leased a 50,000-square-foot building near its plant for wartime production, and received the tax deduction; but it did not take advantage of the plan to build a new plant.

Another incentive was the "pool order," essentially a guarantee to the machine tools industry that the government would buy whatever it made. The pool order was designed to break production bottlenecks by giving companies authority to anticipate its customers' needs and fill them before receiving explicit orders to make a product. If nobody bought the machine tools and cutters produced on spec, the government would buy them anyway. A good example of the success of pool orders was the B-17 Flying Fortress airplane, which required 3,000 anti-friction bearings each. Making bearings required special jigs, fixtures, gages, chucks, and grinding machines. All machine tools companies knew that the Flying Fortress project was going to require all of these items that they could make, so to save time they had carte blanche to make them before receiving orders to make them. Pool orders eventually totaled $1.9 billion, and the government paid $12.3 million for goods never used, a lot of money but just .006 percent of total pool orders, a worthwhile overrun for a program that cut waiting time for critical components.

At Brown & Sharpe, the money came rolling in with the orders. Between 1941 and 1942, the company's assets leaped from $33.9 million to $43.2 million, and the company leased new space, added a parking lot to hold its employees' cars, and bought new tools and equipment. Most of the money, about $660 million today, came directly or indirectly through government contracts, which were a double-edged sword. Government proved to be an excellent stimulator of the economy, a force capable of more than maximizing the nation's industrial capacity. In peacetime, centralized economic planning by government had proven that it could drive the economy to put people back to work and spending money. Government spending proved to be even more proficient at maximizing industrial capacity by making the accouterments of war. But government's central role in stimulating and regulating business came at a price to business owners—in exchange for "undreamed of" contracts the Roosevelt administration demanded concessions: a 40-hour workweek, with overtime pay for increased hours; a "maintenance of union membership" clause, required by the National War Labor Board that further guaranteed the Wagner Act's ironclad rights for employees to hold union elections without fear of reprisal; and resumption of the World War I–era Excess Profits Tax that tied company profits to no more than 10 percent of its prewar profits. With allowable deductions for plant expansion and improvements, industrial assets blossomed as factories expanded and retooled, but companies could not just convert all of their extra war work into cold cash without paying a hefty tax burden.

The Excess Profits Tax bit Brown & Sharpe hard in 1942, when the government hit the firm with a bill for $5.3 million in unpaid excess profits tax; this came on top of $11.9 million the company had already paid in federal, state, and city taxes. In his annual report to stockholders Henry complained that: "The severity of the burden of these heavy levies cannot yet be visualized. Taxation means deprivation of cash. ... [C]orporate taxation has been more than overdone." But really, he did not have much to complain about. Even with the heavy tax burden the company posted a net profit of $2.2 million, and stockholders shared more than $1 million (the current equivalent of $15.25 million) in dividends, 75 percent of which went to Henry and his four sisters.

To the International Association of Machinists, Brown & Sharpe looked like a great white whale. For nearly 50 years the machinists union hoped to union-

15. World War II

A forklift operator hauls scrap metals to the foundry on Promenade Street to be melted down for use in meeting the demand for machine tools during World War II. Responding to the National Association of Manufacturers request for participation in a drive to collect scrap metals, Henry Sharpe, Sr., wrote: "Appreciating fully the need for scrap material, we have been conducting our own collection during past weeks with real success. Having a foundry which produced about 130 tons of castings a day, we of course make use of everything which we find."

ize Brown & Sharpe, the largest open shop, non-union manufacturer of machine tools in the United States, and throughout that half-century Henry Sharpe had avoided the union's harpoon. Their greatest clash had come in 1915, when Henry stultified the union's efforts to organize his shop by refusing to meet with organizers or federal mediators.

By 1941, refusing to meet with the union was no longer an option. Depression-era legislation had ushered in a sea change in union-management relations, and the IAM stood poised to press its advantage. Union officials began a campaign to organize the plant in January of 1941. When Henry got wind of the campaign he sent feelers through the plant to determine what promises the union was making to workers. When he heard that the union pledged to bargain for paid vacation, he offered some paid vacation; he got word that workers wanted a paid wash-up period, he pledged paid wash-ups; night workers wanted a paid lunch hour, they got it; piece workers sought a guaranteed, base hourly rate, and they got that.

The union lodged a complaint with the National Labor Relations Board, charging that Brown & Sharpe offered the new benefits only "in order to induce [workers] not to join the union." In November the NLRB agreed with the union; it ordered the company to hold an election on December 9 to determine whether machinists and foundry workers wanted to join unions. On a chilly, overcast day workers peeled off from the factory floors in small groups in order to cast ballots at supervised polling places throughout the factory, then filed back to work. Like the apprentices of 1896, who agreed that "labor organizations benefit the working class," workers in both the foundry and the factory voted overwhelmingly for union representation.[2]

For the machinists union, the battle for a first contract had only just begun. The International Association of Machinists sought a "closed shop," where every worker covered by contract would have to be a dues-paying member of the union. Henry Sharpe vowed that he would "never willingly sign a contract containing a closed shop clause." The two sides began negotiating in January 1942. Three months and 17 meetings later, they were no closer to signing an agreement. The union appealed to the War Labor Board, which began holding hearings in April.

The IAM, representing 7,600 workers at Brown & Sharpe, complained that Brown & Sharpe was just "going through the motions" and had no real intention of reaching an agreement. The company responded that: "The employer has existed for 109 years with the open shop and sees no warrant for a change at this time"; and: "The union's demand would shift the direction of the working forces from management to the union."

The 12-member panel hearing the case in Washington decided that it wanted to hear directly from Henry, rather than deal with the company representatives he sent to the hearing. When the panel's administrator offered to telephone Henry's office in Providence, company representatives were forced to admit that he had been in Washington advising them all along, he just did not want to appear before the panel in person. So they summoned him to a meeting, where, according to the panel's report, Henry explained that he had "an obligation to the very many old employees who do not want to be members of the union and that he would not do anything which would "sell them down the river."

In a letter to an industry colleague Henry explained the thinking behind his utter opposition to closed shops: "It may be that union is a church," he wrote.

> We readily accede to the right to belong to any religious denomination. It may be the birthright which attaches to anyone belonging to any race or nationality. We accede that right at law to any man, but when it comes to the matter of labor unions our statesmen, lawmakers, and many others seem to think that the closed shop is a proper thing.
> In all of this outcry as to the several freedoms, no one speaks about this fundamental right to work without paying tribute to any union, church, or any other organization whatsoever.

In June, the War Labor Board issued its "Findings of Fact" and made a ruling. "The refusal of the president of the company ... to deal with the union or participate in the negotiations before this panel, lends credence to the misgivings of the union with respect to the willingness of the company to cooperate with it." The company would have to begin negotiating with the union in good faith. As far as the closed shop clause went, the Board issued what it considered a compromise: An "escape clause" that would allow a window of opportunity for workers to resign from the union whenever a contract expired. Once a contract was in place people in the union when it was approved and new workers enjoying its benefits would be union members, only able to resign upon a contract's expiration.

The company loathed the so-called compromise solution. In a protest to the panel, company manager Arthur H. Bainton called the ruling "an infliction of the closed shop feature upon industrial agreements" that was

> thoroughly abhorrent to our company's understanding of the law and the Constitution; it in itself denies freedom to a citizen to work when and where he pleases and for whom he pleases.... As we expressed ourselves to the Panel, "we would not willingly sign a contract containing a closed shop clause or any variation thereof such as the maintenance of membership clause," but, following statements made on behalf of the majority of the National War labor Board, we now feel practically compelled "unwillingly" to comply with this requirement.

Henry Sharpe unwillingly agreed to accept this union maintenance clause, paving the way for negotiations to move forward. But as soon as he agreed he posted the escape clause language on employee bulletin boards to remind workers that there was still time for them to opt out of the union before the first contract was inked. The posting drew an especially stern rebuke from a panel member who wrote to Brown & Sharpe: "You people may feel some pride in your record of 109 years of manufacturing history, but I wish to assure you that you are nearly that long out of date in regard to labor relations with unions in a modern industrial world."

The panel's reprimand stung. Richmond Viall, grandson of the elder Lucian Sharpe's confidant and now himself a company executive, wrote to a friend: "We are very much burned up over the whole thing, but are taking it philosophically, hoping that some day we can get our own back."

But there was some truth to the panel's criticism of Brown & Sharpe being out of date. Henry Sharpe was close to 70 years old. Government responses to two world wars and the Great Depression had profoundly changed labor relations since he stultified the Machinists Union's attempt to organize his factory in 1915. He did not like it, but he had to "most unwillingly" acknowledge unions in his foundry, among his 7,000-plus machinists, and representing his clerical staff. And with men going off to war, Brown & Sharpe once again found it necessary to hire women for traditionally male-dominated jobs, something the company did with great reluctance.

Brown & Sharpe presented a case to the National War Labor Board suggesting that pay rates for women workers should be 20 percent lower than those for men because:

> Women, because of their physical limitations are less versatile than men, and are of less potential value to their employers.
> Women's absenteeism rates are consistently higher than those of men. [This ran counter to the company's own experience with women workers during World War I.]
> Women's working lives are shorter than men. They usually leave to get married after four or five years. [Again, during World War I, Brown & Sharpe actually experienced less turnover with women workers than with men.]"When women are brought in to take men's jobs, it is necessary to make extensive and expensive alterations in plant facilities and layout. It is economically impossible to do this and still pay them at men's rates.

Women require greater supervision than do men.

The going rate in the community (or industry) is lower for women than it is for men. To pay them men's rates would throw the entire existing wage and salary structure out of line.

Men like to think they are worth more than women. It's as much a question of prestige as of productivity.

Most men have to support a family. Most women don't. Men need more money.

In September 1942, the National War Labor Board rejected all of these arguments, ruling that women must receive equal pay for equal work. In self-congratulatory, condescending prose the Board wrote that menstruation renders women inefficient for only "a period of time," so management cannot discriminate against them all of the time:

> If this panel may have a small part in demolishing the fictions and the fallacies which have arisen from certain facts of female physiology it will have served a worthy purpose. The management's figure of 20 percent was drawn from a biological phenomenon which applies only to a period of time and not to relative efficiency and competence at other times. There is no proof, scientific or otherwise, that women are 20 percent less capable than men all the time.

In the Fall of 1942, a teenage girl living on America Street spied a big ad in the *Providence Journal* reading: "Defense Workers Wanted." So Velia (pronounced Val-ya) Costantino walked the few city blocks from her house down Federal Hill to the bustling Brown & Sharpe factory, where she became one of thousands of women who filled out applications.

The employment clerk obviously liked what he saw: A young woman with metal-working experience. Velia's father, an Italian immigrant named Erminio, owned a small blacksmith shop behind his house where he built railings, decorative iron work, fire escapes, and wine presses. Sometimes Velia worked with her father in the blacksmith's shop. She drilled holes, worked on lathes, and watched him twist hot iron on the anvil. In the summer of 1938 they landed a big job connected with constructing the new Mount Pleasant High School: sharpening the picks that day laborers swung in digging out the huge foundation. For that job Velia joined her father working nights when the weather was cooler, heating the steel picks to soften the iron, then honing a keen edge on them.

For Brown & Sharpe, Velia was perfect, a "girl" who was used to the heat, grime, and noise of metal work. There was just one problem: She applied on Columbus Day, 1942, one day before her eighteenth birthday. The clerk told her to come back the next day; she did, and on her birthday Velia lined up for her free, company-issued uniform: two navy blue, short-sleeved shirts with the Brown & Sharpe insignia stamped in red and white on the left shoulders; two thick coveralls; a high hat to cover her pinned-up hair; and a suggestion that she buy her own safety boots. Women with children could sign a child care agreement with the daycare provider of their choice: the U.S. Employment Service, Catholic Charities, or the Children's Friends Society.

The company assigned Velia to a drill press line, drilling holes in parts assembled into tools. Most days Velia liked to don her uniform at home and walk the few blocks to the factory dressed for work, but some of the "girls" as the women called themselves and each other, dressed inside the plant in a huge ladies room with big, round sinks.

At lunch-hour they gathered in circles round these sinks to scrub black grease from their hands.

As their numbers grew, eventually to 2,500, the women of Brown & Sharpe developed a genuine camaraderie. One of the girls from the office, Terry Pelletier, stopped by the sink in a clean white blouse, and looked repulsed by the grime on Vel's hands, prompting the latter to quip: "But the money is very good and clean, and it's three times more than you're making," which was true. Vel earned about $50 a week for drilling those holes. Terry did not hold the dirty hands or the larger paycheck against Velia. "We became friends around that sink," Velia later recalled.

They formed an all-women Brown & Sharpe Glee Club, directed by the women's supervisor, Ivy Whitehead. Ivy liked to say

> It's the women who are winning this war. When they get through doing a man's job at the plant during the day, they have to pick up their regular jobs of managing a home, mothering children, and planning for meals and clothes. While one of our girl workers is washing the dishes, what's her husband doing? He's the reading the newspaper.[3]

The women's Glee Club sang every Tuesday evening in a large break room with a stove in it in Building 5. One day they invited Henry Sharpe, Jr., to attend their spaghetti supper, and Velia made the pasta, sauce, and meatballs. He was just a year older than Velia, still an underclassman at Brown University. He struck Velia as a "nice, conservative person." She looked him over and thought: "I am a factory worker. And you are the owner. No way am I in your league." And then the Glee Club sang for Henry the operetta *Pinafore*.

Velia always nicely pressed her shirt. There was one drill press operator, a man in his 40s, who liked to grab the stiff crease that ran across her shoulder blade, pull her close, and sing into her ear: *A pretty girl is like a mel-o-dy*. "We didn't even think of harassment," she recalled. "I would cringe a little bit. They would do it purposely because they knew my face would get red, all the way up to the tips of my ears."

One day, after many months on the job, a fixture holding down a piece of metal stock loosened. The drill bit kicked up the stock, which sliced into Velia's left index finger. She walked to the dispensary clutching her dirty, bloody hand, where a company doctor stitched her up leaving a permanent scar. Without missing any time she was reassigned to a bench job in Building No. 10, where she filed parts, a job she could do with her good right hand while her left healed. On the filing bench they had a lot of Navy men's wives from the huge naval station down at Quonset Point. A lot of them were from the south, so the Rhode Islanders called them the southern belles.

"The southern belles introduced us to a new life insofar as wearing eyelashes and being beautifully made up," recalled Velia. "Their lifestyle was quite a bit different from us! So it introduced us to new horizons." She liked working on the bench with the belles, filing rifle bolts, mostly because the money was so good. Workers got paid by the piece and as this was a government product, not part of Brown & Sharpe's regular line, the government set the piecework price. Some weeks Velia made as much as $90, the current equivalent of $1,373, which she brought home to help support her 11 brothers and sisters, all of them younger.

The gun bolts contract was an anomaly for Brown & Sharpe. In World War I and through most of World War II, the company did not expand its product line. Factories making munitions, war planes, tanks, and other accouterments of war, consumed all of the machine tools and precision instruments that Brown & Sharpe could make. But in 1943 the Army essentially ordered the company to manufacture the bolt for the M1903 Springfield Rifle. As Wallace Bainton, a company vice-president recalled: "Brown & Sharpe would either do it voluntarily or be ordered to do it. Brown & Sharpe complied."

The U.S. Armory in Springfield, Massachusetts, had built the M1903 since its invention in 1903. Now the Army wanted Springfield to concentrate on the new Garand Automatic Rifle instead, so the old M1903 needed to be made elsewhere. The Ithaca Gun Works was assigned to make the barrels, Brown & Sharpe the bolts; workers at Remington Arms and Smith-Corona Typewriter were assigned to fit all the pieces together, which meant that pieces had to be interchangeable.

Both the Springfield Armory and Brown & Sharpe were pioneers in the business of making interchangeable parts: the armory invented the so-called "American System" to make muskets, and Brown & Sharpe was one of the first to adapt the system to civilian use when it tackled manufacturing sewing machines for Willcox & Gibbs. Bainton and other executives believed that making bolts would be a simple, profitable job—until the machines required to make them arrived from Springfield.

The Armory had been making the M1903 rifle for 40 years, and it still used the machinery it bought at the turn-of-the-century. These machines had turned out more than 800,000 M1903's during World War I alone. Bainton observed: "They were in very poor condition and held together—quite literally—with baling wire."

One of the first machines shipped from Springfield was an old Brown & Sharpe hand screw machine that machinists used to bore a hole in the end of the bolt. The machine was so out of balance that about one out of every five bolts being bored flew out of it and nearly struck the machinist. An instructor sent from Springfield told Bainton that he knew all about this, and after awhile the machinists at Springfield had learned "when to duck."

Brown & Sharpe found it technically impossible to produce bolts on this kind of machinery and have them accurately attach to the barrels made in Ithaca. On further probing, the company determined that the machinists at Springfield had formerly made the complete rifle at the armory, so they filed and doctored each bolt to fit each barrel. In an ironic twist, the armory that had invented the American System of interchangeable parts had not been employing it for decades.

"We placed the junk from Springfield in one corner and went to work," Bainton recalled. "In short order, we threw out the old B&S hand screw machine and found a better way to bore a hole without jeopardizing the safety of our people, and soon tooled up on our own B-000 milling machines. No. 10 Building, where the bolts was made, was transformed."

Once Brown & Sharpe retooled with the proper model, jigs, and fixtures, Velia and the "southern belles" on the bench became superfluous—they did not need to hand-file bolts to fit, so the filers transferred to other departments.

Velia transferred from Building 10 to the sewing machine department, packing Willcox & Gibbs machines for delivery, a contract Brown & Sharpe had been filling since 1858. The work did not pay as well as hand-filing, but she still liked the job, her daily walk down Federal Hill in her blue "Rosie the Riveter" uniform, the gab sessions round the cleanup sink, eating lunch with friends, the Glee Club, even the company-paid swimming lessons in the public pool that Jesse Metcalf built near his Wanskuck Mills.

On Mondays after work she lined up outside the cashier's cage on the fourth floor of the No. 1 Building to receive her pay envelope. Joseph Brown, with his rudimentary bookkeeping system, could never have imagined the complexities in calculating the precise amount of cash stuffed into each one of the more than 11,000 weekly pay envelopes. About half of Brown & Sharpe's employees worked on piece work, so from week to week their pay fluctuated. Supervisors sent time sheets to a group of about 40 women who calculated pay rates by tapping the green and white buttons of Comptometer machines, essentially mechanical adding machines that required a deft touch. A second team of women proofread the data entry, then workers entered the pay information into a Burroughs Posting Machine, which stamped onto each envelope the amount of money required to fill it.

A Brown & Sharpe treasurer named Rockwell Gray recalled:

> Each group of envelopes then went to the "Denominator," a machine which determined the makeup of the cash, with the least number of bills and coins for each envelope. For example, a net pay of $73.33 would consist of three $20 bills, one $10 bill, one $2 bill, one $1 bill, one 25-cent piece, one 5-cent piece, and three pennies.

On Wednesday afternoon or Thursday morning a crew carrying boxes of stamped, empty envelopes headed to the magisterial lobby of the Rhode Island Hospital Trust, where bank employees locked them into the vault to stuff the envelopes with previously ordered bundles of cash. If there were even one cent left over it meant that an error had been made and they had to recheck the entire payroll until they found the mistake. As if this were not complicated enough, in the summer of 1942 the Congress announced legislation requiring companies to deduct 5 percent from each employee's paycheck for income tax, a deduction that would vary from week-to-week for pieceworkers, depending upon their production. Henry tried to stop legislation mandating companies to collect income tax for the government, because he found it burdensome. Brown & Sharpe already spent 16,560 "men days" per year on making up the payroll. In a letter to Rhode Island Senator Peter Gerry (whose great-grandfather, Elbridge Gerry was the eponym for the term Gerrymandering), Henry explained that if the withholding bill passed:

> "[I]it will become necessary for this company to set up an entirely new department, with equipment and personnel to handle this tremendous job. It will not only be necessary to record the marital and family status of every employee at the beginning of the withholding period, but we must keep an accurate record of changes in status of continuing employees, and of all new employees.
>
> "At the present time we have on our payroll just under 11,000 employees, and during the course of a calendar year, due to turnover, there will be approximately 14,000 names appearing on our payroll.

New Deal legislation passed in the 1930s required companies to make deductions from gross wages for Federal Old Age Benefit and State Unemployment Compensation.

Negotiated benefits such as Blue Cross and other group insurance also came out of employee envelopes, and in a successful attempt to push War Bonds, Brown & Sharpe also volunteered to make weekly set-asides for employees wishing to buy the bonds.

"If the deductions for income taxes at the source is thrust upon us, we would require five additional calculating machines and two dual posting machines would have to have substantial alterations made, since all the counter capacity of these machines is used to take care of deductions presently being made," Henry complained to Gerry. New office machines would cost $8,000.00—if you could get them. Factories making adding machines were now largely making products for war. "Add to this the wages of additional personnel, the number being impossible to determine at this time," Henry wrote.

Despite his complaints, the bill passed. And every Monday, Brinks armored trucks arrived at the limestone towers of Rhode Island Hospital Trust to pick up metal trunks holding boxes of pay envelopes stuffed with cash and sorted by department. The Brinks guards delivered the trunks full of cash to the fourth floor of the No. 1 Building for distribution to waiting workers. On Mondays, Velia's father drove the few blocks from Federal Hill and picked her up in his car. She felt safe in her ethnic, Italian neighborhood, but a lone woman wearing the coveralls and blue shirt of Brown & Sharpe on payday would have been an easy mark and she saw no reason to tempt fate.

For Velia and many of the working women, the weeks played out in an agreeable routine: Mondays were paydays; Tuesdays she sang with the Glee Club in Building 5; Thursdays at noon a local band leader, Bill Asturo, led his musicians through a peppy program generally featuring 1940s war music, often with a singer drawn from the factory. Hope, the acrobatic dancer ("a good-looking blonde" in Velia's words) put down her tools to twirl and tumble through the brick courtyard between buildings where workers gathered to eat and listen to the big band sounds. Company trucks parked strategically on Holden Street bearing large placards urging workers to "Buy War Bonds." Velia's mother, Jovanna or "Jenny" made big Italian sandwiches for those Thursday lunch hours when the band was playing. "Oh it was a job even to hold these sandwiches—peppers and meatballs and mushrooms," Velia recalled, wistfully. "It was a job to eat them."[4]

After lunch everyone filed back into the factory, though Hope the acrobatic dancer seemed to have special dispensation to return an hour late, perhaps to give her time to recover from her energetic performances. Soon the milling machines were cutting, the grinding machines grinding, the lathes precisely shaving curlicues of steel. "Oh very noisy," Velia recalled of the factory running full bore. "Very, very, noisy. A lot of action. People coming and going. Everybody seemed to be happy because they had a job, they were making good money."

For both workers and management, the money was indeed good. Inflation drove prices high, but wages rose higher. As Henry observed just after Velia joined the company: "the cost of living in Providence … has increased 13.1% from January 1, 1941 to June 1942, while our wages have increased 21.0% for this same period." Unions collectively bargained new contracts so that before the war's end workers had guaranteed pensions, Blue Cross health insurance, and five paid holidays so that no one had to be "embarrassed" by asking to take a holiday off.

During the war years Brown & Sharpe held war bond drives on Thursdays at noon, with a local big band leader and other entertainment such as comedians, singers, and dancing. Workers watched from the windows and gathered in the brick courtyard to eat and be entertained, while company trucks parked strategically on nearby Holden Street bore large placards urging the workers to "Buy War Bonds".

Despite the company's high excess profits tax, shareholders too were steadily making money, with dividends consistently running between $800,000 and $1 million per year at a time when a dollar bought more than 15 times what it does today. Occasionally friction between workers and management boiled over into small work-place actions—on a 98-degree day foundry workers walked out in protest of the doors being kept shut, and on another day they walked in an unsuccessful attempt to prevent a shop steward's firing—but for the most part labor relations were good.

Economically, the early 1940s were also good times—workers emerged wealthier and with new benefits; factories emerged with more muscularity, upgraded and expanded to work at peak production. But in most other aspects the war years were terrible times. At Brown & Sharpe alone, 3,350 workers entered the military; of these, 65 died. People were making money, but they could not use it to buy basics such as rationed butter, or gasoline, tires, or even sliced bread. And as the wartime letters of Henry Sharpe, Jr., showed, there was the soldier's heartache of enduring time away from home, the ever-present fear of injury or death, and the great grief of the latter.

16. One War Ends, Another Begins

The graduation ceremony for Brown University's Class of 1944 was an incongruously somber affair. Eighteen of the 83 graduates wore the traditional cap and gown—the rest dressed in military uniforms of navy blue, the army's olive drab, or the marine corps' forest green. Instead of marching past the fragrant, flowering bushes of mid-May, the graduates trailed long shadows as they paraded a truncated route through the cold, scentless air of February.

Since the end of the Revolution, Brown graduates had marched in cap and gown through streets lined with cheering spectators down from the campus on the hill to the white-washed First Baptist Church in America; but the Class of 1944 just mingled in the church's basement, then came out a side door. A *Providence Journal* reporter with an eye for detail wrote: "There was a procession into the calm, white severity of the Baptist Church, but it began in the basement. And the elms were black laces against a gray sky, their buds still tight in the February chill."

Only those few in academic regalia had been able to complete all coursework before the rushed graduation day. The rest, including Henry Sharpe, Jr., had been called into active duty before they could finish the final semester. They received certificates of academic achievement; Henry, dressed in the blue uniform of a newly commissioned Naval officer, took his certificate from the hand of his father, the university's longtime chancellor.

Brown University was basically in Henry Sharpe, Jr.'s backyard. He grew up on Prospect Street, which ran straight by the Van Wickle Gates fencing off the university's main quad. He had enjoyed himself as a student there, at the beginning maybe a little too much so. In the first semester of his sophomore year he pulled a prank that resulted in a letter of reprimand written by an assistant dean to his father, the chancellor:

> I am sorry to inform you that your son participated in a dormitory disturbance between midnight an 1 A.M. on Sunday, January 25th. At this time there was a great deal of loud shouting and considerable throwing of water and other things from the windows of Caswell and Brunonia Halls. The entire neighborhood was aroused. … You will realize, of course, that the University cannot tolerate conduct which disturbs our neighbors and brings us into disrepute with the public. Such incidents may easily involve the University with the police and other civil authorities who, particularly at this time, can be expected to have little patience with undergraduate rowdyism.

The assistant dean, Edgar J. Lanpher, had cause to be particularly aggrieved by this stunt. The previous year he had been what Henry later recalled as

> the victim of the most successful joint prank.... I ever pulled off at college. While the Dean was spied innocently waiting for a friend directly below our fourth-floor freshman dormitory window, we managed to completely drench him with a brown paper bag full of hot water, hastily rushed from the fourth floor bathroom.[1]

As a sophomore, Henry Jr. went undercover to interview a Brown & Sharpe executive while researching a term paper titled "A Fair Return to Labor." He met with the works manager, Wallace Bainton, under an assumed name and affected a foreign accent so Bainton would not catch on that he was speaking with Henry Sharpe's son. Bainton struggled to understand Henry Jr.'s muffled accent, and seemed annoyed by the college kid's impertinent questions. The interview did not last long, but it did yield a nice insight into Bainton's thinking: "A fair return to labor," he said, "is what you can get for it."

Henry Jr. rallied from his early "undergraduate rowdyism" to post a stellar academic record at Brown. He won the Hicks Premium in French for the year 1941–1942, an award "based upon an examination set by our Division of Modern Language." He made the Mathematics Department Honors List, awarded to only 2 percent of undergraduates. And when he earned his Phi Beta Kappa pin, the college dean wrote to him:

> The honor is an important one and I am very glad that it has come to you. It seems to me that the fact that you have been able to keep up your high academic standing in spite of the great number of outside duties that you have had makes the award a much more significant one.

The "outside duties" that Henry Jr. had taken on included co-editing the *Brown Herald*, and enlisting in the Brown Naval ROTC program. Henry Jr. loved writing for the *Herald*, and he had a knack for it. For a couple of summers he worked as a copy boy for the *Providence Journal*, where his father sat on the board of directors, and he was intrigued by the bustle and the characters of the Fountain Street newsroom, where the presses roared mornings and nights churning out two daily editions in pungent waves of fresh ink.

In the winter of 1943, Brown suspended classes for a few weeks, taking a Fuel Holiday to save fuel for the war. Rather than spend that time on an early spring break or honing his skills at the *Providence Journal*, Henry Jr. went to work in his father's foundry casting 15-foot beds for Brown & Sharpe's heavy, I-4 grinder. Workers there had just signed their first union contracts, and they were casting 130 tons of steel per day—more than twice the foundry's designed capacity. Victor J. Logan, an industrial engineer, recalled:

"The Brown & Sharpe foundry was not too hospitable to strangers. In the first place, it was a dangerous place to wander around in, especially at pour-off time," when great ladles full of molten metal rolled along monorails to molds awaiting a fresh pour of liquid steel. "The foundry was also extremely hot in the summertime, which meant that the walls were uninsulated, and thus bitterly cold in the dead of winter, except around the core ovens and the newly poured molds."[2]

When school resumed in late February, the foundry supervisor, L.M. Sherman, wrote to his boss: "It has been a pleasure to have your boy work in the Foundry for the past few weeks. He did his work in a diligent and intelligent manner and won praise from the men as well as his Foreman."

That May of 1943, Henry Jr. joined Brown's Naval ROTC program and was whisked away to Harvard Yard for a weekend of training aboard a converted yacht in Boston Harbor. The old boat leaked gasoline vapors that ignited in a harmless but startling explosion. The night before he set off for naval training his parents bid good-bye in a stoic, stiff-upper-lip kind of way that left him feeling rankled. So as he sat in his Harvard dorm converted to a barracks, he wrote a scorching letter to his mother, Mary Elizabeth:

> As the years go by, it has occurred to me that it is our economic position which has given us the excuse for our stupid outlook on family relations. Somehow we have come to believe that the possession of an abundance of wealth, a beautiful home and an easy life places us above all those common things of life—the beauties of close, loving and intimate living. In college especially, I have seen this doctrine disproven. I have seen it disproven by the part which family life plays in the lives of all my associates and by the fact that, to my deep-felt shame, I myself have never sensed that feeling of intimate understanding. The family, in one's youth and a wife, in one's later life, should provide a reference point on which all the good and constructive acts of one's life are to be constructed. Neither you nor I—and to a certain extent father—have ever been wise enough to base our life on such an underlying "reference point." We have gone our merry way from year to year thinking that all life can run on insincerities and false show of emotion. We have thought our money enough to keep us afloat morally and felt ourselves superior to those "lower beings" who must resort to weakling show of emotion for true enjoyment of life.

Then Henry hit his mother with an admission: "I for one am a radical—always trying new schemes—always hoping perhaps that one will succeed. This is something new—to my knowledge it is the first time that admissions such as the above have ever been made within our family." His "radical" bent, his willingness to try new ways of doing things would serve Henry well in his future career.

All of that June day in 1943, Mary Elizabeth walked about "in a haze—with a positive and real pain" in her heart. Her son's criticism had stung. She wrote back to Henry Jr. explaining that she had hoped to "create a climate in which each shall feel loved and wanted, yet untrammelled—welcome yet uncoerced with freedom for individual tastes and growth with room to avoid treading on one anothers corns!" Apparently her attempt to give Henry space had made her appear distant, unemotional. Well, she welcomed his frankness. "Pour it on," she wrote him,

> give me your heart searching and let nothing be withheld. I can take it. I can take it joyfully, offering as it sincerely does—a chance for better understanding, happier relationships¬—In a year you really join the Navy—to be gone perhaps for years—perhaps never to return—I utter frankness—such thoughts never leave me.

Before really joining the Navy, Henry Jr. drilled with the ROTC on Brown's Thayer Field. He tried his hand at leading a platoon on the march and "nearly tumbled the men into a sandpit" before bellowing: "To the rear-march!"

"As it was," he wrote, "the command came on the wrong foot—the whole platoon lurched [—] on its haunches for several seconds before 'coming about.' This business of telling people what to do is not as easy as it looks!"

16. One War Ends, Another Begins

Henry Jr. did not possess his father's natural ease in ordering men about. For him leadership was like an equation, a problem he could solve through study. He worked to earn his commission as a naval officer, then, within weeks of his somber "graduation" from Brown he found himself ensconced in a crowded Quonset hut at the hastily-built naval training base in Little Creek, Virginia. This time Mary Elizabeth did not let her son get away without a proper sendoff, as evidenced from the letter he wrote her in March 1944 from the barracks: "Perhaps I forgot to thank you for the swell time you gave us in N.Y. … It was really swell."

Little Creek was a ramshackle amphibious training base not far from Norfolk. There was little to do on base for entertainment, and Norfolk was not much better—a city teeming with sailors prowling the streets in search of a good time, and finding only each other.

Henry Jr. arrived at Little Creek thinking he would receive "small boat" training on the Higgins Boats also known as "Landing Craft Vehicle Personnel" (VP's in Navy lingo) used to ferry infantry platoons ashore. To his surprise he was assigned to a "Landing Ship Medium," 200 feet long and 5 stories high, large enough to carry 5 medium-weight tanks and 59 men, officers and enlisted. Henry Jr. described his ship as "a great tin cigar box with two beautiful diesel engines in her."

While he trained to ship out either across the Atlantic or into the Pacific, nobody yet knew where, Henry Jr. staved off requests from his mother and aunts to shower him with things—a phonograph machine would be cumbersome, he told his mother, and it was foolhardy to keep large sums of money in a barracks, he wrote to an aunt. Rather than having them send him items, perhaps he could send some to them?

"It has occurred to me that certain commodities are available here on the base that it may be impossible—or hard—for you to get at home," Henry Jr. wrote to his mother. "If I can be of any help to you, let me know and I will try to pass a 'shipment' along to you. Two items which come to mind immediately are Kleenex and alarm clocks. Lots of other things like candy bars etc. are also available in more than peacetime profusion."

Henry Jr. turned 21 in camp, and Mary Elizabeth determined not to let her son go off to war without one last party to show that his family was not the coldly stoic unit he had felt that it was. While he was on leave she rented a block of rooms and imported some friends for a heart-felt sendoff. Her son appreciated it, writing her: "I don't quite know how to go about thanking you for the two most-memorable days which have just passed. Naturally, they will *never* be forgotten, for they were the result not only of extraordinary happenstance but also of warm sympathy and touching thoughtfulness on your part together."

In fact, in his father's eyes, Henry Jr. may have appreciated the party a little too much. Henry sent his son a letter criticizing his "recent indiscretions" at his birthday party; young Henry took it in stride, responding:

> Your words of advice on this score, however, offer me a chance to say a few things I have long wanted to say and lacked opportunity.
> It's often occurred to me that you have wondered much about how your well-taken "lectures" to me on conduct were received. You, as well as I, no doubt have noted little immediate result—

if not utter relapse, as in this "posture question." And it must have seemed to you as though countless talks on more study and less "week-end party" were in vain.

The thing I want to say here is simply this: No, these talks have not been in vain at all. For from them I have gained an acute sense of my character weaknesses—and that is half the battle toward making oneself a respected (and respectable) human being.

In June the fleet of amphibious ships swept out of Chesapeake Bay en route to the Panama Canal, destination: the South Pacific. En route, Henry wrote from the tropics:

> The air is oppressive the day long, and sticky clouds, scudding not far from the sea, are wont to gather again and again through the day and drench us head to foot. Moving as they do so low, they seem to squash us and all the tepid air together, to make one feel humid and unhealthy. The sea teams with life unheard of. Giant jellyfish; phosphorescence that flashes big and powerful as though somewhere in the depths a searchlight was pointed upward; flying fish skitter left and right the day long....

At Panama he switched ships to the LSM 312; for some reason, the Navy assigned him to the engine room as chief engineer, a job usually held by men who trained at Cornell Diesel at General Motors for six months before being assigned. Henry Jr. had worked in a newsroom and for a few weeks in a foundry; he had a keen mind for math, but no clue as to how run the engine room.

"I hope the old B&S [Brown & Sharpe] instinct comes in handy!" he wrote to his father. "In the meantime, I think I'll race back to Brown for an engineering degree."

The old B&S instinct did not kick in. When the fleet pulled into San Diego, the LSM 312 plugged into electric power at the wharf—and everything went haywire. The cook came to Ensign Sharpe to complain that his galley was nearly frozen out—and all this with the stoves going full tilt. The ice box was getting hot instead of cold and the air compressors were evacuating the high-pressure air tanks instead of filling them. Somehow the ship's engineer had plugged everything in backward and everything on the ship was running full speed reverse.

When the ship stepped off for Pearl Harbor, Henry Jr. wrote: "Well, it's a long way home now—and every minute makes it further. Would that I were back there too! This far away stuff is not so hot. One gets a terrible sense of loneliness being so distant. It is an awesome sense of emptiness at the bottom of my heart. The day of eventual return seems as a dream in the far-off mist. And the world ahead is strange, cold and unfriendly....

"We sit nights in the wardroom hearing first broadcast stations, then short wave stations growing fainter—fainter behind us. We stand on deck, and all about us is sea and sky—and we realize we are *alone*. The idea of the 'great adventure' leaves one entirely when the last promontories of 'good ol' USA' fall behind and the open ocean rolls out ahead. It is a sensation I hope never to pass through again."

At Pearl Harbor, Henry Jr. was relieved of his engineering duties and first made a deck officer, ensuring that men and material moved smoothly along the LSM's long steel deck, before he switched to navigation. From Hawaii the fleet sailed to the Russell Islands, two small islands near New Guinea. "There is MORE water out here than you could ever imagine," Henry wrote home. "And if you ever do see land, it's no job to see the other side of it. The 'Romance of the Pacific' is also a nonexistent myth. There is nothing but hot sun, dry palms and blue salty water."

One Sunday morning he was chosen to lead the deck-side service, an occasion for which Ensign Henry Sharpe, Jr., wrote an original prayer:

> O God, as our small ship sails on into the waters of danger and human strife, help us to keep our sense of proportion. Lead us to see in thy great natural beauty, which surpasses all that is manmade, the Glory of thy presence. Show us, in the infinity of the stars, in the vastness of the sea and the mass of the land the immensity of thy presence.

On **the day** before the attack, the fleet slipped past Dinagat Island in the Philippines, en route to its ground zero: the u-shaped, 20-mile beach lining the Lingayen Gulf. Peering from the silo-like conning tower of the LSM 312, Henry Sharpe, Jr., watched the sunset over Dinagat, perhaps the most beautiful sunset he had ever seen. The air was filled with a violet glow. On the deck the speaker blared: "I'll Be Glad When You're Dead, You Rascal You!"

As an ensign in the U.S. Navy, Henry Sharpe, Jr., served in the Pacific aboard the LSM 312, a landing craft that delivered tanks to the Lingayen Gulf, and missiles to Okinawa, where he was targeted by sniper fire. Ensign Sharpe earned two battle stars during the Pacific campaign, returning home through a typhoon that buffeted his ship with 30-foot waves (courtesy Peggy Sharpe).

During the fleet's two-month layover in the Russell Islands, Henry Jr. had had plenty of time to mull over where he had been and where, god willing, he might like his life to go. In a letter to his father he had broached the idea of "not following in the footsteps" and working at Brown & Sharpe:

> Whatever the future may hold for me, one thing will always be understood, Brown University shall hold a place of closest affection in my heart. I want to see her become re-established as one of America's truly great colleges. Hence the desire, possibly, to take up the profession of teaching. History, I believe, would be the field of my attention.
>
> In addition to the fact that Brown so holds my interest (and history in general) there comes the added factor of, "what am I, as a person, qualified to do." Frankly, I'm no great shakes as a leader of men—strange as the admission may seem. My successes, small as they have been, have been as an individual—Phi Beta Kappa, the Sphinx, the Herald. ... Hence it might seem more logical to build a career as an individual rather than as an individual working in close cooperation with hundreds of others.

But for the present, in January 1945, Henry Jr. was working in close cooperation with thousands of others, and as an officer. His LSM 312 was one of nearly 500 Navy ships plowing toward the Lingayan Gulf, where they planned to deposit more than 200,000 soldiers to take back from the Japanese the island of Luzon and its capital, Manila. S-Day—the Pacific equivalent of D-Day, fell on January 9. In the pre-dawn darkness, the fleet formed a line broadside to the beach; the big battleships fired their guns.

"Of a sudden a yellow flash filled the semi-dark to port," Henry Jr. wrote. "Two

spots lept from the turrets of the *Colorado* and began a graceful arc toward shore....
Whap! ... The air about us whipped around the ship and the bombardment was on.
The heavies all down the line began belching yellow bursts and soon columns of smoke
began to appear on shore as well."

The small "VPs" or Higgins Boats, roared full throttle toward the beach, ferrying
infantry to secure a beachhead. "Then ... the 1st reports so tensely awaited," Henry
Jr. wrote. "10:00 [a.m.] ... no resistance whatsoever to the first 7 waves ... 25,000 men
on the shore. A smile and a sigh flitted through the ship." Japanese troops had withdrawn
to the hills of Luzon to do their fighting there. The beach was open for landing. The
LSM 312 crunched ashore, the first LSM to land, disgorging its cargo of five tanks, a
bulldozer, and dozens of men.

Early the next morning the ship idled near a large "attack cargo ship," or AK, planning to take on more equipment to deliver to the beach. Then a voice on the radio
barked a warning, a "Flash Red," and the air raid was on. Ships fired up their "smoke
pots" to cloak the fleet, octopus-like, in an inky, protean black cloud.

"We started our own smoke generator ... and waited," Henry Jr. recalled.

> There was the usual mid-air-raid tense calm. The lively putput of our own smoke generator ...
> the hissing of the "pots" on the "VP"s as they swished past in the murk unseen. Occasionally we
> could see the stars through the smoke ... a clear early dawn. Suddenly, as Carbone and myself
> were standing aft by the fantail ladder the calm split ... "C-R-R-OCK."—once on the reverse
> side of the KA ... then again between us and the ship. A bare 150 feet away looking out of the
> corner of my eye I could see spray 100 feet in the air. Then I was at the bottom of the ladder,
> huddled together with Ekenberg and Carbone.[3] We looked at each other stunned ... and knew
> then and there the war was on.

In May of 1946, Henry Sharpe, Jr., came home from the war. For a few months
he took a summer vacation, to decompress. In his time in the Pacific he had
come under fire in the Lingayan Gulf and was targeted by a sniper on Okinawa, where
the LSM 312 had disgorged tractors and men for the bloodiest invasion of the war; he
had convalesced in a South pacific hospital, suffering "jungle rot" in his hands and
feet; and on the homeward leg after the war, his narrow, top-heavy ship had been nearly
swamped by the 30-foot seas of a typhoon.

As Henry Jr. pondered his career options that summer—news reporter, history
professor, or businessman—his father strongly suggested that he apply to Brown &
Sharpe's two-year apprenticeship program for college graduates. Mindful of duty,
Henry Jr. sent his application in July, and received a formal acceptance letter from company vice-president Arnold K. Brown.

Brown could not have been too pleased to see Henry Jr.'s application come across
his desk. As vice-president, Brown was the logical choice to succeed Henry Sharpe, Sr.,
now in his 74th year. But if his father held onto the reins for a few more years, Henry
Jr. might be the heir-apparent to the presidency. Nonetheless, Brown extended a courteous welcome to Henry Jr. in his acceptance letter: "In making these arrangements it
is our hope and expectation that you will develop in such a way that we will wish to
have you remain in this organization, and that your experience at the same time will
make you wish to do so."

Henry Jr. punched in on September 23, 1946, put in a now-standard 40-hour week, and took home $25.46, the current equivalent of $330.98. Over the next two years he learned the business from the shop floor up, taking his turn on the screw machines, grinders, millers, and lathes; near the end of his term he travelled to Europe to report on post-war conditions there, then returned to where he had begun in his college years: the foundry, this time to learn patternmaking. In November 1948, his last full week in the program, Henry Jr. netted $31.50, a 23 percent increase over week one.

The year 1948 had been a bad one for Brown & Sharpe, and for the machine tool business in general; as after the last global war, the world was awash in surplus machine tools. Between 1940 and 1945, U.S. companies had built more machine tools than they had made in the previous 40 years; the government owned half of these, and was only too willing to dispose of them at 20 cents on the dollar, or even as lows as 5 cents on the dollar to "real and pseudoeducational" trade schools that promised to train the next generation of machinists.[4] In his trip to Europe, which drew intense interest from the newly created Central Intelligence Agency, Henry Jr. found the prospects there even worse than at home.[5] As his father reported to stock holders: "Following the war unstable, and in many countries nonexistent, foreign exchange destroyed the market for our product despite the continuing demand of foreign customers for goods of our manufacture."

Post-war inflation also hurt the bottom line by driving up the costs of goods and labor. In his annual report to stockholders Henry Sr. explained:

> Net profit for 1948 is less than half that reported in 1947, notwithstanding a drop of only 20% in sales. This considerable decline in net profit is attributable in large measure to a further rise in the so-called "break-even point." ... Labor and material costs reached an all-time high during 1948. Prices were increased modestly, but not sufficiently to offset cost increases.

To ease the transition to a peacetime economy the U.S. government awarded Brown & Sharpe a small, classified contract to build gear cases for the radar-controlled anti-aircraft guns made by the Sperry Gyroscope Company, but it added little new work. With high labor costs and low sales, Brown & Sharpe began weekly layoffs, eventually dropping thousands of people from the payroll. Velia Costantino, who had begun work as a drill press operator during the war years, was one of the last of the woman floor workers to be let go. Brown & Sharpe had received a big order for screws, and Velia ran a small lathe to smooth the burrs off finished screw heads. That lasted into 1948. "And I was laid off," Velia recalled. "I did feel a little bit down. You made friends. Their husbands came home. Everybody went onto their own lives.... No we didn't protest. We were lucky to have had those jobs for so many years. It wouldn't last forever. It was a man's job."

In 1949, a job for anybody became a scarce commodity at Brown & Sharpe. At one point the company pared its workforce to just 500, a cut of nearly 12,000 since peak employment and a figure not seen since the Depression. The company finished the year $315,000 in the red.

Like his father, Lucian, Henry Sharpe, Sr., liked to take advantage of slack times to develop new products, and the late 1940s were no exception. The company began making a line of electronic gauges for measuring pieces, and in 1948 it bought the Ford Motor Company's exclusive rights to make and sell Johannson Blocks in the Americas. The blocks came in sets and could be stacked to produce gauges as precise as 4 millionths of an inch. Along with Vernier calipers, micrometers, and six-inch rules, the "Jo Blocks" had become a staple of machine shops; they had even proved useful in high-precision plants such as Pratt & Whitney, where mechanics making jet engines demonstrated their use to Henry Sharpe, Jr., during a tour he took of their plant.

There were a few hopeful signs for economic recovery: U.S. military spending was again on the upswing, driven by interservice rivalries within the new Department of Defense. Rather than argue about receiving a larger share of pie, the branches of the military pressured Congress for a larger pie.[6] The aircraft industry, a heavy user of machine tools, had grown exponentially during the war, and its leaders pushed for subsidies to maintain their technological know-how until a lagging civilian industry could put new technology to use. Future sales to the aircraft and defense industries looked promising. And, hoping to stave off Soviet influence in Europe and in China, the Truman administration had pushed through "The Marshall Plan," a $17 billion plan for rebuilding 16 European nations and propping up the government of Chiang Kai-sek.

On a June day in 1950, Henry Sharpe, Sr., felt ambivalent about the company's immediate future. On the one hand, he wrote to a Wall Street firm, the first four months had seen a "considerable increase" in orders; on the other, demand had suddenly dropped off, creating "a varying opinion in the trade as to what the remainder of the year will develop."

Ten days later, 75,000 North Korean troops crossed the 38th Parallel dividing communist North Korea from the Republic of South Korea. The 5-year-old United Nations responded with a *Complaint of aggression upon the Republic of Korea,* and dispatched a coalition of UN forces to Korea, with 88 percent of the soldiers drawn from the United States. Though Congress made no formal declaration, the United States was again at war, and war, as always, was very good for the machine tool business.

At the war's outbreak the government still held thousands of machine tools in reserve—the Joint Army-Navy Machine Tool Control Committee counted 131 Brown & Sharpe machines stashed in a single airplane hangar in Georgia—but nearly half of the reserves were more than a decade old. A Senate Committee determined: "The tools of 1940 were unable to manufacture the armament of 1950, particularly in the aircraft field, where jet engines were supplanting the piston type."[7]

In October 1950, the Chinese forces that had deposed Chiang Kai-sek flooded onto the Korean Peninsula, driving UN forces into retreat, and creating a bloody war of attrition. Once again the federal government instituted a "pool order" program, guaranteeing to buy machines made without contracts, to expedite production. While fighting the war, General Dwight Eisenhower complained:

> There seems to be a bad shortage of machine tools. When we get over this emergency I am going to take as one element of my personal ambitions, that of preaching the need for machine tools as part of military preparation until some d—administration will take the necessary measures. I've heard the same story time after time and it seems to me we should learn.[8]

16. One War Ends, Another Begins

Henry Sharpe, Sr., did not like the looks of this. In late March 1951, he received a letter from a Mrs. Sanford who claimed that she and Arnold K. Brown, the company's No. 2 man, had had an affair. Now she wanted financial support. Henry responded promptly,

> Your favor of the 28th is at hand suggesting my interest in seeking to alleviate your needs, and possible complications as relating to one of our employees. Your communication, however, in no way states what amount is owed you.
>
> We have interviewed the party in question and he says he does not owe you any funds and that you have been apprised of this fact.

To paraphrase James Joyce, Henry Sharpe, Sr., "dealt with moral issues the way a cleaver deals with meat." Soon, Arnold K. Brown severed his relationship with the company and was out of the picture, leaving Henry Sr. without a reliable right-hand man. Henry was now 78 years old; he moved in a stiff, arthritic fashion, the result of an ice skating fall in his mid-sixties that had knocked him unconscious. A recent X-ray showed an old, undiagnosed, poorly-healed spinal fracture from that fall, and it was causing him great pain. In March of 1951, Henry's doctor had said: "Within another month, Mr. Sharpe, you will no longer be able to walk. What are your plans for the business?"

The company's annual meeting, held like clockwork in mid–April, was just weeks away. Henry Sr. needed someone to take over the day-to-day grind of running Brown & Sharpe, and he had just lost his senior vice-president. "It was under these conditions that he took me aside late one morning and invited me to join him (as he often did) for lunch," recalled his son, Henry Sharpe, Jr.

Father and son rode the elevator 16 stories to the top floor of the Turk's Head Building, a U-shaped skyscraper that stood as downtown's tallest building until the Biltmore Hotel eclipsed it. In the early 1800s a shopkeeper hung a ship's figurehead of an ottoman warrior as a sign, so customers could know his store as the place with the Turk's head. When the 16-story building went up on the spot in 1913, an architect incorporated that history into the building's façade with a replica of the Turk's head cast in concrete. The entire top floor of the building was leased by The Turk's Head Club, a men's weekday social club comprised of 600 bankers, manufacturers, doctors, and lawyers. The club's dining rooms and reading library commanded a sweeping view of Narragansett Bay.

The Sharpes stepped into the top-floor lobby, bound for the club's prestigious "Table #1," where Henry Sr. traditionally dined with old acquaintances. "For some reason my father saw fit to divert to a long couch in the club lounge before going directly, as he usually did, to his accustomed table," Henry Jr. recalled. "And it was there, sitting on that couch within public earshot of his business friends and acquaintances, as they read their *New York Times*' or *Wall Street Journals*, that he suddenly turned to me.

"By the way," he said casually, and all too loudly, "next week is the Brown & Sharpe Annual Meeting. I think you should be elected President."

PART THREE: A PEOPLE'S CAPITALISM

17. Flying into the Jet Age

On Henry, Jr.'s first morning in the president's chair, the door to his office creaked open; his father peeked in with a smile upon his face, gently closed the door and sat down. "Now I know how you feel about flying," he said, an oblique reference to the secret father and son shared, but never discussed: Henry Jr. had been taking flying lessons from a neighbor up the street, Mary Ann Lippitt. "Now I know how you feel about flying, but I don't want the company rule that our people are never to fly on company business repealed."

Henry Jr. had in fact planned to repeal the prohibition on flying his father had imposed nearly 30 years before when a group of his business acquaintances flying from Paris to London went down in the English Channel. The flying ban may have been prudent in aviation's pioneering days, but it was, after all, 1951. Only the previous week an important customer, the Douglas Aircraft Company of Santa Monica, California, had sent Brown & Sharpe an emergency call for a repairman post-haste. And Henry Sr. cleaving to his long-held company policy, had assured the customer that, yes, help would be on its way to the West Coast immediately: leaving aboard the Merchant's Limited at five o'clock that night.

Henry Jr. had his own ideas on how the company needed to run in the second half of the twentieth-century, but in many ways his basic business philosophy was not much different than his father's. On his first day as president he issued a memo to his managers saying:

> I think we should review together the things which have made Brown & Sharpe so successful under my father's 52 years of leadership. The rules are simple—there are really only three, but they leave a rich inheritance on which to build our future.
> Make certain that machines you create are of really good design.
> Make certain the integrity of your workmanship is of the highest.
> Make certain your reputation for fairness and honesty is never blemished.
> There is no "modern touch" that can change these rules on which Brown & Sharpe will continue to operate.

Father and son agreed that the company had outgrown its warren of buildings clambering up Smith Hill from the Woonasquatucket River, some of which dated into the nineteenth-century. The company had recently bought the old Greystone linen mill in nearby North Providence, and an additional 160 acres in suburban North Kingstown. There were no concrete plans for the North Kingstown land, which the company bought with an eye toward the distant future; but shortly after its purchase the Greystone plant was working at maximum capacity with 600 workers dedicated solely to a top-secret project, making the gear cases for Sperry Gyroscope's radio-controlled anti-aircraft guns. With the Korean War, a small contract that had originally been an attempt to buffer a downturn soon morphed into a commitment too large for the Promenade Street plant, requiring the purchase of Greystone.[1]

"Young Henry," as the old-timers called Henry Sharpe, Jr., took over the company at a challenging time. Business was booming with the war, but problems loomed: in June the Department of Justice named Brown & Sharpe as conspirators in an anti-trust suit, and three union contracts were expiring.[2] Union leaders knew that business was good; they wanted in on it, and in negotiations they were prepared to test Young Henry's mettle.

On July 12, machinists in the union voted 3,041 to 384 to authorize their leaders to call a strike. By a similar margin they also voted for a Union Shop, requiring all machinists to be union members. Henry Jr. responded to the strike vote by sending a letter directly to workers:

> In the first place, all of you know this company's traditional opposition to all forms of union security. This opposition has continued with good reason. The management's concern has been that union security means a weak individual.

At 28, Henry Sharpe, Jr., followed in the footsteps of his father by assuming the presidency of Brown & Sharpe, one of the country's largest manufacturers of machine tools. He had earned a degree in economics at Brown University, served as a naval officer in the Pacific theater, and completed a two-year apprenticeship in the plant. As president, he received on-the-job mentoring from company treasurer Frederick Austin (courtesy Peggy Sharpe).

> Our system of working together is founded on the ability of the individual to stand up for himself. Insofar as the changes now proposed by the machinists' union would weaken the powers of the individual, we feel that they threaten the very basis of our society.

The union was not buying it. Joseph Kane, president of a local 1142, shot back: "The whole purpose of collective bargaining is to accomplish the greatest good for the greatest number. Our whole nation is founded on this truism."

Kane's response revealed the crux of the matter: rugged individualism vs. the collective good. The Henry Sharpes, junior and senior, descended from Lucian Sharpe, who rose above childhood poverty through the twin virtues of intelligence and ambition. From the perspective of the managing class, the Republic of the United States of America was a place where virtuous individuals could find opportunities to succeed and lead. Successful individuals ensured a strong republic by building strong communities through philanthropy and by leading the less able.

But the machinists union represented a class of people who had obtained better working conditions, the right to vote, and political clout only through collective action. For them, America was a place of equals where people built a strong nation by selflessly sticking together to "accomplish the greatest good for the greatest number."

A week after Henry Sharpe, Jr., sent his letter, 4,000 workers walked off the job in a one-day protest. When they staged a similar wildcat walkout one week after that, Henry Jr. closed the plant. As the strike dragged on into September, it caught the attention of the Truman administration, which was trying to keep planes flying and soldiers fighting in Korea. As Henry Jr. recalled:

> Korean-War supply tensions had mounted sufficiently even to place the question of solving the Brown & Sharpe Strike on the agenda of President Truman's Cabinet in Washington. Both the company and the Union were to come immediately to Washington and have a Federal Mediator adjudicate a solution. The union, in a public statement, had immediately responded that they were flying to Washington in their effort to break the deadlock.

Union reps were winging their way to Washington, and company executives were going to be riding the rails? "That did it," Henry Jr. recalled. For perhaps the first time in his life, Henry Sharpe, Jr., defied his father: the famous "No Flying on Company Business" rule finally came to its end, as Brown & Sharpe executives flew to Washington.

But the strike ground on.

The **strike** took most everybody by surprise, even the workers who had authorized it. Albert Q. Perry, a union machinist who also served as a pastor in the First Universalist Church, recalled the strike's beginnings this way:

S. Day, August 1, S for Strike, and for a good many of us, S for Surprise. Perry worked the night shift and was part of a car pool that drove down to Providence from Burrillville, the state's northernmost town. On August 1, 1951, the men in the car pool arrived for second-shift only to find a deserted parking lot. As their driver drove around the block he spied a union steward.

"Is the strike on?" the driver asked.

"I honestly don't know. I'm on my way on to the plant too, but it looks like it."[3]

The steward jumped onto the car's running board. Coming down Promenade Street the workers could see pickets in every doorway, wearing sandwich board signs and speaking to clusters of men in work clothes carrying lunch pails. The pickets directed them to a schoolyard, where they gathered around a sound truck. The voice of a union spokesman squawked from the truck's speakers: The company had accepted the union's demand to expand insurance coverage to cover surgery; but in exchange management demanded a corresponding cut from their wage offer. A sound of derision, what Perry described as an "angry growl" spread through the crowd.

The company, the spokesman said, would not pay out any vacation pay until the contract was signed. Those vacation checks were supposed to go out on August 10, when the plant closed for its annual two-week maintenance period. The union read the threat to withhold that money as an ultimatum: accept the contract within the next week, or there would be no vacation pay. "We were in a rather ugly mood," Perry recalled.

Around him, Perry heard snatches of conversation: "I've got a rush job on my machine. The foreman begged me to get it done. Maybe 'Henry' will do it if he wants it so badly."

"I hope my whole damn machine rusts out," another man said. "I'll never use it again. I'm off for Detroit where a miller can make some money."

A second-shift Supervisor reported that strikers broke his car's windows and flattened a tire in the Promenade Street parking lot; tacks strewn across the driveway leading into Greystone resulted in several flats. Later, a worker claimed a car driven by supervisor struck him and broke his leg.

"We knew the strike was going to cost us money"; Perry wrote, "we hoped it would cost the company as much. Anger was mingled with humor as we drifted away. Above all else, we were worried." Driving back to Burrillville, Perry thought about how much things had changed from the night before, when a friend at work showed him with some pride a case of special micrometers he was making for Pratt & Whitney. They looked as though they were custom-made for special work in building jet aircraft engines.

"Gracie Fields used to sing a song about a girl who made a 'thingamabob' that won the war," Perry wrote.

> I expect that every working man feels that he is doing exactly that in time of national emergency. In a tool factory, this attitude was very easily held, and we took great pride in the importance of our work. Not much was ever said, but you could see that pride as a man smoothed off a burr with a file, stopped to look into a perfectly reamed out hole, or smiled when the micrometer reading agreed exactly with that called for on the Blue Print.

As the strike wore into its sixth week, Perry probed to hear the reaction of men who had felt a personal stake in building tools for the war in Korea. One man told him: "Why should I be more patriotic than the company? I'll bet they didn't take the contract until they were sure they were making money on it. I won't work on it until I'm sure of making both ends meet."

A second said: "Look, my own kid is headed for Korea or Germany or some other miserable hole, and while he's fighting for me I'm fighting for him. He'll probably work in this damn mill, and he won't get paid a penny more than the union demands."

A third worker recalled the hardship of his layoff after the Second World War. "I listened to all that hokum during the last war and worked my fool head off but I couldn't live on patriotism when work got slack," he said. "Nobody worried about me not working a couple of years ago, what are they yapping for now?" Perry observed: "Very few are influenced by the effect of the strike on the defense effort. Most seem to think the company must accept at least half the responsibility for any delays and many that it ought to receive all the blame since it can better afford to give in."

As far as who could afford what, Perry noted,

> It must be admitted that most of us were making pretty good money, but only because most families had two wages. With all the talk about the desirability of the 40-hour week, little is said about the fact that a family can't live at an average level on the wages from one, and either both husband and wife work or the husband works two jobs. In my department half the men had other jobs and the rest had household duties because of a working wife. Between our jobs at the plant and those outside, we were working 70, and 90-hour weeks and were tired. Perhaps that explains some of the willingness to vote for a strike.

The average wage of Brown & Sharpe worker was $62 a week; machinists earned above average pay, closer to $100, the equivalent of $960 a week in current currency. In Perry's opinion, workers were not striking to avoid abject poverty. They were trying to keep pace with the ever-escalating temptations of the marketplace.

> We would certainly seem to an outsider to be enjoying the strangest poverty in history: carrying fried chicken because we couldn't afford beef, buying television sets in order to save on going to the movies, hoping that prices would be frozen as soon as ours rose a little bit more and driving ourselves because a man couldn't live on 100 dollars a week.

At first, workers feared the losses of their Brown & Sharpe paycheck. But as August and September rolled by without them, the pay cuts were actually not as bad as many expected: a man with a small farm had work that really required him to take off several weeks, but he had been reluctant to do so; a door-to-door salesman took on extra lines and made more money at his second job; some skilled workers moved out of state. "One thing only that I have not heard," Perry wrote, "and that is any willingness to 'scab' on the strike."

After 13 weeks, labor and management finally hammered out a deal. More than 2,000 union machinists filed into Veterans Memorial Hall to hear their leaders explain the proposal. Then they voted 1,340–183 to accept the contract, sending union leaders hustling down the hill to meet company officials in the Sheraton-Biltmore to sign the contract, ending the strike after 13 weeks.

The union did not win a modified union shop, but it did erode management rights by winning the right to arbitrate disputes affecting changes in jobs or the establishment of new jobs, and the rates of pay appropriate for them. The company also agreed to give the union an office in the main plant, where union reps could meet with new workers and explain to them the benefits of joining the union. All of the plant's 8,100 workers also received one new paid holiday, Armistice Day, making 8 total.

The 13-week strike struck a blow to the finances of workers, the company, and the merchants of Providence. The Union Electric Railway calculated that it had lost 800 to 1,000 riders each work day, and the company announced a net loss of $1.4 million, an inauspicious beginning for Young Henry Sharpe, who still lived with his parents in the place he jokingly referred to as "The Castle."

As a girl, Maybury Viall loved the garden at 84 Prospect Street, where the Sharpe family lived. Her father, Richmond, a former World War I pilot, worked at Brown & Sharpe, and sometimes Mary Elizabeth invited her in behind the walls of her mansion.

"She knew I loved her garden," Maybury recalled in her 80s, when she was known as Maybury Fraser. "You wouldn't think you were on Prospect and Cushing Street. The garden was just so—it was all trees and shade. The foliage of the place, the brick walks and everything, made you feel as though you were somewhere else"—no specific place, just somewhere magical. "I mean it was an oasis in a city."

By the 1950s, Mary Elizabeth had been nurturing that garden for more than 20 years. She had always had a good eye for art, and on Prospect Street she used that eye to create art of her own, using the three-dimensional world and all of the senses—even the tastes of quince, pear, and grape, as her canvas.

At first Mary Elizabeth thought that she would hire Frederick Law Olmsted to design her grounds, but she wanted a French garden, a series of "outdoor living rooms" as she put it, to complement a French manor. It would take some ingenuity to transform this long, narrow, barren western side of their property into a stately green space. With the help of her English gardener, trained at Kew Gardens, she created a formal French garden near the house, stepping down to a middle informal garden that flowed into a naturalized wild garden below. By making each of the side-beds and stepping stones that edged the rectangular lawn in the upper garden narrower and narrower as the eye moved away from the house, it made each seem farther away than it really was, extending the sense of distance. She created perspective the way a painter does to make a scene of beauty.

In planting, Mary Elizabeth looked to create what Japanese gardeners called the "stolen view"; by spacing plants so that they drew the eye to pleasant sights and blocked ugly or unnatural elements. Beauty, she believed, "doesn't have any nationality or belong to any date in time. It can be a rock, or it can be an African carving. It can be by someone who doesn't have any intellectual interest in things, or it can be by a very great master of the middle ages. It can be anything. The point is that it communicates something to you."

The president of nearby Brown University, Henry Wriston, felt the connection to her garden, and he asked her advice on landscaping at Brown. She thought there should be a plan that united disparate buildings together with "pretty big bulky things" such as laurel trees that would bloom and fragrance the air at graduation time. She set about a years-long plan, bankrolled by her husband and sisters-in-law, to unite the patchwork of campuses on the East Side.

When President Wriston bought up and tore down a city block to create a new quadrangle lined with fraternity houses and a dining hall, Mary Elizabeth advised him on how to create a stolen view by lowering walls and allowing paths to meander rather than run linearly. Henry Sr. led a fundraising campaign for the Wriston Quadrangle, getting it off to a good start by writing a check for more than $200,000, a debt that Wriston repaid by naming the dining hall The Sharpe Refectory. Students soon renamed it the Rat Factory, and generations of Brown undergrads imply called it The Ratty.

After Mary Elizabeth redesigned the Brown campus from the Wriston Quad to Pembroke, she became a sought-after designer. She did not charge for her services and chose only projects that interested her, such as planting a courtyard in front of the new Superior Court Building on South Main Street, and landscaping the neighborhood's African-American Church, likely the Congdon Street Baptist Church.

In 1950, Wriston surprised Mary Elizabeth by granting her an honorary Ph.d., not for her accomplishments as a businesswoman, but in recognition of her second career as a designer who shaped the landscape of the city's East Side. She had not ever earned a high school degree because as a teen, she was too busy rescuing her family from poverty to go to school. "I just thought if I could have three square meals a day and go on a trip once in a while it would be wonderful," Mary Elizabeth recalled.... "My interest, when I was free, was to try and make more beauty in the world."

Peggy Boyd strolled along Cushing Street on her way downtown to the movies with Russ, her current "beau." The sun was still high in the sky in late spring, and fragrant blossoms of flowering cherry trees spilled over a high garden wall, brightening the streetscape. Peggy was close to earning a degree in Landscape Architecture at the Rhode Island School of Design, and she could not help but remark on the beauty of these trees. Russ allowed that he knew the family who lived behind that wall, that he was in fact a fraternity brother of Henry Sharpe, Jr.'s; perhaps to impress his date, he took her behind the wall and knocked on the door of "the Castle."

As it happened, Henry Jr. was just bustling out the door; he was the 29-year-old President of a large machine tool manufacturing company, a busy man heading off on a business trip. If he noticed the thin, dark-haired girl in a kerchief and tennis shoes, she did not set his heart aflutter—this time.

That November they met again, at the home of Maybury Viall on a nearby street. It was Armistice Day, November 11, 1952; the factory was closed because the union had negotiated the day as a paid holiday. It was also Maybury's birthday and her parents had invited young un-marrieds to a party for her on that date. Henry Jr. came, because he had always kind of been there in Maybury's life, sort of a distant, big-brother figure, son of her father's boss and seven years her senior. Peggy, a native Floridian, came because she had gotten to know Maybury through a mutual friend at RISD. Henry walked through the door of the Library, where he spotted Peggy in a red dress, sitting on the floor among a group of friends next to a vase of flowers. Peggy looked up at him and smiled. He came to sit next to her. And then Maybury recalled, "It was love at first sight, it really was."

They found they had a lot in common. Both were only children; both were interested in exploring ideas about the meaning of life and how the world began. Henry had read a book called *The Art of Thinking* by a French priest and family friend, the Abbe Ernest Dimnet. Peggy was reading a book given to her by her grandmother called *Human Destiny* by Le Compt du Nouy. Peggy suggested that he might like to borrow it. And a few days later, Henry stopped by her apartment to borrow *Human Destiny*.

She felt thrilled when he met her train at Union Station when she returned to Providence after her Christmas vacation. He asked his mother if he could bring "some friends" to dinner at Prospect Street. For the winter months "some "friends" always turned out to be just Peggy. Mary Elizabeth put her at ease. They shared a passion for landscape design. Peggy had also studied set design while earning a degree in theater at Sarah Lawrence College, and she appreciated the artistry that Mary Elizabeth had put into designing her home's interior.

As winter melted into spring, Henry and Peggy realized that they both loved nature and enjoyed taking walks in the countryside together. After a spring vacation to meet her parents in Florida, Henry Sharpe, Jr., and Peggy Boyd planned an August wedding. They married high on the hills of Onteora, New York, a summer community where everyone opened their houses to visiting wedding guests. Mary Elizabeth and Henry Sharpe, Sr., enjoyed watching their only child wed, though Henry Sr. was by then using a wheelchair. It would be his last adventure away from home. The bride and groom went on what Peggy termed a month's "Bunnymoon" in Europe, a honeymoon mixed with business.

All that August morning in 1954 the wind blew, and about a half-hour before noon the factory's power cut out. The winds of Hurricane Carol funneled a surge of seawater from the wide mouth of Narragansett Bay up into the narrow headwater of the Woonasquatucket River, which continued to rise till it spilled over its banks and washed across Promenade Street. When the surging water topped the curb and stove in the factory's tall windows, the men in the basement grinding room beat a hasty retreat to a higher floor.

Forge workers found themselves wading through thigh-deep water washing through in their building; workers in the shipping room frantically stacked finished tool orders on a high counter as flood waters swirled around their knees. The bench caved beneath the weight, and finished precision tools tumbled into the dark and dirty water, becoming scrap. The No. 34 shed, packed with finished pumps, flooded nearly to the ceiling.

Hurricane Carol struck on August 31, 1954, killing 17 people in Rhode Island, and shutting the factory down for a week. The storm was just the latest in a series of blows to wrack the newly-wed Henry Sharpe, Jr., as he tried to follow in the large footsteps of his father and grandfather. The Korean War armistice, signed the previous year, was good news for the nation but bad for business. In the first three months of 1954, orders plummeted by 65 percent, forcing Henry Jr. to layoff 2,500 workers.

In April, the flu laid Henry Jr. low, forcing him to send a proxy to deliver a speech in Rochester, New York, about the importance of harnessing technology in the atomic age to a management group at the Rochester Institute of Technology.

And on May 17, at 1:30 in the morning, his father died. Henry Sr.'s death must have been stop-the-presses news, because that morning's *Providence Journal* reported it on page one, alongside news of the McCarthy hearings, perennial presidential candidate Harold Stassen's alarums about the Soviets building up a "war machine," and U.S. opposition to any Indochina settlement that handed a part of Vietnam to communists.

The news story of Henry Sr.'s death contained all of the predictable encomiums: it noted that he had served as chancellor and trustee of Brown University; vice president and director of the Providence Journal Company; trustee of the Rhode Island School of Design; director of the Rhode Island Hospital Trust Co., and the Rhode Island Hospital National Bank; vice president and a director of the Providence Gas Company; president of the Puritan Life Insurance Co; originator and "guiding genius" of the

Henry Sharpe, Sr., and Mary Elizabeth Evans Sharpe dressed up for the marriage of their son, Henry Sharpe, Jr., to Florida native Peggy Boyd. The two wed in August 1953; Henry Sr. died the following May, prompting the *Providence Journal* to comment: "there has been an increasing understanding that Mr. Sharpe acted out of a deep love of Providence—deeper than the general public will ever know" (courtesy Peggy Sharpe).

Providence-Cranston Community Fund, precursor to the United Way; twice served as president of the New England Council, a regional group of business and academic leaders. "And his lasting desire to see his own city governed well bore fruit in the organization of the Providence Governmental Research Bureau," and the Rhode Island Public Expenditure Council.

But the editorial board of the *Providence Journal* knew Henry Sharpe, Sr., very well, and in a full-page, six-column spread its members took the time to limn the man not just as an icon of industry but as the flesh-and-blood human being whom they had known:

> Many saw him as a distant, austere and severe man and interpreted this front as aloofness when in truth he was the captive of a shy, reticent and introspective personality whose chains he was incapable of breaking. Even some of his closest friends at times found it a strain to be with him when he would withdraw within himself beyond penetration. Mr. Sharpe had no defect of character; he did have an impediment of temperament which shut him off from people, who looked upon him with a sort of awe and therefore as a man difficult to approach. His tremendous power of concentration kept him preoccupied and a man who had smoked a cigar and chatted pleasantly with him the night before when he was in a relaxed mood would find Mr. Sharpe looking through him on the street the next day. That was a difficulty he could not overcome and, with his thoughts centered on some other matter, he often appeared curt and brusque. When he was cantankerous, as he could be, that was his New England inheritance.
>
> Because of this personality impediment and because he was the largest employer of labor in Providence with positive economic and political views, he became a symbol of the Old Guard and as such drew upon himself the fire and abuse of politicians of differing persuasion and demagogues with their own personal ends to play. He met these onslaughts with dignified silence and if he were hurt inside that fact was known to but a few. Sometimes he would turn off mention of a personal attack upon himself from the hustings with a wry smile and a humorous remark.
>
> Yet his character was so massive and his sense of responsibility so compelling that disillusionment did not set in to swerve him from his duties toward the community and cynicism never bit into his spirit. A symbol is hard to break down and a myth hard to catch up with; still there has been an increasing understanding the Mr. Sharpe acted out of a deep love of Providence—deeper than the general public will ever know—and sought to make it a better city. It was the city of his parents and of his birth. He was never far from it for long.
>
> Over the years it has come to be realized that the inspiration of Mr. Sharpe was not in a magnetic personality, which he did not possess, but the inspiration of integrity, respect, ability and leadership with purpose.

18. Retooling

There was little money to be made in manufacturing hair clippers anymore, so after his father's death Henry Sharpe, Jr., dropped them from Brown & Sharpe's product line. The story was the same for sewing machines: After 96 years, Brown & Sharpe severed its relationship with Willcox & Gibbs, the company founded in 1858 by an Appalachian mountaineer with a woodcut and a dream.

The manufacturing plant that Henry Sr. had handed over to his son, a warren of brick buildings punctuated with smokestacks ambling up and across Smith Hill, was a throwback to another era. Even the tools inside of it were old, in need of replacement. But "Young Henry" had big plans for the old plant. He had once written to his mother that he was "a radical—always trying new schemes—always hoping perhaps that one will succeed," and the time had come to put his new schemes to the test.

All of 31 years old, Henry Sharpe, Jr., wanted to cut production costs by changing the old-fashioned way of making machine tools in small batches. A week after his father's death in May 1954, Henry Jr. issued a stockholder's report announcing plans for a "general modernization and retooling" of Building 10 that included a $2 million expense for streamlining production of the company's automatic screw machines.

Brown & Sharpe had been a pioneer in civilian applications for the Armory System, the use of a model, jigs, fixtures, and gauges to manufacture precise, interchangeable parts. This system, and the machine tools that it produced, made possible the mass production techniques developed by Henry Ford and others, yet Brown & Sharpe and other machine tool companies had never themselves adopted the techniques of mass production—they still relied on batch production, churning out small batches of machines at a time. Mass production depended on high-volume sales; now the company's backlog for its various models of screw machines stood at 450, enough to tempt Henry Jr. to risk the investment in retooling in hopes that quick turnaround times would reduce manufacturing costs and draw enough new customers to keep the production lines flowing.

He decided to experiment with the most popular model of the company's automatic screw machine. Brown & Sharpe designers did not invent the automatic screw machine, a designation best applied to Christopher Spencer, who in the 1870s was the first to apply a blank cam cylinder to a screw machine. The cam, an oddly shaped wheel attached to a rotating shaft, tripped parts of a machine to kick in at carefully prescribed moments, obviating the need for a human operator to manipulate the machine. Spencer's machine borrowed the automatic feed and self-revolving turret that Joseph Brown first applied to Frederick Howe's screw machines in 1861, but the cam's automatic control of cross slide tools running at right angles to the rotating stock made it a truly automatic machine that advanced stock, switched tools, and performed all operations required to complete a part.

The first screw machines produced screws and studs, but it soon became clear that they could make any small, cylindrical part used in clocks, guns, sewing machines, typewriters, valves, motors, and all kinds of everyday products. In the late 1890s, Brown & Sharpe designer William Davenport won a company contest to develop a relatively easy-to-use, inexpensive automatic screw machine that quickly became a best seller. By the mid–1950s, Brown & Sharpe's 00G screw machine ran a single spindle at speeds of 6,000 rpm; carbide-tipped tools cut, drilled, reamed, and shaped pieces from the spinning steel stock. The machines completed eight separate operations within 8 seconds.

By the 1950s, Brown & Sharpe's plant on Promenade Street was becoming an anachronism. Frederick Howe had designed the No. 1 building at the lower right in 1872, followed by the addition of the foundry, the low-slung building just to its left 1880. By 1913 the plant had grown to 967,000 square feet sprawling across 28 acres near downtown Providence. During the Korean War the plant employed 9,000 people working three shifts, but the haphazard layout required workers to push parts from building to building and up and down floors to assemble machinery, an expensive and cumbersome process.

18. Retooling

To a layman, watching one of these screw machines at work was like looking at a magic box: steel bars or tool wire flowed into the box; the box whirred, clucked, buzzed, and within seconds small, complex parts popped like hail into a hopper which quickly filled with shiny steel parts as the machine chewed through raw stock.

Under the current way of doing things, a gear made for use in a screw machine had to be turned in the turret-lathe department, drilled in the broaching department, shaped in the grinding department, and finally hardened in the foundry. Along the way laborers pushed pallets heavily laden with unfinished gears from building to building, up and down floors; occasionally the gears sat in a pile until a machine operator had finished previously scheduled jobs on other components before sending them on to screw machine assembly. All told, a production batch of 100 gears would be on the move up, down, and through the factory for 14 weeks.

Henry Jr. spent $600,000 on special-purpose machines to set up screw machine assembly lines in Building No. 10, a flat-roofed building 90 yards long and almost as

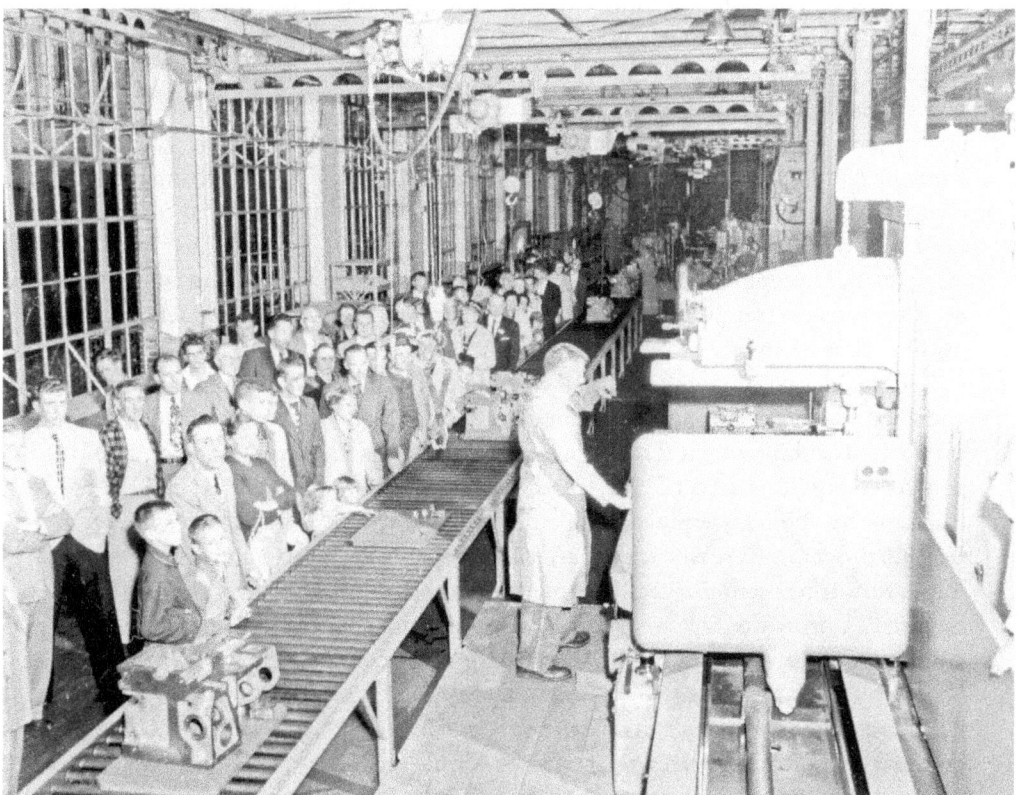

With a backlog of 450 screw machine orders, Henry Sharpe, Jr., decided to take a chance by investing millions of dollars in retooling Building No. 10 to mass produce the machines. Though Brown & Sharpe had been a pioneer in the system that made mass production possible, it had never itself used assembly line production until the retooling in 1954. When the production lines were ready to go, the company held an open house for employees and their families, who observe a cigarette-smoking machine operator demonstrate the new production techniques.

wide. There were no guarantees this would be money well-spent. "Maybe it will blow up in our face," one anonymous executive told *Business Week* magazine.

In retooling Building No. 10 for screw machine production, the company set up 12 parallel lines—5 for making base parts, 6 for other parts (gears, spindles, turrets,) and one for final hand assembly. Each conveyor line featured newly designed special tools that made many cuts on a single component at one shot: A rough-cast gear emerged finished, with teeth cut precisely, in an estimated four weeks instead of 14; the large base castings, formerly machined on huge, antiquated planer mills that filled Building 5 with a roar like a bellowing elephant, now ran through multi-blade cutters which machined the surfaces in one or two passes; rather than bore a series of like-sized holes in a screw-machine bed, then re-tooling to drill a different-sized hole, a forest of 10 boring tools with different diameters all worked at once, boring the bed in about 15 minutes instead of 8 hours. Henry Jr. predicted that by 1958, Brown & Sharpe would halve the manufacturing time of a screw machine from foundry castings to final assembly, enabling the firm to cut 25 percent off the machine's list price.

One Thursday in January of 1955, Brown & Sharpe axed 140 executives and department heads. At the height of the Korean War, the firm employed more than 9,000 people; now it needed just 2,500 workers, and fewer workers meant fewer bosses. Among those let go on "Bloody Thursday" was John Machon, the legendary Director of Apprentices, known for his sartorial style—he always wore starched, white shirts and a bowtie—his bass voice, and an insistence on probity from his teenage charges. Machon was expendable, because at the time of his firing the company had only 26 apprentices left in its last class. The previous year, Brown & Sharpe had shuttered the two apprentice dormitories, let the housekeepers and their assistants go, laid off 2 secretaries, and 20 full and part-time instructors, leaving just Machon to teach his final class of 26.

Like the hair clippers and the sewing machine, the four-year apprenticeship program had become an anachronism. For decades the federal government had been funding apprenticeship programs through legislation such as the Smith-Hughes Act of 1917, and the George-Dean Act of 1937, which appropriated $4 million for teachers of industrial subjects.[1] Both government regulators and union officials feared that some apprenticeship programs were ripe for abuse, with companies using them to avoid paying full wages to workers without really giving them the type of rigorous training that they received at Brown & Sharpe. With government funding the bill, almost no manufacturing company ran an elaborate program of worker training anymore. Brown & Sharpe retained a two-year apprenticeship for training managers and supervisors, but it trained only 55 workers at its peak. During and after World War II, the federal government encouraged the growth of new vocational secondary schools such as the New England Institute of Technology, and Wentworth Institute of Technology, in some cases offering them machine tools for pennies on the dollar. The onus for education had shifted from companies to unions, the government, and to workers themselves, who could now enroll in an array of new programs. Machon had begun his own Brown & Sharpe apprenticeship in 1916, and had put in nearly 40 years with the company when he was unceremoniously dumped. He became so depressed that he spent the

Apprentices in Brown & Sharpe's rigorous, four-year training program wore numbered patches on their shirts signifying which year of training they were completing. At right in this 1952 photograph is fourth-year trainee Bill Nixon, who recalled, "There were so many overhead belts you could not see the other end of the building." Archives identify others in the picture from left to right by last name only: Ellsworth; Swanson; Alcott. Wary that some companies used apprentice programs to underpay young workers, unions and the federal government worked for decades to create government and private training programs, taking apprenticeship programs out of company hands. Brown & Sharpe severely curtailed its apprenticeship program in 1955.

month of March in bed, before rallying to become a vice president of a Wall Street commodities firm.

After the Korean War, Brown & Sharpe's orders plummeted from $53 million in 1953, to just over $29 million in each of the next two years. Through layoffs and cost-cutting, the company managed to turn profits of $1 million and $1.8 million (currently equivalent to $15.9 million) in those post-war years. The slump following Korea was not as harsh as the slack times following the World Wars because President Dwight D. Eisenhower felt that national security required a healthy machine tool industry. He made sure that the federal government did not repeat mistakes, such as flooding the market with surplus machine tools. To help the industry bridge the transition to a civilian economy, the Government Office of Defense Mobilization bought $100 million of new machine tools from U.S. builders in 1954, and ordered another $100 million worth in 1955.

While the economy struggled to find its post–Korea equilibrium, Henry Sharpe, Jr., began looking to move some manufacturing overseas. An arm of the United States government, the International Cooperation Administration, agreed to provide economic protection to U.S. manufacturers that built plants overseas as part of a Cold War program designed to boost economic development in 37 non–Communist nations.

In May 1955, Brown & Sharpe registered a new company in England, Brown & Sharpe Ltd., "Manufacturers of and dealers in screw, milling, grinding and other machines." After World War II, England banned machine tool imports in hopes of resurrecting its bombed-out factories. The ban removed Brown & Sharpe's best export market. Henry Jr. decided to build a manufacturing plant there. The International Cooperation Administration guaranteed to cover losses of up to $1.12 million on the new plant attributed to the exchange of British pounds for U.S. dollars. With this guarantee, later upped to $3.9 million, Brown & Sharpe bought 17 acres overlooking the Tamar River outside of Plymouth, which had been flattened by the Luftwaffe in World War II. Here the company built a 20,000-square-foot plant designed for easy expansion.

Over the next few years, Brown & Sharpe embarked upon a coordinated buying spree of domestic companies, designed to complement its core products: First it bought Nelco Tool Company, which primarily made cutting tools in Manchester, Connecticut; it also obtained the American Twist Drill Company; Double A Products Company and the Rosaen Co., manufacturers of hydraulic control valves, a key to automated machine tools; it also bought Howe & Fant Inc. of East Norwalk, Connecticut, maker of turret-drilling machines to speed drilling operations; and parts of several smaller companies. All of these purchases gave Brown & Sharpe better in-house control of the milling cutters, drills, and hydraulic controls installed in its own machine tools.

After consulting with the management firm of Booz-Allen & Hamilton, Henry Jr. reorganized the company into four distinct groups, each one running like an independent business, responsible for its own profit and loss: Machine Tools, Industrial Products (precision gauging, small tools); Cutting Tools, which included the new Nelco Division; and Hydraulics, which subsumed Double A Products and pump manufacturing. In reorganizing the company, Henry Sharpe, Jr., transformed it from a traditional American business firm run by a small number of owners, often family members, into a truly "modern business enterprise," defined by business historian Alfred Chandler as a business comprised of distinct operating units managed by a hierarchy of salaried executives.[2]

In 1957 one of the four divisions, the Industrial Products sector, faced a critical problem: the profit it made by selling tool holders was shrinking; the manufacturing price for some tool holders actually exceeded the price that Brown & Sharpe could charge for them. A price raise was out, so the manufacturing engineers examined three cost-reduction possibilities: design change, lower-cost raw materials, and different production methods.

Engineers found the current designs were good, and the cost of raw materials was out of the company's hands. They had to find a way to make tool holders of the same design and materials in a way that was less expensive than the current methods of machining them, casting them in sand, or forging parts from heated steel. They found a solution in "investment casting," a process in which designers created parts made of wax, a malleable material that they could mold into complex shapes. Workers covered the wax patterns with a ceramic coating, which hardened around the wax. They then fired the ceramic to melt out the wax, leaving a cavity inside the ceramic which they filled with molten metal. After the metal solidified within the ceramic mold they broke out the casting from its ceramic shell.

After a months-long learning curve, the new production method paid off handsomely. The process worked well for more than 80 stock-tool parts, with an average saving of 30 percent on each tool. The machine tools division also found dozens of applications for investment casting in making high-volume parts used in milling, grinding and screw machines. Besides being cheaper to produce, investment castings emerged from their ceramic shells with particularly smooth surfaces and sharper contours. The method allowed a great flexibility in shape, giving engineers greater design freedom, which helped them in designing new, more precise machines such as the "Cedasize" grinding machine that hit the market in April 1958.

With the Cedasize machine, Brown & Sharpe became the first company to make a machine tool that could shave metal to the millionths-of-an-inch. The Cedasize machine could grind parts to tolerances as fine as 30-millionths-of-an-inch. The machine married electronics to mechanical engineering. The Cedasize grinder was essentially a plain grinding machine with an electronic comparator caliper that rode on the work; as stock fed into the grinding wheel an amplifier electronically registered changes in work diameter in increments of a few millionths-of-an-inch. *Iron Age* magazine reported that such fine tolerances were "aimed directly at the production of touchy guidance system components for missiles and supersonic planes," and "the new grinder is already in use in key defense plants" to grind parts for Inter-Continental Ballistic Missiles.

Though he accepted the president's job with some reluctance, Henry Sharpe, Jr., quickly immersed himself in it. He dressed formally, always wearing a bow-tie, and maintained a formal decorum around the plant, where workers referred to him as Mister Sharpe. But unlike his father, Henry Sharpe, Jr., was at ease in social situations, a good storyteller and sought-after public speaker. In meetings he was more of a listener than a talker, but when he did speak he had a way of condensing a discussion's main points, often while leavening the dialogue with humor. Within five years of his father's death in 1954, Henry Jr. had opened a manufacturing plant overseas, bought several machine tool-related subsidies, improved parts production, streamlined screw machine assembly, developed new lines of grinding machines, and created a modern business enterprise by reorganizing management into four distinct divisions. Yet all of these improvements were not enough to maintain profitably through the recession of 1958.

Six months into 1958, Henry Sharpe, Jr., sent an urgent memo to all divisions announcing "Operation Updraft," a sales campaign. In order to "spur salesmen to make every possible extraordinary effort between now and year's end," Brown & Sharpe announced a bonus plan to reward salesmen who could reach even the minimum sales goals it had set for the year.

Collecting those bonuses would be tough. Shipments during the first quarter were 54 percent below what they been the previous year. The company had already pared the workforce as much as it felt that it could without jeopardizing the future—

when good times returned, and history has shown Henry Jr. that they invariably did—the firm would need a core of workers who could churn out new orders.

The primary goal of the sales push was "keeping the organization intact for a better day, ready to provide good jobs for the company's many laid-off and short-time employees," the memo said. The method of achieving that goal "had to come through the only avenue left open, through increased sales of products the company is now in a position to make."

At the end of the year, Brown & Sharpe managed to pay a dividend of 60 cents a share, better than nothing, but an 87 percent earnings cut from the previous year. And 1959 started no better: sales were up, but overall profitability was down. One of the reasons profitability dropped while sales increased was Henry Jr.'s decision to boost spending on research and development.

"Although we're working hard to cut costs, we're budgeting more for research and development in 1959," Henry Jr. told the *Boston Herald Traveler*. "We've recently placed an order for our first computer, which is indicative of the kind of thing we're spending money on."

The new computer helped Brown & Sharpe design its Numericam Series of "tape-controlled" machine tools that it rolled out the following year. To cue a machine to cut a specific shape, operators entered data into the computer which then spit out machine-readable cards, strips of punched, paper tape, or magnetic tape. The tape fed into a scanner that read the numerical code and directed movement of the tool and the work piece to cut cams for screw machines in any mathematically defined contour. Numerical controls dramatically cut screw machine set up time, and reduced scrap rates.

Victor Logan, manufacturing engineer at Brown & Sharpe, had first spied a numerical control milling machine four years earlier in 1955, while touring the Servomechanisms Lab at MIT. He had not been impressed. The machine managed to push its cutter through an elliptical motion, a simple cut. "All that impressed us, really, was the amount of hardware with hundreds of glowing electron tubes, that was required to perform that minor operation," Logan observed.

Engineers in the Servomechanisms Lab had spent six years trying to coax that cutter into describing a mathematically precise arc through numerical controls. John C. Parsons, who owned a company that made helicopter rotor blades, first approached MIT in 1949, seeking help with a $200,000 contract he had won from the Air Force to develop numerical control of machining operations. MIT took over the contract, and within a few years its engineers managed to build the ungainly machine that Logan saw. When no machine tool company sought to develop the machine beyond that primitive phase, the Air Force offered to subsidize development by buying 100 numerical controlled milling machines, pushing the government's investment to $62 million in hopes of transferring the technology to a civilian market.

In 1957, Burg Tool of Los Angeles introduced a numeric controlled turret drill that was affordable and easy to use. A turret drill was a sensible choice for the application of numerical controls, as drilling required fewer complex calculations than milling. Hughes Aircraft, Convair, and General Electric's Atomic Power Equipment Department all reported huge savings with Burg's "NC" turret drills, and despite a recession that lingered into 1959, Brown & Sharpe invested in developing its own line of NC turret drills.

By December of 1959, Henry Sharpe, Jr., felt optimistic about the coming decade. In his December "Season's Greetings" to stock holders he wrote: "1960 FIRST OF SEVERAL BOOM YEARS ... It is difficult to recall any year in more than two decades when the indicators are so unanimous in pointing toward a high level year. In our mind, the only thing that can prevent [a] substantial boom in 1960 would be widespread strikes in the mass production industries."

The Defense Department was still buying tens of millions of dollars worth of machine tools annually, building a stockpile of 300,000 units. Aircraft manufacturers were retooling to build new fleets of commercial jet liners. And the Federal-Aid Highway Act of 1956—which had had pledged $26 billion to build 41,000 miles of interstate highway—was now in full-swing: bulldozers, trucks, compressors, pneumatic drills, and all manner of gear-requiring machine tools were rolling north and south, laying down pavement for more and more cars.

"Automobile production should range between 6.5 million and 7 million units," Henry Jr. wrote. "Big orders ahead.... Orders from the automobile industry are expected to provide a growing volume of business for both metal-cutting and metal-forming machine builders. As soon as the market for the new small cars is evaluated, Detroit will be making plans for expansion and new tooling for '61 models to be introduced in the fall of 1960. ... Ford has a tooling program for tractors that is reported to be placed."

Already, Henry Jr. was anticipating the summer's 1960 Machine Tool Exposition in Chicago. "All major builders have developed new machines for this show," he reported, Brown & Sharpe among them. "Many of the new machines are designed to increase production efficiency ... and really do. We believe that the metalworking and other industries will be forced to re-equip with the most modern machinery available to compete with European and Japanese suppliers."

For 20 years after World War II, American machine tool companies enjoyed a huge advantage over companies based in Europe and Asia. While companies overseas were struggling just to rebuild the capacity they had had before the war, American companies were pushing ahead with new technologies. About 95 percent of the machine tools used in America were built in America, and exports routinely ran 5 to 7 times higher than imports.[3]

Still, Henry Jr. could see competition looming from European and Japanese builders. Some West German manufacturers were already capable of building precise, if expensive, machine tools. And the Japan Machine Tools Builders Association had taken a novel strategy for getting its industry back on its feet: instead of encouraging foreign companies to build inside the country, the way England had, Japan offered to buy licenses from U.S. firms to duplicate their technology. Faced with a choice of laying out money to fight expensive patent infringement suits against firms that copied their technology or making money by selling licenses, many U.S. firms took the latter course. By 1959, *American Machinist* noted: "Japanese machines for the first time appear to merit recognition and to be competitive with machines of the most advanced industrial nations."[4]

At Booth No. 452 at the Machine Tool Exposition in Chicago, Brown & Sharpe rolled out all of its latest advances: the Numericam, tape-controlled cam milling machine; the newly designed No. 5 plain grinding machine with Cedasize equipment, capable of grinding missile parts to within 30-millionths-of-an-inch; a Model 618 Micromaster slicing machine that cut silicon "wafers of the highest quality from semiconductor materials with extreme accuracy"; a Model B automatic turret drilling machine; an automatic screw machine that could produce 3,600 pieces per hour, detecting and rejecting pieces not cut to tolerance. Henry Jr. was concerned about advances made by overseas competitors, but the Chicago Exposition showed foreign firms had a long way to go to best Brown & Sharpe.

On an October night in 1960, Henry Sharpe, Jr., stepped off an airplane and crossed a dark road near the Hillsgrove Airport in Warwick, Rhode Island. Henry was walking toward the overflow parking lot, probably looking down as he pulled his car keys out of his pocket, ready to head home. He never saw the pickup truck that hit him.

19. An Industrial Eden

A police officer sent to notify Peggy Sharpe about her husband's accident knocked on the door of the Sharpe home, a new house of a prize-winning modern design set by a pond in the woods of North Kingstown. Peggy, home with the couple's three children, opened the door and heard what little the policeman knew: Her husband, Henry, had been hit by a truck. He was taken by ambulance to the Jane Brown Hospital up in Providence.

Peggy drove up the Post Road into Providence to be there when he came out of surgery. The blow from the truck had spread his pelvis "like a woman's in child-birth," Peggy recalled; his right leg snapped, and a gash in his forehead required 23 stitches to close.

News of the accident spread fast. Company vice-president Frederick Austin telegraphed Hugh Calkins, head of President Eisenhower's Commission on National Goals:

"HENRY SUFFERED HEAD INJURIES DISLOCATED PELVIS WHEN HIT BY TRUCK LATE LAST NIGHT CONDITION SATISFACTORY...."

By morning's light the truck owner, Roland Onorato, found Henry Jr.'s keys imbedded in his truck's grille. Onorato, a Marine veteran and electrician who had once worked at Brown & Sharpe, was horrified by the accident. Henry Jr. spent seven weeks bedridden in the hospital, much of that time silently ruminating about the company, its past performance and future potential. When he returned home in December he struggled step-by-step to learn how to walk again. At Christmas, he received a card from Onorato, a tradition that lasted 50 years.

Three months after his December discharge, in March 1961, Henry Jr. walked unaided into the annual Foremen's Club Meeting, where he addressed the 90-or-so foremen gathered there. He thanked the foremen for "the hundreds and hundreds of sympathetic and cheerful greetings" he received from so many Brown & Sharpe workers. The Club secretary reported:

> With customary good humor, he advised against anyone duplicating his experience just to determine the extent of the *esprit de corps* that exists in this Brown & Sharpe family, but for him it was a most heart-warming and exhilarating experience, "as I lay there with lots of time to think and count the flies on the ceiling—and there *are* flies on the ceiling in a hospital."

As he launched into his Foremen's Club speech, it became obvious that Henry Jr. had spent his convalescence thinking about more than the flies on the ceiling. He rattled off a string of statistics, some of them bad: between the time of his accident and the first of the year, incoming orders across all of Brown & Sharpe's divisions had dropped by half, perhaps because of the uncertainty of his condition.

But as Henry had predicted in the waning days of 1959, the machine tool industry stood on the cusp of a boom: besides increased orders from car companies, Brown & Sharpe stood to gain from the Bomarck and Atlas missile systems, which used the company's grinders to grind for exceedingly fine tolerances; from precision gun drilling, sales of products cast in its foundry, and from its new line of numeric controlled turret drills that used an IBM computer to punch tape directly, removing the need for a keypunch operator. Henry Sharpe, Jr., was back on his feet, and both he and his company were growing stronger.

May 10, 1963: The news broke at mid-day, and was soon stripped across all six columns of the *Evening Bulletin*. "Brown & Sharpe Will Move to N. Kingstown." Some 130 years after David and Joseph Brown opened their watch and clockmakers shop on South Main Street, Brown & Sharpe would have no presence in the state's capitol city. For Providence, this meant the loss of its second-highest taxpayer, behind only the Narragansett Electric Company. "There is something both inspiring and nostalgically depressing in the announcement that big Brown & Sharpe is moving out of Providence," the *Providence Journal* opined.

Brown & Sharpe planned to spend $10.5 million building and equipping the new manufacturing plant. At 707,000 square feet, the plant of pressed concrete and glass would cover more than 15 of the 160 acres that Henry Jr. and his father had purchased more than a decade before. Pavement for 1,500 cars would cover dozens more.

In moving to the suburbs, Brown & Sharpe was following its largely white, middle class workers out of the city. With miles of new highways catering to cars, the second and third generation descendants of ethnic immigrants were on the move from Federal Hill, Olneyville, and Smith Hill out into suburban areas that were no longer the white-collar enclaves of John Cheever's stories. By the 1960s, blue collar workers outnumbered white collared in suburbia by 9 million to 8.8 million. As census director Richard Scammon told newspaper columnist Sylvia Porter: "Away back, the suburbs consisted of the native wealthy families and the newcomers, who were upper-middle class white collar people. But today, the suburb is less class and more mass."

The 1950s saw the population of Warwick, a city sandwiched between the new plant in North Kingstown and the old one in Providence, boom by 59 percent. In a letter to its workers, Brown & Sharpe stressed the importance of the new Routes 95 and 195, the former still an unpaved gash in the ground, as an aid to commuters:

> Unbelievable as it may sound, it will probably take little more time and far less energy, for example, to reach the new site from Fall River by fast moving freeway, than it now takes to buck one's way through the traffic of downtown Providence to the parking lot and three block's walk to work.

Once he announced the move, Henry Sharpe, Jr., moved quickly. Just five days later he held a groundbreaking ceremony at Precision Park, a name chosen through a company-wide naming contest.[1] A newspaper photographer captured his mother, Mary Elizabeth, wearing an ankle-length, leopard print coat against the chill of a May morning;

On a May morning in 1963, Henry Sharpe, Jr., presided over groundbreaking ceremonies for a new manufacturing plant in the suburban town of North Kingstown, where Brown & Sharpe planned to build a 707,000-square-foot plant sprawling across 15 acres. Henry Jr. in bow tie is joined at the groundbreaking by his mother, Mary Elizabeth Evans Sharpe, and his son, Henry Sharpe III, wearing a crested blazer. Also standing next to Mary Elizabeth is then–Governor John Chafee, Jr., later a U.S. senator and secretary of the Navy, a great-grandson of company founder, Lucian Sharpe.

she churned a spadeful of sandy soil with a gold-plated shovel under the watchful eye of her 9-year-old grandson, Henry Sharpe III.

Throughout that summer of 1963, Henry Jr. watched the construction on Sunday mornings from his seat in the Unitarian Church on a hill in East Greenwich. The new plant was less than three miles south-southwest from his church pew, and during sermons he could keep an eye on swiveling cranes, and on the progress of the new water tower that grew into a North Kingstown landmark.[2]

Work on the new building was something of a family affair for the Sharpes. Mary Elizabeth designed the landscaping, covering six acres with woodchips to retain moisture in the sandy soil, with red bearberries forming a natural ground cover. Henry Jr. wanted an open courtyard surrounded by walls of glass to create a friendly, visual connection between the factory floor and the corporate offices; his mother filled the courtyard with something soothing to look at: a small mound evoking a Japanese garden, studded with pines, rocks, and benches for breaks.

Peggy Sharpe also had a talent for design—she held a bachelor's degree in set-design and theater from Sarah Lawrence College, as well as her degree in Landscape Architecture from the Rhode Island School of Design. Designers of all types were then feeling the influence of ingenious architects and furniture designers immigrating from Europe after World War II. These designers championed the notion that "good design is good business" in that it signaled a company that was facing the future with innovation and cutting-edge technology. Schooled in this belief, Peggy eagerly saw to it that the interiors of Brown & Sharpe's new factory reflected such "good design," carefully helping to harmonize a variety of paint colors with fabric colors and textures of elegant, modern furniture that a design team chose for the building.

"I asked my wife to come pick a few colors," Henry Jr. told the *Pawtucket Times*. "She spent eight months choosing practically everything."

In July 1964, workers began moving the first of 1,400 machines into the new plant, working on a carefully planned schedule. Each machine trucked from Promenade Street had to make the move within 36 hours so its operator would miss just one shift on his usual machine. On July 20, a Pawtucket machinist named Edmond J. Perreault turned out the first chips of metal in the new plant while drilling into an unfinished, universal grinder. These first chips were ceremoniously swept up and imbedded into a clear, plastic paperweight presented to the governor, John Chafee, Jr., a great grandson of Lucian Sharpe's.

Within days the 11,000-pound grinder, the first machine finished in Precision Park, was crated and ready for shipping to the company's agent in Stockholm. It was slated to sail aboard the NS *Savannah*, a futuristic looking, nuclear-powered ship. U.S. government agencies funded the merchant ship's construction in the 1950s to show the potential, peaceful uses of atomic power. The interior of the ship, with its 30 state-rooms, a swimming pool, and 75-person lounge with sleek, modern furniture, looked like a set for the new cartoon *The Jetsons*. The *Savannah* arrived in Providence on July 31 and slipped from port four days later, bound for Stockholm with the $25,000 grinder. The temperature of the ship's reactor control rods was regulated by valves produced at the Double A Products Company in Manchester, Michigan, a subsidiary of Brown & Sharpe.

Throughout that summer and into the fall, employees arrived at the new plant in waves: first came the office workers with 400 new desks, 200 tables, and 1,500 chairs; the machine tool division made the move in 60 days, bringing with it 825 pieces of machinery and 1,050 workers. The final group to move in was the Pump Group, which set up shop in the Hydraulics Division in mid-November. The entire move, with the exception of one plating room and the foundry workers, unfolded over 19 weeks. Brown & Sharpe sold the old Providence plant for $3 million to a group that planned to lease its multiple buildings and stories to dozens of different industries. The new plant, Henry Sharpe, Jr., told *American Machinist*, "should project the whole tone of Brown & Sharpe bodily out of the Victorian Age and into the Nuclear World."

By January 14, 1965, Precision Park was ready for a coming out party. On a chilly morning more than five-dozen guests from industry, newspapers, and trade publications gathered outside the manufacturing plant, its front doors symbolically sealed off

In January 1965, Brown & Sharpe held a ceremonial "tape cutting" to officially open its new plant, symbolically replacing the usual ribbon with a taut length of tape punched by a numerical control machine. After holding a company-wide naming contest Brown & Sharpe christened the new plant as Precision Park. The plant's water tower, at right, added a new landmark to suburban North Kingstown.

by a taut length of tape punched by a numerical control machine. Henry Jr. said, "In the spirit of Almighty God give us this morning the wisdom, the resourcefulness, the patience, the courage and the leadership to accomplish in the service of mankind not only what tradition demands but 100-fold more." He then snipped the punched tape, welcoming the assembled guests into the warmth of the new, brightly accented factory.

"It strikes you as a symphony of light and color," wrote Mike D'ambra of the *Pawtucket Times*. "Bright reds, blues, yellows, etc., easy on the eyes, lilting to the spirit."

"If the word 'factory' up to now has evoked visions of grease and grime and dark cavernous areas, dusty, noisy and smelly, you're in for a shocker on your first look at Brown & Sharpe's Precision Park. ... You have room to move, to enjoy the feeling of space." Precision Park was, D'ambra concluded, "an Industrial Eden."

In November of 1965, a machinist named Ed Green heard some good news: Brown & Sharpe had accepted his application to work at Precision Park as a grinding machine operator. Ed had worked a few years as a grinder at Federal Products in Providence, which did some sub-contracting for Brown & Sharpe. He had learned the trade through machine shop and drafting classes that all of Rhode Island's public high schools now offered.

At first, his job at Federal Products had seemed pretty good; but now that he was married with a third child on the way, machinist's work seemed like it might be a dead end. Then he got hired by Brown & Sharpe, where his pay nearly tripled from $1.80 to more than $5 per hour, the current equivalent of $37.80 for an hour's work, more than $78,000 per year and a high school graduate's ticket into middle class America.

"Working the piece work, once you got the hang of it, you could make some decent money," Green recalled. "Five or six bucks an hour.... You had all your benefits, you had your pension plan, medical benefits, they were all included." Eventually he fathered four children raised by a stay-at-home mother in a home of their own. Weekdays he punched in at 7 a.m., and at 3:30 p.m. he punched out.[3]

Green was just one of 953 new hires in 1965. About 300 replaced people lost to attrition, but more than 600 were new jobs created by a boom in business. Green came aboard in the same month that Brown & Sharpe Manufacturing Co. won a coveted listing on the New York Stock Exchange with the assigned ticker symbol BNS. In December the company reported annual sales of $53 million, about $400 million by current standards and the best since 1953, the height of the Korean War.

Green began his Brown & Sharpe career on the J-5 cylindrical grinder, a small grinder with which he finished parts for the Industrial Products Division—small parts for micrometers, Verniers, gauge indicators. "If you were the junior man, you got the lousier jobs," Green recalled. "You had to put in your time, pay your dues. I had some jobs called a two-machine run, 180 to 200 pieces an hour to make up your money." On a two-machine job he fed metal stock into one machine, spun around and fed the other machine, spun around and fed the first. "We used to do the micrometer spindles. I had one machine that was a centerless grinder. I used to do lots of 1,000 or 2,000 micrometer screws. You could be on this job for a week. When we were busy, we put through a lot of work."

And in 1965, they were busy. The Industrial Products Division alone made 5,000 different kinds of precision measuring and machinists tools, each one requiring dozens of different parts. The Machine Tools Division accounted for more than half of the company's sales, thanks in part to a big order from the National Cash Register Company for 330 screw machines to churn out parts for its cash registers. In 1967, *American Machinist* ranked Brown & Sharpe the nation's ninth largest maker of machine tools. The news that year remained upbeat, as Brown & Sharpe again set new records with $75.9 million in sales, and an unprecedented profit of $5.3 million, adjusting for inflation nearly $38 million. "The war in Vietnam, of course, was instrumental in fueling demand," noted Max Holland in his book, *When the Machine Stopped*. "By some estimates, 50 percent of all [machine tool industry] production was devoted to war and war-related industries."[4]

All of this success made Brown & Sharpe a plump takeover target, particularly in the "merger mania" of the late 1960s, when corporate giants such as ITT, Gulf & Western, and Providence's own Textron Corporation began accreting smaller companies into their ever-growing conglomerates. Holland cites several reasons for the boom in conglomerates: a desire to grow without running afoul of anti-trust laws by buying up

In November 1965 Brown & Sharpe Manufacturing Co. won a coveted listing on the New York Stock Exchange with the assigned ticker symbol BNS. Celebrating the achievement are, from left: Exchange President Keith Funston; Brown & Sharpe President Henry D. Sharpe, Jr.; company Vice-President Frederick P. Austin, Jr.; and Robert J. Jacobson of Benjamin Jacobson & Sons, specialists in stock. Mary Elizabeth Evans Sharpe symbolically bought the first issue at 34½.

assets that had nothing in common with each other; spreading investment risks across different industries in the hope that if one sector were down, another might be up; and deferring taxes by acquiring new assets through tax-free stock swaps.

Buying up assets also helped corporations deal with the ever-growing problem of inflation. By the mid–1960s, America was spending money on war ($80 billion in 1967) and on expanding social programs ($15 billion) at home. The United States was buying more goods and services from other countries than it was selling, creating a trade balance deficit. The U.S. was flooding the world with its dollars and had been guaranteeing them, since the postwar Bretton Woods conference, at the rate of an ounce of gold for 35 U.S. dollars. More dollars abroad in the global economy meant that the dollar bought less, driving up the cost of goods, which required higher wages, leading to an inflationary spiral. The Federal Reserve tried to put the brakes on inflation by raising interest rates to 7 percent, which discouraged corporate borrowing but did not erode inflation. The consumer price index, which measures changes in a market basket of consumer goods, was steadily rising; rather than hang onto cash that was rapidly losing its value, corporations bought up other corporations for stock, investing money that otherwise would have lost value, while also avoiding tax payments on capital.

The year 1966 saw 22 machine tool mergers or acquisitions; 1967 saw even more, with half of *American Machinist's* top 10 machine tool companies merging with conglomerates that year. The industry's postwar record of six consecutive years of growth made it a prime target for conglomerates that historically had never been in that business.

Both Henry, Jr.'s father and grandfather had been adamantly opposed to yielding Brown & Sharpe's independence through a merger. They had resisted previous eras of merger mania, in the late nineteenth-century and in the 1920s. As recently as the 1950s, Henry Sharpe, Sr., had rejected a merger offer: "When younger voices in the company, including mine, urged that he consider what was seen as an advantageous merger, he announced firmly to his Board that he did not believe in mergers," his son recalled. "And I'm absolutely certain that he did not. He preferred the values that one created directly by one's own technical and material progress, not by manipulation."

Taking a page from his father, Henry Jr. resolved to protect Brown & Sharpe's independence. To avoid being eaten by a bigger fish, Brown & Sharpe had to act like one. In December 1968 the company restructured by boosting the number of shares of common stock authorized from 3 million to 4.5 million, and by creating 1 million shares of a new, preferred stock. At the same time, directors modernized their century-old Rhode Island charter by incorporating Brown & Sharpe in Delaware. Half of America's biggest conglomerates incorporated in Delaware, in order to take advantage of that state's business-friendly corporation laws. For example, Delaware did not require companies to set a par value on its stock, which eliminated a theoretical liability and an actual bookkeeping headache. The primary reason for restructuring, Henry Jr. told the *Providence Journal*, was to "make it more difficult for other persons or corporations to gain control of the company."

Even before the stock restructuring, Brown & Sharpe began buying firms that had some connection to its core business of machine and precision tools. Between 1967 and 1970, the company bought: Cleveland Grinding Machine Company and Clev-Co. Jig Boring Company, both of Ohio; Olmsted Products of Ann Arbor, a manufacturer of hydraulic valves; Anocut Engineering, which in lieu of cutting metal removed particularly hard metal through electrochemistry; 90 percent of the stock of Airoyal Hydraulics of Maplewood, New Jersey; $3 million worth of stock in Chamberlain Industries of London, makers of the Staffa hydraulic motor; 40 percent interest in a new Company, Herramentacion Industrial, S.A. of Mexico City; Petz Technical Instruments, Inc., which made dial indicators for Brown & Sharpe's micrometers; and Tesa, S.A. in Lausanne, Switzerland, a manufacturer of coordinate measuring machines.

Some of these acquisitions paid dividends, while others ended, such as Anocut, disastrously. Anocut's method of removing metal through electrochemistry was tailor-made for the aero-space industry, which required machining extremely hard materials to very precise tolerances. The technology had little practical application outside of the aerospace industry; in 1971 the U.S. Congress abandoned funding for a new class of supersonic transport planes, the SST Concorde, leaving Anocut without its sole major customer. Only 20 Concordes ever were built, and those by an unlikely government partnership of Great Britain and France. Brown & Sharpe paid $8 million for Anocut in February 1969, and voted to shutter it in 1971.

The two Cleveland companies turned out to be a good fit for the company—with technology pioneered by the Cleveland Grinding Company, Brown & Sharpe was able to replace the optical spotting scopes on its most precise grinders with a television screen that allowed machinists to watch operations on a magnified screen rather than on the machine itself as they shaved metal to the 30 millionths of an inch. Brown & Sharpe merged the two Ohio firms into a group called Cleveland Precision Instruments, which also manufactured products for the exacting demands of the nuclear power industry.

The purchase of Tesa, the Swiss manufacturer of coordinate measuring machines, or CMMs, looked particularly promising. These machines used electronic probes sliding on an x, y, z axis—that is, away from and toward an operator, left and right, up and down—to precisely measure any object. They were very useful as a kind of gauge for the quick and accurate testing of any part or assembly of parts against the ideal model. For example, if a design called for a sphere with screw threads tapped, or drilled into it, a CMM could measure the diameter of the finished ball, the depth of the hole, and the width of the screw threads in one quick operation. Brown & Sharpe bought all of Tesa's shares in 1967, and two years later released the Validator CMM.

Since its earliest days Brown & Sharpe had been involved in metrology, the science of measurement. Joseph Brown's first linear dividing engine of 1855 precisely graduated rulers, protractors, and Vernier calipers to help machinists achieve precision in their measurements. Brown & Sharpe had added electronics to some of its precision tools, but all of these tools were still small, hand-held measuring devices. The coordinate measuring machine was something like a machine tool—it was a powered machine, large, not portable; but instead of being designed to remove metal it was designed to precisely measure it, and other material. The Validator represented Brown & Sharpe's first foray into the machine metrology business, and its applications in aerospace, automotive, and heavy industry businesses appeared promising.

In the spring of 1968, rumors of a "reduction in force" began sweeping through the shop floor at Precision Park. Workers anxiously eyed the seniority lists to determine who would be the first out the door in event of a layoff. In May, the rumors proved true: Brown & Sharpe was letting go 300 production workers, drawing employment at Precision Park to 2,700 from a peak of 3,250. Second quarter income had dropped by 70 percent from the halcyon year of 1967, prompting a *Providence Journal* business writer to pen a column titled: "When Machine Tools Began to Soften." About the only thing prescient about this article was its title: 1967 really had been the high-water mark for Brown & Sharpe, and to some extent for the American machine tool industry.

The article's writer, Donald M. Houghton, blamed the downturn on: a 10 percent income tax-surcharge recently imposed by Congress to pay for the war; high interest rates for borrowing; race riots in Detroit, Newark, and Chicago; "the war in Vietnam, and the change of administration [from Lyndon Johnson to Richard Nixon] this November." Industry actually favored the 10 percent income tax surcharge that Houghton lambasted, in lieu of the excess profit taxes enacted during the World Wars and Korea. Houghton did not explain how race riots and the war softened the industry. But high interest rates did curtail corporate borrowing, which slowed orders for machine tools. The consumer price index continued to grow, fueling a demand for higher wages to pay for more expensive goods; higher wages and higher materials costs cut corporate profits, and Brown & Sharpe was no exception.

Besides inflation, another challenge facing machine tool businesses was changing technology. In 1968 the National Cash Register flooded the resale market with all of its barely used Brown & Sharpe screw machines, as it switched its new cash registers from mechanical to electronic, requiring fewer metal parts and more plastic ones.

Technological change within the machine tool industry itself also proved challenging. At the Philadelphia Tool Show of 1968, a Buffalo-based firm called Strippit unveiled a numerical control machine that linked directly into a computer. The computer numerical control, or CNC machine, allowed a machinist to program a tool's movements into the machine, a step beyond the numerical control machine that could only read instructions punched onto a tape. Brown & Sharpe had been an early adopter of the NC technology, but it lagged by two years in developing a CNC product, partly because it was so successful with its existing lines of screw and grinding machines, and partly because the big conglomerates had more money to spend on development.

Brown & Sharpe's string of profitable years ended at 22 in 1970, when sales fell by $15 million, and the company posted a loss of $5.6 million, the current equivalent of $35 million, more than $100,000 per workday. "I just hope and pray that this pulls out," Henry Sharpe, Jr., told the *Providence Journal* in July. "I hope the rest of the economy doesn't do what the so-called 'lead-indicator' machine tool industry is doing." President Nixon was "tinkering with the economy," Henry Jr. noted, and "should know the effect of the buttons he's pressing."

Another wave of layoffs followed in December: the firm cut employment at Precision Park from 2,700 to 1,000, including office help and managers. Ed Green, the grinding machine operator, recalled once again "checking out the seniority list to see who would survive." This time, he did not make the cut. "The bad part was, anybody working at Brown & Sharpe, you put in an application somewhere else, they saw you had a few years at Brown & Sharpe they knew you were going to go back," if conditions there improved. "The pay was better, the benefits were better. They knew you were going to go back to Brown & Sharpe because they couldn't compete." With four children and a mortgage, Green took a job working third shift in a Coventry mill, winding thread; then he worked under the table at a friend's father's casting shop, awaiting the return of good times and a call back. He would have a long wait.

In 1971, Brown & Sharpe did not lose as much money as it had the year before thanks to Russia's order for 132 automatic screw machines. Still, the economy remained stuck in the doldrums, prompting President Nixon to take dramatic action. On a Sunday evening in August, Nixon pre-empted the popular *Bonanza* TV show in order to deliver an important, primetime announcement: "The time has come for a new economic policy for the United States. Its targets are unemployment, inflation, and international speculation. And this is how we are going to attack those targets."

Nixon issued an executive order freezing wages and prices. He also repealed a 7 percent federal excise tax, a move that cut an average of $200 from the price of each car, the intent being to stimulate sales and manufacturing of domestic cars. He offered companies a 10 percent tax credit for buying new machinery; a $50 income tax credit for all tax payers; and a 10 percent tax on all imports.

To world markets, Nixon's most important news was that for the first time since 1944 the United States would no longer peg the value of a dollar to the price of gold. Instead of being guaranteed at the rate of an ounce of gold for every $35 redeemed, the U.S. dollar would "float" against the value of other currencies. This would lessen the dollar's value on the international market, making it more expensive to buy foreign goods, which could act as a stimulus for domestic production.

In his *Bonanza* address Nixon explained:

> At the end of World War II the economies of the major industrial nations of Europe and Asia were shattered. To help them get on their feet and to protect their freedom, the United States has provided over the past 25 years $143 billion in foreign aid. That was the right thing for us to do.
>
> Today, largely with our help, they have regained their vitality. They have become our strong competitors, and we welcome their success. But now that other nations are economically strong, the time has come for them to bear their fair share of the burden of defending freedom around the world. The time has come for exchange rates to be set straight and for the major nations to compete as equals.

Nixon was correct, the economies of western Europe and Japan had rebounded, and in the machine tools industry they had become particularly strong. By 1971, the United States was no longer the world's largest producer of machine tools—that distinction had been claimed by West Germany. Japanese builders were also making significant inroads, with Yamazaki Machinery opening a plant to manufacture numeric control lathes practically in Brown & Sharpe's backyard on Long Island, New York.

For Brown & Sharpe, Nixon's plan worked well. In July 1973 company directors learned from management that: "Inventories and receivables were over Plan mainly because of foreign exchange rate changes resulting from dollar devaluation ... estimated machine [tool] orders for the year will, with the exception of 1965 and 1966, exceed any year since 1956."

A strike by the machinist's union at the Double A plant in Michigan plant dampened profits the following year. The three-week strike in March of 1974 was the first labor shutdown at a Brown & Sharpe plant since Henry Jr.'s first year at the helm in 1951. The strikers won a 30–cent per hour pay increase in the first year and a 25–cent per hour additional pay raise on the anniversary date of the contract. But Donald Roach, recently promoted to executive vice-president, observed in a company newsletter:

> Our calculations indicate it will take each employee of the bargaining unit 253 weeks to recover the monies lost during the three week strike. In addition, each employee will probably lose approximately $400 profit-sharing contribution to their retirement plan, something that will never be recovered.

This strike was a tragic thing for Double A employees and Brown & Sharpe as well, and bears out the fact that "nobody wins as the result of a strike."

That was a lesson that Roach would come to learn all to well, too late.

Part Four: Managerial Capitalism and the Global Corporation

20. Fenced In

Brown & Sharpe stood on the verge of losing money for a second consecutive year, and Henry Sharpe, Jr., knew that he needed professional help. He had methodically built the firm into what he termed "a complex, worldwide company" that had grown into something more than he could oversee on his own. By 1971 rapid technological change combined with competitive pressure from West Germany and Japan required more managerial talent than he had in the offices of Precision Park, the company's global headquarters.

Brown & Sharpe, long a leader in mechanical engineering, was falling behind the curve in electronic controls for numeric and computer numeric machining; the firm had also lost millions by betting on the Wankel engine and the Concorde supersonic jet, and was struggling to keep pace in developing "machining centers," big machines that performed multiple functions, drilling, boring, turning, milling, all in one station.

To meet these technological and global challenges, Henry Jr. hired an executive vice-president with academic training in management skills, Donald A. Roach. At age 40, Roach was just 7 years younger than Henry Jr.; the two men shared some similarities in their backgrounds: both had been naval officers, Henry Jr. in World War II, Roach in Korea; both held Ivy League degrees, Henry Jr.'s undergraduate economics degree came from Brown, while Roach had earned his Master's in Business Administration from Harvard.

Roach was a charismatic young man, thin, meticulous in dress, with bushy eyebrows and an expressive face. He had distinguished himself at Brown & Sharpe's Double A subsidiary, makers of valves, pumps, and hydraulic power units. Roach's reports to headquarters from the plant in Manchester, Michigan were almost always upbeat. Even in 1970, when the machine tools division lost $1.7 million and industrial products was in the red by $1.89 million, Don Roach's hydraulics division actually posted a profit of nearly $1 million.

Roach moved into the Precision Park headquarters in March 1971, taking an office on the big plant's small second story, right next to Henry Jr.'s. Roach was part of a growing movement of corporate managers who felt that the relationship between government and labor had become too strong at the expense of business. In a "Statement of Operating Strategy" he wrote that "this country still has not taken the necessary steps to redress the fundamental problem of the balance of power between management and organized labor."[1]

By the early 1970s, Brown & Sharpe had grown into a complex, worldwide company with production plants in Europe and in multiple states. For generations, most Brown & Sharpe executives began their careers in the company's apprenticeship program, but President Henry Sharpe, Jr., left, decided that the time had come to hire an executive vice-president with academic training in management skills: Donald Roach, at right, a 40-year-old manager from the Company's Double-A subsidiary, with a master's in Business Administration from Harvard.

Naturally, Roach wanted to recruit managers who shared his beliefs—Brown & Sharpe had hired him to put modern, professional management techniques in place, and to do that he needed to hire people trained in those skills. Before Roach could hire a new team the company had to shed some of its older managers. Roach dubbed these older managers "the Apprentice Mafia," because most of them had come up through the company's apprenticeship program. The board of directors encouraged older managers to retire through a Special Pension Supplement that boosted benefits for early retirement. The board expanded the generous pension deal to include older union workers too, thus cutting the average age of its union members.

After shedding some old-timers, Roach began assembling his own managerial team. One of Roach's first hires, a new Director of Design named Richard Hook, immediately stepped on the toes of the union that represented engineers, the American Federation of Technical Engineers. Hook hired a number of top-tier mechanical engineers, then held them out of the bargaining unit claiming they were managers working with proprietary information. The union objected; an editorial in the union's "Straight Line" newsletter claimed that the company was seeking to destroy the union. There was no room for compromise: either the designers were in the union, or they were out. The case worked its way through arbitration, and the company won.

20. Fenced In

Henry Sharpe, Jr., possessed one skill that his father had never displayed while dealing with unions: tact. Unions had been entrenched in Brown & Sharpe for all of the younger Henry's professional life; as a young apprentice he worked side-by-side with union workers, unlike his father, who had completed his apprenticeship in the 1890s then spent nearly 50 years trying to shut the unions out. From his second year as company president in 1952, Henry Jr. always reached agreements with union negotiators without either side resorting to lockouts or strikes. Generous wages accounted for part of his string of successful contracts—Brown & Sharpe was not a pushover at the negotiating table, but the firm always tried to pay at or better than the "community standards" set by other firms for similar work. As long as the company was making good money, it paid good money.

Henry Jr. had also won over union officials by soliciting their opinions on proposed changes in working conditions. For example, the company's decision to mass-produce a line of automatic screw machines in the 1950s dramatically changed working conditions for people working in that department. Before the change, a machinist building screw machines worked on either a drill press, a milling machine, or a boring machine. New equipment installed along the assembly line combined all of these steps in one machine, which meant that a machinist formerly assigned to milling would now have to bore and drill as well. The union agreed to create a new class of machinist called Miscellaneous Machines Operator that allowed a worker to perform multiple tasks. Most of the 15 people assigned to this group received a pay raise as the "labor grade" describing their skill level got bumped up a notch or two from grade 7 or 6 to 5. The company reaped its reward through a dramatic production increase that more than paid for the pay raises.

In 1967, Henry Jr. had approached the union about scrapping the piecework system that rewarded workers for speed. That system had been in place since Henry Leland and Richmond Viall introduced it in the 1880s. Henry Jr. did not like the idea because it created constant conflict between managers seeking to minimize labor costs, and workers trying to maximize wages. If managers set the price for a piece too low, demoralized workers complained that the price was too "tight" and they could not make any money; if the piece price were too "loose" or generous, workers would "rob the job" by producing just enough pieces to earn a good wage but not so many that management would suspect that the rate was too loose. Piecework generally put a lot of pressure on workers to produce; it also resulted in uneven paychecks, making it hard on household budgets from week-to-week. The union did not mind giving up the piecework system, as long as those machinists who had been making good money from it were not penalized.

To phase out the system, the company agreed to pay higher wages to workers who had been able to earn more than the base wage through piecework, which was almost everyone. New workers would receive an improved base rate, but existing workers each received a "personal rate" of pay that was based on how much he or she had historically exceeded a percentage of the base. In exchange for paying a higher personal rate, the company won the right to monitor every machinist's performance—how many parts each made per hour. A machinist who failed to make his rate could be disciplined, even fired.

By the mid–1960s the union and the company were pretty good at working things out, even on an issue as complex as scrapping an 80-year-old pay system. One of the reasons for this was the people involved. Both the company and the union had long-standing executives in place, people who worked together for years and came to trust each other. On the company side, a former apprentice named James Rigney had worked his way up the ladder to apprentice instructor, personnel director, and finally in 1960 to Director of Industrial Relations.

On the union side, J.J. Turbitt, enjoyed popular support among members and handily won reelection throughout the 1960s. The business agent for the machinists union, Joseph Kane, had been a union president at Brown & Sharpe before Turbitt, and was thoroughly versed in the inner workings of the company.

In settling the 1951 strike, Brown & Sharpe agreed to provide the International Association of Machinists with an in-plant office where union officials could meet with new workers in an attempt to convince them of the benefits of union membership. The union's in-plant office offered an unintended benefit: whenever problems arose on the shop floor, stewards wandered into the office at lunch break to talk things over with Turbitt, the union president. He then called up Rigney, and the two of them hashed things out before small problems escalated. Rigney believed that management had the right to do anything that the contract did not expressly forbid. Therefore he found it best for the company to settle disputes on a case-by-case basis, without codifying the language—almost always these informal settlements worked out between Turbitt and Rigney bore the disclaimer that they were not setting precedent.

The team of new managers assembled by Donald Roach saw first-hand how this approach could work. In the last quarter of 1974, the company faced the challenge of shipping 333 machine tools; failure to meet the quota would result in disgruntled customers, and loss of a bonus offered to managers through their Profit Plan. Lower level managers told Rigney that machinists would not meet the target, because they did not like a new, Midwestern manager whom Roach had recently hired to run the shop floor of the machine tools department. This manager "did not understand their way of getting things done," Rigney recalled.[2]

Rigney held a dinner meeting with the foremen and their managers to see if they could smooth things out between the Midwestern manager and the Rhode Island machinists in time to ship the 333 machines by year's end. Rigney told them, "we can make fun out of this effort if we try." Rigney assigned Dick Jocelyn, then a worker in the industrial relations department, to make a kind of game out of the effort to ship the machines. Jocelyn talked it over with union's new president who had recently succeeded Turbitt, Robert Thayer, and the union bought into "Operation 333."

For a couple of months, no one paid any mind to daily production quotas, or who worked on which machine, or whether non-union supervisors worked on union jobs. Everyone pulled together. At one point, Jocelyn promised a case of beer to the team that contributed the most to the effort, but on the day before Christmas, as workers affixed a Christmas tree to the 333rd machine, he changed his mind: Everyone would receive a case of beer. Workers greeted this news by marching around the plant's wide perimeter, trailing musical colleagues playing trombone and trumpets. Tractor-trailers hauling more than a thousand cases of locally brewed Narragansett Beer backed up to the loading docks, sending everyone home happy before the Christmas break.

Still, the new managerial team did not like the idea of Rigney and the union president quietly ironing out conflicts that had arisen in their departments. It did not seem right that a problem could come and go without them ever having been apprised of it, or having a say in the solution. "To these individuals ... the Brown & Sharpe labor management style presented a problem," Rigney wrote.

By the mid–1970s, Rigney observed "the new production executives imposing their own professional management style on the company, which was in the traditional style of direct authoritarian action and decision making...." The union did not like the new management style. Union-management disputes about seemingly petty things such as "horseplay" that Rigney and the union might have informally worked out in the past, now metastasized into formal grievances.

"Lodges 1088 & 1142 protested the discharge on May 2, 1975, of Mr. Mark Muller, a Burrer & Filer on the night shift," read one complaint.

> The Company maintained that he was responsible for the explosion of firecrackers in the plant on two occasions that day which had resulted in the collapse of another employee, Mr. Jessup, in the area and led to hospitalization for several days for what was thought to be and may have been a heart attack. [Jessup had cause to be nervous: This happened just months after a fire erupted in the assembly department, nearly killing a worker who suffered deep burns over most of his body.] During the time Mr. Jessup was being attended to, he stated to several individuals that Muller was the individual responsible for this condition. Later, on June 13, 1975, Mr. Jessup advised the Company, in the grievance hearing, that he was no longer certain that Muller was responsible.

The Arbitrator ruled that Jessup's first impression was probably accurate, but since no one saw him throw the firecrackers, "the Company's case was not substantial enough to sustain discharge."

Another complaint alleged that a worker was throwing iron filings over his shoulder into a metal locker, "horseplay" intended to make a disruptive noise. "Horseplay is a very general word including a wide variety of activities," the arbitrator wrote, "some of which are so dangerous as to be treated as very serious, perhaps justifying immediate discharge. I would not rank Mr. Beauregard's action on the afternoon of January 14 very high on the scale." He cut Beauregard's suspension from five days to two.

This contentious atmosphere between new managers and younger union workers continued into the Fall of 1975, when the contracts of three unions—first the machinists, then the clerks and then engineers, came up for renewal.

One day before the contract with the machinists union was set to expire, Brown & Sharpe announced that due to a lack of new orders it would soon shut down its Rhode Island properties for a two-week furlough. It was true that incoming orders had plummeted since mid-summer, but the October 1975 announcement was probably just posturing—in a memo marked "Confidential," a financial manager wrote that he had run the numbers and determined that a short-term layoff of production workers would actually cost the company more money than it would save.

The shot across the union's bow did not scare negotiators into submission. On

Sunday, October 22, more than 1,000 machinists and maintenance workers gathered in a large hall at the Rocky Point amusement park overlooking Narragansett Bay. By a large majority they voted to strike, the first strike at Brown & Sharpe in a generation. The union president, Robert Thayer, told picketers at the plants to remain peaceful: "For anybody to destroy property won't do anything for this labor organization. We hope someday there will be a settlement, and we want a place to go back to work."

The issues were pay and pensions. Workers wanted hourly increases of 50 cents in each year of the two-year contract; the company stood firm at a formula that would raise rates by 70 cents an hour at the end of two years. (The average wage of a union member was $4.61 an hour, the current equivalent of $21 an hour, plus $1.90 worth of benefits.) Since the first union contracts, pension plans had been one of those informal agreements: the company managed them and the union was happy to have them. Now the company wanted to negotiate a reduction in pension payments.

The day after the strike vote Brown & Sharpe shut down production in Rhode Island, locking out members of the other unions and idling 2,000 workers. "None of the production workers showed up, so you don't operate," Rigney told the *Providence Journal*. Strikers slowed cars driven by managers driving into Precision Park, but state police quickly got them moving.

For three weeks the two sides did not even meet. Picketers milled about outside the plants, and managers understood that they meant nothing personal: picketers during strikes were as ubiquitous as hot dogs at baseball games. At Precision Park's carpenter's shop, non-union foremen built a small shed and erected it just outside the main entrance so the strikers would have a place to warm up in November's chill rains. Rigney recalled that managers even sent out coffee and donuts "to keep peace in the family."[3]

Negotiators resumed meetings on a Friday morning, convening in the Sheraton Motor Inn off Route 95 in Warwick. They reached an agreement in the wee hours of Saturday morning, and on Sunday, machinists and maintenance workers gathered inside the Cranston Portuguese Club to vote on a proposal that had won the backing of union leaders. They rejected it by 693–279—more than a 2-to-1 margin. The company was offering 40 cents per hour in the first year and 30 cents in the second. This was better than the initial offer in that it front-loaded raises in the beginning of each year, rather than phasing them in through six-month increments. But the rank-and-file did not see it that way: it was still just 70 cents after two years, and they wanted $1 more per hour.

When he got word of the rejection, Henry Sharpe, Jr., could not believe it. Brown & Sharpe had just closed its twist drill plant in Denton, Texas, laying off 100. And on Halloween 1975, just a few weeks before the vote, the 95-year-old Providence foundry fired for the final time. The only time Henry Sharpe III had ever seen tears in his father's eyes was the day he closed the foundry.[4] The plant and foundry closings "are growing illustrations of the fact that operations which do not contribute to a company's profitability cannot be indefinitely supported: they must either carry their own weight or face the inevitable," Henry Jr. warned workers.

Rather than deal through the union leadership, Henry Jr. mailed a letter to every worker's home: "The vote to continue the strike just does not come to grips with reality." He called the rejection "one of the most unfortunate and surprising judgments ever made by a responsible group of our employees." Naturally the union leaders did not like the company's direct dealing with their members. They shot back a reply, mimicking the tone of Henry Jr.'s letter; the company offer was "the most unfortunate judgment ever made by a responsible management since it was obviously designed and did, in fact, precipitate the current strike action."

The clerical workers contract expired on November 20 and they also voted to strike, rejecting increases of 33 cents and 25 cents per year, atop their average wage of $3.76 an hour. The day after that, the engineers union officially went on strike.

By December 9, the machinists and clerical unions had hammered out agreements and were ready to go back to work—but the engineers were not. The company had offered the engineers an extra 80 cents an hour over the two-year term, the same increase accepted by machinists; but the engineers made more money so the raise was not as high, percentage-wise, and they were insulted. Their leadership declined to even bring the offer to the members, much to consternation of machinists union President Robert Thayer.

Finally after a marathon bargaining session, the engineers voted to ratify a contract—they negotiated a 13.88 percent wage increase that gave their lowest-paid workers an hourly raise of 80 cents; all of the unions agreed to pension cuts. After 55 days, the first strike in a generation came to an amicable end.

A notice went up on the employee bulletin board in February 1976, reminding employees to wear their photo identification badges at all times. The company began issuing photo ID badges before the strike, and now that everyone was back management intended to enforce the new rule requiring workers to wear them.

"It was our purpose in establishing this procedure to tighten up our plant security situation which was an area of great concern and, secondly," the memo read, "to provide a visible demonstration that we were a disciplined Company by carrying out this requirement in an orderly manner as in other successful companies."

The idea for photo ID badges came from a colonel in the U.S. Army, who conducted security surveys of plants that did business with the Department of Defense. As part of his Industrial Defense Survey, the colonel also recommended that Brown & Sharpe install a chain-link fence "eight feet high with triple strand barbed wire topping" around the entire perimeter of Precision Park. He also wanted No Trespassing signs posted along the length of the fence, and more lighting installed above the employee parking lot.

Brown & Sharpe received the colonel's confidential report in 1973, and began a piecemeal approach of complying with its requests—first the company installed the brighter lights, then it required the photo ID badges; in the summer of 1976, a bulletin board memo announced that new fencing and gatehouses were going up around Precision Park to control access to the plant.

In the memo, James Rigney wrote that the reason for installing the fence and gatehouses was a rash of employee theft.[5] Management used a similar explanation, theft from employee cars, when it installed the parking lot lights—and thefts had played a role in the company's decisions. As a company that dealt in high volumes of expensive metals and small, easily-smuggled tools, Brown & Sharpe had always had problems with employee theft. Whether theft in the mid–1970s had grown beyond the historical norm, or whether managers used theft as an excuse to cover the Defense Department's confidential security concerns, the fences, floodlight, signs, and gatehouses went up around the perimeter of Precision Park.

ID badges, gatehouses, floodlights and fences were not the only changes inside Precision Park. In the spring of 1976, Henry Sharpe, Jr., announced that after 25 years as Brown & Sharpe's president he was stepping aside. Donald Roach would take his place. In some ways the change was merely titular: Henry Jr. retained his position as chief executive officer, and planned to report to work every day to oversee operations. Still, the announcement was a planned step toward the separation of the company's management from its ownership, and it augured well for Roach's career at Brown & Sharpe. Henry Jr. was now in his mid–50s, with retirement on the horizon; Roach clearly held the inside edge on one day running the company.

They rattled along the left side of the road in a rented car, Henry Sharpe, Jr., and Donald Roach, en route to Wotten-Under-Edge, a market town in Gloucestershire. Here they were scheduled to meet an ingenious English inventor, David McMurtry, who had developed a product that Brown & Sharpe very much wanted to buy.

McMurtry called his invention the Renishaw Probe, and it provided a remarkably simple solution to the problem of quickly measuring three-dimensional objects. The probe was tipped with a tiny sapphire or ruby ball, not chosen for gem quality but for hardness. A probe tip was wired to three switches; whenever the probe sensed the slightest pressure it would trip one of these switches, signaling to a computer that it had touched something beneath it, in front, behind, or to its left or right. The computer logged this data, and in this way it could precisely measure anything that the probe scanned, up and down, in and out, left and right. With a probe attached to the fast-moving arms of a coordinate measuring machine (CMM), the possibilities for precise, real-time measuring were vast. Producers such as car companies would never have to wait until a large and expensive piece was done before finding out whether it met specifications— they could make corrections to products such as car bodies while they were being made.

McMurtry was something of a latter-day Samuel Darling—his insatiably curious mind would not be pinned down to a single product. By Roach's recollection, McMurtry explained in the October 1978 meeting "he could well tire of working in the metrology area in another five years, and he thus wants to be free to do something else." He would only commit to a partnership with Brown & Sharpe for five years; after that, he needed to be free to go. The three men eventually struck a deal, giving Brown & Sharpe an almost exclusive right to use and sell the Renishaw Probe in its expanding line of CMMs.

As 1978 wound down, Brown & Sharpe was on a roll. Besides negotiating with McMurtry for exclusive rights to his probe, the company had developed an invention of its own: the DigitCal Micrometer. Brown & Sharpe trumpeted the DigitCal as the first hand tool to embrace microchip technology. The new tool was a Vernier caliper

In 1978, Brown & Sharpe introduced the DigitCal, billed as the first hand tool to embrace microchip technology. At a glance, a machinist could read a Vernier caliper's measurement on a small, electronic panel affixed to the tool which digitally displayed measurements. With a push of a button, a user could switch the reading from English to metric, or could determine whether a part was within tolerance. Sales of the DigitCal helped the company achieve then-record profits.

similar to the one that Joseph Brown had introduced to America in the 1850s; instead of reading measurements through tiny numbers etched into the steel, a machinist could glance at a small electronic panel affixed to the tool that digitally displayed measurements. With a push of a button, a user could switch the reading from English to metric, or could determine whether a part was within tolerance.

Not everything that Brown & Sharpe was trying was a success. Designers working on a Programmed Turning Center, essentially a computer numeric controlled screw machine, were having a devil of a time trying to keep up with the marriage of electronics to screw machine technology. Big customers, such as car companies, still liked Brown & Sharpe's cam-operated automatic screw machines—they were tough to set up, but once they were running they reliably churned out parts forever. But smaller customers needed to retool more frequently; they sought the flexibility offered by electronic controls, and Brown & Sharpe's designers lagged in that field.

Every year, Chicago hosted an important International Machine Tool Exposition where manufacturers showed off their wares. At the 1976 show, Brown & Sharpe had unveiled its "System 1500," a second-generation of numerically controlled machining centers that automatically grabbed a wide variety of tools and brought them down to cut, drill, or bore metal fixed vertically beneath the tools. The company promised to make horizontal models—machining centers that could cut into stock from all sides. Generally, customers wanted both vertical and horizontal machining centers, and they wanted both made by the same company. Brown & Sharpe made a good vertical machining center, but again its design engineers seemed bedeviled by the challenge of getting tools to index, or line up properly, for horizontal machining.

Overall, 1978 had been a great year, the most profitable in Brown & Sharpe's history; and the next one was shaping up to be even better. The twin energy crises of the 1970s had been a double-edged sword for Brown & Sharpe: on the one hand, the 1973 crisis had destroyed the company's hopes of being a leader in the development of the Wankel engine, which generated power through a lop-sided rotating disc rather than through the hammering of pistons. Several Detroit car companies had planned to build

Wankel engines, and Brown & Sharpe had specially developed a machine to grind the engine's shamrock-shaped housing. Then the Organization of Arab Petroleum Exporting Countries (OPEC) placed an embargo on oil exports, boosting gasoline prices, and forcing President Nixon to call for a 55 mph speed limit to curb fuel use. Early Wankel engines were less energy-efficient than piston-driven models, so Detroit pulled the plug on the Wankel leaving Brown & Sharpe with a surplus of specially-designed grinding machines. On the other hand, the 1973 crisis forced American car companies to retool in order to build smaller, more energy efficient cars.

The Iranian Revolution of early 1979 forced the U.S.–backed Shah of Iran into exile, causing a panic in oil markets that again drove up prices, exacerbating the trend for car companies and builders of jet engines to scrap old models and retool to make new, fuel-efficient ones. Military spending had dipped to 5 percent of gross national production in the wake of Vietnam, but in response to the Shah's fall President Jimmy Carter and the Congress boosted defense spending by the inflation rate plus 3 percent—a big raise, and big raises in military spending historically meant big orders for machine tools. With oil prices high, oil companies also retooled in order to develop new ways of extracting oil from previously tapped or unprofitable drill sites. With the car, aerospace, defense, and energy sectors all tooling up, 1979 promised to be better than 1978's record-breaking year at Brown & Sharpe.

Members of the machinists union knew that their work was generating record profits for Brown & Sharpe; their two-year contract was set to expire in October 1979, and they wanted in on some of the spoils.

On October 14, more than 1,000 union workers gathered in the Rocky Point Palladium to hear the company's final offer: wage increases of 9 percent per year, which averaged 55 cents per hour per worker. Jim Rigney, the company's longtime director of Industrial Relations, had resigned to take a job with the state, but he still kept in touch with Dave Waterman, his former protégé, on the company's bargaining team. As Rigney understood it, the company was confident that its offer was strong enough to win union acceptance.

Union negotiators sought across-the-board increases of $1 per hour each year. They agreed to bring the company's final offer to their members, but they declined to endorse it. After all ballots were cast and counted, only 78 of the 1,151 workers present voted to accept the contract. And after rejecting the contract, they voted overwhelmingly to go on strike. As one grinding machine operator told the *Providence Journal*, "This company is doing a record business, and they can afford to do better."

The vote caught the company by surprise. Through the first half of the year, Brown & Sharpe was on pace to smash its records of the previous year, with a net profit of $6.4 million on sales of $94 million—the current equivalent of $348 million—in just the first two quarters. Shutting down the plant in a boom time like this would be disastrous. On the first day of the strike, police and picketers got into a shoving match. A North Kingstown patrolman hurt his leg after falling over a guardrail and into the patch of a car as it slowly drove through a gatehouse. State Police also arrested two strikers for allegedly assaulting a trooper.

On day two of the strike, the company's top managers met inside Don Roach's house on Providence's East Side. The finance director, Richard Duncan, laid out numbers that showed the company had made a huge mistake in taking the strike. Henry Sharpe, Jr., still the chief operating officer and chairman of the board, told his team to get back to the table and do what they had to do to get workers back into the plant. The company's negotiating team, led by Dave Waterman, needed something to save face, but the union did not give them much. After a 12-hour session in a mediator's office, the two sides agreed to pay raises of 11.8 percent for all employees, plus increased pension, insurance, and health benefits.

Presented with this package, workers voted to end the strike after just three days. Bob Thayer, former union president and now a Providence-based business agent for the International Association of Machinists and Aerospace Workers, crowed: "This is the best contract settlement that's ever come across at Brown & Sharpe."

But in the late 1970s there was no reason for labor to be cocky. From the 1950s into the mid–1970s, labor and business had enjoyed a kind of détente that was now under attack from businessmen who, like Don Roach, felt that the balance of power between the two forces had swung too far toward labor. Business felt that competition from other capitalist economies, Nixon-era regulations such as the Occupational Health and Safety Administration and the Equal Employment Commission, and domestic political groups advancing a liberal agenda required a push back. New and emboldened "Think Tanks" such as the Heritage Foundation implemented plans to reset the balance of power, and in 1978 they successfully shot down the federal Labor Reform Bill favored by unions.

Union leaders had felt that the Labor Reform Bill was a sure bet to pass. A few Fortune 500 Companies had been flouting labor law, and unions had hoped to restore the original intent of labor laws by requiring judges to make quicker decisions, to make judges' decisions binding during appeals, and to slightly increase penalties for violations. With Democrats holding the White House and majorities in both the House and Senate, the bill's passage seemed certain—until conservative groups, coordinated by the Business Roundtable, surprised complacent labor organizers by spending more than $5 million lobbying to defeat the Labor Reform Bill, a bill of relatively minor consequence but fraught with symbolic value. Business's concerted effort to reduce taxes on the wealthy, increase military spending, and change the balance of power between labor and business in Washington was well underway for a few years before Ronald Reagan loaned his name to the "Reagan Revolution," and became its public face.[6]

In quickly settling the 1979 strike, Brown & Sharpe averted disaster and again set records for profitability and sales. But company negotiators had been forced to eat humble pie, and the experience left a bad taste in Don Roach's mouth. By Rigney's recollection, Roach would not even talk to Thayer. "Clearly," Rigney wrote, "this was a return to the conflict of the pre–1951 era. As that famous Yankee philosopher, Yogi Berra, is quoted as saying, 'Déjà vu, all over again.'"

The Board of Directors urged stockholders to attend the 1980 Annual Meeting, because they had big news to share. So on the appointed day, April 25, 1980, curious investors filled the tiers of seats inside Precision Park's Auditorium. From the podium Henry Sharpe, Jr., wearing his trademark bowtie, told stockholders that in its 147-year history Brown & Sharpe had never made as much money as it had in

1979, a net profit of $12.68 million on sales of $227.5 million, nearly $800 million in today's currency.

"There is no question that *right now* is the most technically exciting time to be in the metalworking capital goods business since the nineteenth-century—in the days when precision measurement and metalworking production machinery were truly in their infancy," Henry Jr. said.

> The *reason* for all this excitement is also very simple: the availability of reliable, cheap electronics and their booming successful application to machine control and precision measurement. No across-the-board technical development since the first machine tools themselves were widely introduced in the 1800s can remotely compare to the present revolutionary turn of events.

The way Henry Jr. saw it, Brown & Sharpe stood in a good position as the 1980s began. General Motors and Chrysler planned to spend $65 billion within the next 5 years in order to build smaller, more fuel efficient engines. More than half of the 6,000 commercial airplanes flying would need replacement within the decade. "And,'" Henry Jr. told investors,

> now we hear of a growing conviction that America must re-equip its outmoded arsenal of conventional military hardware. Already in recent weeks we have begun to receive high-priority DX military procurement ratings on certain machines. Some civilian business is actually being displaced for armament needs.

Looking over the company's progress and its prospectus, Henry Sharpe, Jr., figured that this would be a good time to turn over control of its day-to-day operations. He and Peggy had three children, and the eldest, Henry Sharpe III, was by inclination and training more qualified to understand machine tools than his father, grand-father, or even his great-grandfather. Lucian Sharpe had found bookkeeping suited his talents better than machine design; Henry Sr.'s passions had been horses and the law, while Henry Jr. might have preferred journalism to the precision tools industry. But Henry Sharpe III was never happier than when he was tinkering with the inner-workings of machines. He had earned a bachelor's degree in arts and engineering from Brown University, and was then enrolled in a graduate program at Stanford, learning how to marry electronics to mechanical engineering—the very challenges Brown & Sharpe then faced.

In 1980, Henry Sharpe III was just 26 years old. His father and grandfather had taken the company's reins at 27, but with a global sales force and manufacturing plants in three states and four countries making 4,000 types of precision tools, machine tools, pumps, valves, probes, and coordinate measuring machines, Brown & Sharpe was now way too complex of an organization to be run by a 26-year-old with a head for engineering. Besides, Henry Sharpe III was something like David McMurtry, the restlessly ingenious inventor of the Renishaw Probe. When Henry III did eventually speak with Don Roach about joining the company, Roach told him that if a Sharpe worked at Brown & Sharpe he could never leave the company, for the departure of a founding family member might signal that something was internally amiss.

For Henry III, the notion of a design engineer dedicating a career to a company was an anachronistic approach to business. He had spent some time at Stanford and in the Silicon Valley, where designers were dedicated to ideas, not to companies; and when ideas came to fruition, their progenitors moved on to the next big idea. In a sense, the Silicon Valley of the 1980s was a lot like the Connecticut Valley of the mid–1800s, a place where designers roamed from shop to shop, spreading ideas and creating a culture of creativity. Henry Sharpe III could not be tethered to a desk, looking out at the wide world of product design from the limited peephole of one company. He eventually opened his own product design firm, helping clients create everything from mobile toys to complex medical devices.[7]

"The choice of succession at Brown & Sharpe thus involves the imposition of a unique trust," Henry Jr. told the packed auditorium on that spring day, "one which I have surely felt keenly over the years and have always hoped would be equally keenly felt by each in succession who followed on."

> Now, time is passing; today's meeting is my 29th anniversary as your chief executive officer, and I am gratified to say a successor who fully deserves the unique trust I described a moment ago, is more than ready and deserving. I have recommended to your directors, and they have unanimously concurred, that Donald A. Roach be elected president and chief executive officer. ... Brown & Sharpe is very lucky indeed to claim his talents. The tradition is worthy of the man and the man is worthy of the tradition.

21. The Longest Strike

> I swear to it that this strike will never go away. We intend to keep their feet to the fire on this issue for as long as God gives us the strength to do so and as long as we're on the face of this earth.—*Ed McElroy, President Rhode Island AFL-CIO*[1]

At sunrise Don Roach peered through an airplane window at the flat, truncated peak of Mount Fuji poking above a thick blanket of clouds. His plane climbed from Tokyo and banked toward South Korea, the next stop on his tour of the Far East. The trip made a deep impression on Roach, particularly the visits to Japanese factories where he saw workers begin their days with group calisthenics led by plant managers. After this 1977 trip to Japan, Roach felt the need to cultivate a more cooperative relationship between union and management; he decided to take the union into his confidence—to share with the leadership the kind of information he used in making decisions as president. He believed that with the knowledge and camaraderie of confidences shared, he could build a better, more responsible attitude in the leadership.

The strategy did not work. Not only did the union go on strike in 1979, Roach felt that its business agent, Bob Thayer had used shared information to gain a negotiating advantage. The strike struck Roach as a public slap in the face. "This was a turning point in the company's relationship with the union," wrote Jim Rigney, who had recently retired as the company's industrial relations manager. "Feeling completely betrayed by Thayer, having shared all possible data with the union, Roach was dismayed to say the least. Totally frustrated, he did not speak with Thayer for a year."[2]

Roach published a story about his Far East visit in the December 1979 issue of *The Brown & Sharpe News*, a four-page tabloid the company began publishing after that year's brief but divisive strike. Union leaders did not like *The News*—they viewed its publication as a company attempt to bypass them and directly communicate with their members. "The most startling aspects of the visit to Japan," Roach reported, were the employee production figures: Japanese plants shipped $100,000 worth of product per employee, more than twice Brown & Sharpe's per-worker rate. The story quoted Roach as saying "we may have won the fighting war, but the Japanese are certainly winning today's economic battle in the market place on almost every front."

From Roach's viewpoint, Brown & Sharpe's workers had to become more productive for the company to compete on the global stage. He consistently sounded this theme in the "Telling It Like It Is" column that he wrote for the *Brown & Sharpe News*: "the way we can best help ourselves right now is to produce more product at lower cost and higher quality," he wrote in June 1980; "The only way our company will survive and grow is by raising performance standards," he wrote that fall.

In January 1981, Roach asked four department heads to publish pieces addressing the company's "objectives" for that year, and all of them kept to the script, citing a need for better speed and efficiency. In his published piece Dave Waterman, the new director of industrial relations who succeeded Jim Rigney, observed:

> 1981 is also the year in which we will negotiate new collective bargaining agreements with our Unions. A successful negotiation is one in which each party is able to peacefully respond to concerns of the other to their mutual benefit. I am extremely hopeful that both the Company and those employees represented by the Unions will be able to describe the 1981 Negotiations as a success.

In the spring of 1981, Brown & Sharpe prepared for contract negotiations in an unprecedented way. The manager driving the new method, John Gordon, was a 34-year-old whiz-kid with an English accent and degrees from the London School of Economics and Columbia. Don Roach had hired Gordon a few years earlier, wooing him away from the nearby Ciba Geigy chemical company where Gordon had served as personnel director.

In reviewing Brown & Sharpe's past performances at the negotiations table, Gordon was not impressed. In the last round of talks the company had caved into union demands after a two-day strike; the time before that, the company had not offered a single change in contract language. Gordon believed that collective bargaining presented an opportunity to tackle bothersome issues. He wanted to handle negotiations the way that Ciba Geigy did—aggressively, and with a well-established plan. Roach agreed, and for the first time in 40 years of negotiations the company set up a Steering Committee that supervised all aspects of negotiations from behind the scenes. A Negotiating Committee would still meet face-to-face with union negotiators, but that committee would be subservient to the unseen Steering Committee.

As the company prepped for negotiations that spring, two federal mediators warned the company's lead negotiator, Dick Jocelyn, that because of "frustrated union leadership" the union would likely strike this year. The union's business agent, Bob Thayer, could not abide Don Roach, and vice-versa; also, union stewards did not like the increasingly formal relationship between them and departmental managers, which made it harder to resolve differences on the shop floor. Company managers, too, felt some frustrations. Time and again the sub-committee drafting a position paper on the upcoming negotiations heard managers complain about their inability to assign workers to jobs that needed to be done.

The expansive floor inside Precision Park was an extremely Balkanized place. By past practice, the floor workers were split up into 118 different groups, an average of 14 workers per group. Within each grouping, seniority held sway—the most senior worker of a group could decide which machine he wanted to use, right on down the line so that the most junior employee in a group got stuck with whatever machine nobody else had chosen.

The practice of "machine seniority" was a hangover from the pre-1967 days, when workers got paid by piecework and the less desirable jobs paid less money; since then, a worker would make the same hourly rate no matter what he was doing, but some jobs were more physically taxing than others and senior workers could pass on those harder, dirtier jobs. A manager could tell a senior worker that he had to work on another machine, but if the senior employee did do work on another machine, his usual machine had to remain idle; this rankled managers trying to meet production quotas.

Another unwritten rule that annoyed the company was known as job transfers. Managers felt they should be able to temporarily swap workers from one department to another that had a need for his skills—for example, if the machine tools line received a big order they might want to move a good grinder in the industrial products department to a grinding machine in the machine tools line; by an unwritten rule, a senior worker could veto a transfer.

The committee drafting the position paper concluded that the company should not address the machine preference and job transfer issues at the negotiating table. If negotiators tried to repeal those practices, the union would certainly strike; and if the company failed to win its points then it would be stuck with bad contract language rather than unwritten rules that it was trying to address through subtle changes. And ultimately, the committee wrote, though there was some "psychological benefit" in giving managers the right to determine who did which jobs where, there was no real cost savings in eliminating machine preference and job transfers.

Dave Waterman held a seat on the all-powerful Steering Committee by virtue of his job as director of Industrial Relations. When the Committee met on June 1, he recommended that the company maintain silence on the two contentious issues that annoyed managers but did not seem to have a real cost. Two of his committee members, vice presidents William Masser and Howard Geyer, withheld their assent. They felt that machine preference cut down on productivity, which resulted in a real cost to the company. They wanted to keep the issues on the table pending further research.

The next day, Don Roach fired Dave Waterman. The Steering Committee felt that Waterman was not a team player. They believed that Waterman had prompted the union to file a complaint with the National Labor Relations Board when managers bypassed his department to deal directly with workers; they also felt he was not committed to the new methods of negotiating espoused by the young John Gordon. Gordon, in turn, leapfrogged over Dick Jocelyn, a veteran negotiator, to take Waterman's job at the top of Industrial Relations. Jocelyn would still be the face of the company during negotiations—he had been part of the company's team for every negotiation since 1973, and he would act as principle spokesman for the Brown & Sharpe team as it prepped for contract talks that everyone felt would be particularly contentious.

In early September 1981 the two negotiating teams sat at long tables facing one another like two warring ships drawn up to fire broadsides. Bob Thayer did the talking for the International Association of Machinists and Aerospace Workers. Thayer was now the union's business agent in Rhode Island, though he had worked at Brown & Sharpe and served briefly as the local lodge's president.

21. The Longest Strike

Dick Jocelyn, a veteran of four negotiations, acted as the principle spokesman for the Brown & Sharpe team. Jocelyn vowed that at least one aspect of these negotiations would change from previous ones: the way that the company communicated its message to employees. For years, the union insisted on a negotiations blackout; union leaders would not reveal the company's proposals until the final offer was on the table. Jocelyn called this "crisis bargaining," as employees scrambled for information on the day of a contract vote. This year the company announced its intention to defy union leaders and publish information about its latest contract proposals.

The two sides began negotiations on September 4, 1981, the Friday before Labor Day; that day in Washington, Ronald Reagan (who had just fired the nation's striking air traffic controllers) was recording a Labor Day speech in which he extolled the benefits to workers of federal budget cuts, tax cuts, and replacing pension plans with individual retirement plants. The teams sat down to face each other inside a carpeted conference room of a suburban Sheraton Inn. As usual, the union opened talks by presenting its proposals, 29 of them—a two-year contract with annual raises of 13 percent, more holidays and sick days, better insurance, et cetera. Company negotiators listened, but did not respond that day—they were still looking to their Steering Committee for guidance on the issues of job preference and job transfer.

After the Labor Day weekend three company executives held a transatlantic conference call. Gordon and Jocelyn spoke from Precision Park to Roach, who was in Germany at the Hanover Machine Tool Show. They discussed a new study of the machine preference issue that showed there was indeed a cost savings if managers could assign workers to jobs that needed to be done, regardless of seniority. The elimination of machine preference would boost productivity by 3 to 5 percent—a savings of about $1.2 million a year. Everyone agreed that if the company tried to eliminate machine preference, the union would strike. The Negotiating Committee recommended that if the company intended to stay in Rhode Island, it should eliminate machine preference and be willing to take a predicted six-month strike ending with a rich settlement. But if Brown & Sharpe planned to leave Rhode Island within a few years—a distinct possibility given that it had 11 manufacturing plants in 4 countries, including a recently built plant in North Carolina—the company should not be willing to take a strike. It should just tolerate machine preference for one more contract, and then move on.

Although Henry Sharpe, Jr., was not part of the negotiations he remained chairman of the board of directors, and his presence hovered over the talks. Henry Jr. held a keen appreciation of history. The day he closed the old foundry, he cried; the company's Rhode Island roots were important to him, and he let that be known. High workers compensation costs, unemployment insurance, and New England's cold weather energy demands made running a plant in Rhode Island cost $5 million more per year than operating a similar plant in North Carolina. But for Henry Jr. moving headquarters away from the company's Rhode Island roots was not a possibility. At the end of their overseas phone call, Roach, Gordon, and Jocelyn agreed: the company was staying, but machine preference and worker ability to veto temporary, in-plant transfers would have to go. The Steering Committee was willing to take a six-month strike and to pay handsomely to achieve those ends.

The company made its first proposal to the Production and Maintenance Negotiating Committee on September 18, 1981, exactly one month before the contract's expiration date. In his opening statement company negotiator Dick Jocelyn said that the company had more than modest wage increases in mind and that there would be a difference in the company's negotiating style this time around. There would be no fight on economics. He said that the company's goal was to get rid of inefficiencies and noted the decline in business at the Precision Park and Greystone plants. Through three quarters the machine tool division had lost $2.1 million. Jocelyn attributed the decline to foreign competition, noting that the Japanese paid labor approximately $3 less per hour.[3]

In its initial proposal the company sought expanded rights to establish working schedules, and the right to require junior employees to work on holidays and during vacations, the right to internally transfer people without notice, and a three-year agreement instead of the usual two. Then Jocelyn dropped the bomb: abolition of machine seniority, job preference, and permanent job assignments.

According to the Brown & Sharpe Negotiating Committee's minutes,

> Thayer and [union President Billy] Martin reacted bitterly, asserting that if the Company's real problems were with imports, it made no sense to attack the union through unnecessary language changes in the Collective Bargaining Agreement. Thayer clearly saw the proposal at that time as a personal attack on his leadership of the union. Martin characterized the proposal as a continuation of an effort to eliminate union involvement, which in his eyes was an indication of a collapse in management.

On October 16, with just two days left before the contract expiration the company laid its final offer on the table. Brown & Sharpe would grant annual raises of 11 percent, 10 percent, and 9 percent. After three years, with compounding percentage increases, workers would be earning 33 percent more than their 1981 base. Even by the inflationary standards of the time, the money was staggering.

When Jocelyn put that money on the table he saw "a committee which was wreathed in smiles." The only person present who was not smiling, in Jocelyn's view, was Bob Thayer; he looked "kind of uncertain," confused.[4] This was not conventional Brown & Sharpe negotiating style; usually the company offered lower raises than it intended to pay, then did some horse trading to obtain agreements it wanted before settling higher. This time the company was putting it all on the table up front to show that there was no more money to be had, that this was it, an extraordinarily fat financial package, the biggest the company had ever offered the IAM in 40 years of negotiations.

Jocelyn said that the company wanted an answer to its final proposal by 9 p.m. Thayer countered that there was a strong rumor afloat that the company planned to mail its final proposal to the employees and that the mailing was the reason for the 9 p.m. deadline. Jocelyn, who wanted avoid another session of "crisis bargaining" replied that the company was indeed planning to overnight its proposal directly to the homes of its employees. Thayer protested that there was no benefit to the company in mailing the proposal and that it would simply result in mass confusion.

Clearly, the money was not an issue. But there was still no agreement on mandatory transfers, machine preference, and, a new wrinkle, pensions. The company offered a late change in pensions that it said would actually pay employees more, but the union said it liked the current pension plan. For the union, all three of these issues were strike issues. For the company, machine preference and mandatory transfers, were strike issues.

At 9:45 p.m., past the company's self-imposed deadline, the two sides took a break so union negotiators could mull over the fat financial package versus the company's demands. Negotiations reconvened at 2:45 a.m. on October 17, one day before the expiration date. Thayer said he could not give a simple answer. The union was in a box—issues identified as strike issues, the pension, the three-year contract, and the clause regarding machine preference were just unobtainable at any price. Thayer said, "We're calling it quits. We accept that it is the Company's final offer. We don't doubt your sincerity, but believe the Company is on a suicide mission."

Before the two sides drifted off into the chill of a fall night, Jocelyn offered a one-week contract extension—current pay and conditions would stay in effect for a final, week-long push at reaching a settlement. Thayer responded that the "membership must agree" to an extension through a union-wide vote. "We (the negotiating committee) can't do it." Jocelyn reiterated that the offer of a one-week extension was on the table—but he formally withdrew the Company's final offer, including the hefty pay raises, so the membership could not have a chance to vote it down.

At dawn of the 17th, a Saturday, the company sent private couriers scrambling around Rhode Island to deliver its proposal to voters prior to their Sunday afternoon vote. The gambit backfired. "I'm the one who works at Brown & Sharpe—not my wife, not my kids" a grinder named William Cahroff told the *Providence Journal*. "She says, 'Look at the money. That's beautiful. She was already spending it. Then I read down further and saw the stuff about the seniority."[5]

When Ed Green heard that Brown & Sharpe wanted to strip away machine preference, he knew that the union would strike. Green had worked as a grinder at Brown & Sharpe from 1965 through 1979, when he left to take a similar job at a nearby company. In his years at Precision Park he had gone from a job that required churning out 200 pieces an hour, to one that required five pieces an hour; this had resulted in less "physical time" for him to be actually loading stock, and more "machine time" spent watching the machine do its work, a benefit of his senior status. When friends showed Green the company-delivered proposal, he knew that there was no way workers would approve the elimination of machine preference, not even for pay raises of 33 percent.

"What they tried to implement was a flexibility clause, where they could move a senior person out of his job and put a junior person in there," Green said. "It wasn't flexibility. It was a weapon. If they didn't like you they could move you off your job and put you back to doing 200 pieces an hour. ... That was the crux of the whole thing. The union decided not to buy into this flexibility. They called it flexibility. It was a weapon."[6]

On Sunday, October 18, most of the 1,597 union workers gathered in the Rocky Point Palladium to decide whether they should grant negotiators a one-week extension, and whether they should strike. Thayer saw Brown & Sharpe as a company intent on busting the union by making it irrelevant. The company seemed to be trying to weaken the union's role in a variety of ways: by plopping a big pay raise on the table without the traditional tradeoffs of negotiations; through offering a three-year deal instead of the traditional two years, thus reducing the union's opportunities to negotiate; and by doing an end run around union leaders to communicate directly with members. He accused company negotiators of "Boulwarism," a term coined after Lemuel R. Boulware, a former executive at General Electric. Boulware negotiated by laying the company's final offer on the table at the first negotiating session, and by communicating the offer directly to employees. Unions loathed this take it or leave it strategy, as it undermined their authority; and in 1969, the National Labor Relations Board ruled that "Boulwarism" was an illegal breach of a company's duty to bargain in good faith. Inside the large hall of the Rocky Point Palladium, Thayer and union President Billy Martin urged workers to reject Brown & Sharpe's offer to extend talks, and to authorize a strike. Workers agreed, authorizing the strike by a vote of 1,177 to 40.

In a departure from previous strikes, Brown & Sharpe decided that this time managers would keep the plant open and attempt to fill orders themselves. A hard rain fell after midnight on Monday, when the first pickets gathered outside Precision Park's main gatehouse. By 6 a.m. nearly 1,000 strikers massed outside the plant's fenced perimeter, ready for a faceoff when plant managers arrived for work. A platoon of 40 state and local police officers dressed in riot gear that included crash helmets and long riot sticks showed up before the managers did; one of the policemen controlled two attack dogs, and the company's own force of Wackenhut guards also patrolled with attack dogs and video cameras.

The show of force incensed picketers, who felt that the helmets, sticks, and dogs were overkill. At 7 a.m. a caravan of managers driving their own cars began rolling toward the gatehouse; picketers swarmed cars and jeered at managers as they slowed to pass through the gate. As one police lieutenant recalled it: "Pushing became quite vibrant." Dick Jocelyn saw one picketer toss a full cup of coffee across his windshield, briefly blocking his vision. Police officers trying to clear the way in front of his car became irate, and tackled the coffee-thrower, touching off what newspapers described as a "scuffle" between police and picketers. When it was over, one 27-year-old striker bled from the scalp, the result of a blow from a riot stick; another 25-year-old man blew out his knee in a takedown.

In ensuing days, violence abated along the picket line. The company complained about picketers throwing bottles at the Wackenhut guards; the union countered that the guards deliberately provoked picketers. And after some initial posturing, the two sides returned to the negotiating table with an important deadline looming: Rhode Island and New York were the only states that paid unemployment benefits to striking workers. In the first week of December, Brown & Sharpe's striking workers would begin receiving tax free unemployment checks of $143 per week; combined with the union's weekly payments of $40, most picketers would earn just $25 per week less than they earned when working. Company negotiators believed that if they did not settle the strike by the time extended benefits kicked in then they were in for a long haul.

On November 23 the company mailed employees its final offer—formally submitting the raises of 11 percent, 10 percent, and 9 percent, and modifying the transfer language to give the union veto power over temporary transfers. The company still insisted on eliminating machine preference so it could assign workers within production groups to any machine they were qualified to run, but it agreed to limit such assignments to the worker's usual shift. The day before Thanksgiving workers rejected the contract by 968 to 113. As a company lawyer later said, "the perception was that we were in trouble."

After the Christmas break, Don Roach convened a meeting of the top-level executives on Brown & Sharpe's Steering Committee in order to plan the company's response to the strike. They decided to hold out for a settlement on their terms. A company brief on the subject read:

> The advantages of settling on the Company's terms were seen as demonstration of the Company's resolve for change, improvement of the Company's self-image because it would be perceived as 'the winner,' maintenance of high management morale, weakening of the union's position in future negotiations, and productivity gains resulting from improved flexibility.

To "win" the strike, committee members embraced a strategy that they called "an escalation approach." A company brief described the plan as "an attempt to escalate the conflict by communicating through the media and directly to homes before negotiating. If this did not result in a resolution of the strike, the Company would then hire replacement workers." Dick Jocelyn, a native of East Providence, recalled "I thought it was insane to hire replacement workers in Rhode Island," a densely-packed, pro-labor state. Many workers were second and third generation immigrants, desperately scrabbling to maintain a toe-hold in the middle class. They would not go down without a fight.

On Valentine's Day, 1982, Brown & Sharpe published an open letter in the *Providence Journal*:

> Given the prospect of a never-ending strike, we have under study:
> A. The transfer of certain product lines to other locations; and
> B. The hiring of a new workforce for our Rhode Island operations. ...You will see our [help-wanted] advertisements soon, and we did not want this to happen without first sending you this letter of explanation.

The letter was signed by Donald A. Roach, president, and Henry D. Sharpe, chairman of the board.

"Part of the thought was the threat to hire replacements might end the strike," Jocelyn recalled. "Usually you hire on the QT, but we were putting ads in the paper saying we were going to hire replacements." Company executives had no idea whether they could make good on their threat to hire new workers. They opened up a hiring center inside a Howard Johnson's Motel off Route 95, and were astounded by the response—nearly 2,000 people applied, many of them machinists displaced by a deepening recession.

"We lovingly called them scabs," recalled Chris Jocelyn, a manager in the human resources division who later married Dick Jocelyn; "they loved to be called scabs." She and other managers operated a hiring mill in the motel, two managers to a room for security and efficiency.[7] A doctor performed on-site physicals. As the end of each shift she swapped her list of new hires for a pre-vetted list of qualified candidates. Thus

Brown & Sharpe joined a growing list of companies such as Phelps Dodge and International Paper that followed Reagan's lead by replacing strikers rather than concluding successful negotiations with them.

When union leaders suspected that replacement workers were rolling into the plant they called for a union "Solidarity Day"—February 22, 1982, a Monday. As the sun rose shortly after 6:30, hundreds, perhaps a thousand picketers milled outside the main gate, awaiting the arrival of managers and "scabs." Winter-pale workers, mostly men in their 20s and 30s dressed in heavy coats, knit hats, and ball caps huddled around wood fires burning in metal barrels.

Around 7 a.m., the cars came. Picketers swarmed like angry bees, kicking at car doors and spitting on windshields. Someone smashed a pickup truck's windshield with a baseball bat; again, someone smashed a car's side window. Police moved in to clear a path through picketers, and a North Kingstown police officer went down, knocked off his feet and kicked in the back. Another officer broadcast over his police radio: "Can you send somebody out here? We're getting our butts kicked!"

When reinforcements arrived, police formed a phalanx of 50 state troopers and town policemen. With arms linked they spanned the width of Frenchtown Road and pushed through the jeering crowd, cleaving it in two till they reached the main gate. Police dismantled a barricade of tree limbs from outside the gate while strikers hooted and jeered.

Bob Thayer stepped before the crowd, joined by the union's lawyer, Raul Lovett, a flamboyant man who had arrived that morning in a Rolls Royce bearing a license plate reading: M-MOUSE. He had earned a fortune through worker's compensation and personal injury cases. A client had once called Lovett a "Mickey Mouse lawyer," and he had embraced the term, posting a five-foot tall neon Mickey Mouse outside his downtown office. On this morning he sported a thick sheepskin coat, and his eyes were barely visible between his broad beard and the low brim of a gray cowboy hat stuffed with a plume of ostrich feathers. Lovett and Thayer asked the crowd to clear out. By 9 a.m. the road to the plant was opened, and a few cars carrying managers and non-striking workers rolled slowly through a gauntlet of shouting strikers. Police logged seven arrests, and two injured policemen.

When the negotiating teams met again, four days later, Dick Jocelyn was livid. He told the union that the company's November proposal was no longer on the table. This was now a 19-week strike that had cost the company $12 million in lost business, security, and hiring expenses. He said that the union had better make a proposal that offered some return on that $12 million.

The union countered with six new demands, including full payment for the Christmas 1981 vacation.

Every workday morning managers and replacement workers ran the gamut of picketers lined up along the road to hurl insults, and sometimes eggs or rocks, as cars drove through the main gate. Some replacement workers and non-strikers carpooled with managers; others arrived in rented vans or in yellow school buses.

Merrick Leach, a sales engineer whose union was not on strike, recalled those bus rides. Every rider received a sheet of cardboard with instructions to press it against the window when they neared the plant, to contain shards of breaking glass. "The people on the picket line were banging on the sides of the buses," Leach said. "Rocks, baseball bats—whatever they could reach the windows with. You were in fear of your well-being crossing that line."[8]

Bus riders met at a Park and Ride lot near the state airport. Sometimes at the end of the day they returned to find their cars with slashed tires or broken windshields. Jocelyn kept a running tally of the broken windshields, more than 80.

In early March a bomb placed in a rented van sitting in a supervisor's driveway exploded. Three hubcaps blew off, the hood crumpled, and the engine sat cockeyed, its engine mount blown ajar. The next day a woman who had been crossing the picket line at the Greystone plant reported three shotgun slugs fired at her house in rural Glocester, Rhode Island. And an arsonist burned a wooden shed outside the main gate (Rigney believed it was the shed that Brown & Sharpe had built to shield picketers from the weather in the 1975 strike) by sticking a burning box beneath it.

On March 22, a group comprised mostly of workers' wives and a few sympathetic Brown University students staged a sit-in, dropping to the ground in the driveway to the factory entrance. A State Police lieutenant, Frank McGee, told them to move. They stayed put. A North Kingstown patrolman, TJ Varone, stood by with a motorized tear gas fogger that looked like a leaf blower; instead of blowing air, it blew a cloud comprised of an acrid chemical called Chloroacetophenone.

McKee again asked the picketers to get up and move. Again they did not. He gave the signal to Vadrone to wade among them in a helmet and gas mask, waving a "pepper fogger" that sprayed a foggy stream of chloroacetophenone gas. The Center for Disease Control warns that exposure to the chemical may cause skin blisters and fluid accumulation in the eyes and lungs.

As the gas engulfed them, the picketers bowed their heads and sang, "We shall not be moved."

Again, Vadrone passed among them closer this time. "It was like somebody threw beach sand in your eyes and then rubbed it in," one striker named Bobby told the *Providence Journal*. "It burns. It smells like sulfur from a match, but a lot stronger." Picketers buried their heads in each other's clothes and held their ground. They endured the fogging for as long as they could, and then, choking, began to stumble away. A knot of three people hung on, then they too stumbled, zombie-like, through the fog. Newspaper photographs and television footage of the event spread across the country and overseas, leaving lasting, indelible images of the violent strike at Brown & Sharpe.

The fogging did not stop picket line violence—and the violence came from both sides. In mid–April, two men seeking jobs at Brown & Sharpe took umbrage when picketers kicked the doors of their Oldsmobile. They returned that night, brandishing a home-made pistol that fired .22-caliber bullets. They fired at picketers massed around the plant's west gate. One bullet struck a 58-year-old picketer in his torso, sending him to the hospital with a wound in his left side.

For a couple of weeks after the shooting an uneasy détente lingered over the picket lines. At the annual stockholders meeting about two-dozen strikers gathered outside a Providence hotel to peaceably leaflet the stockholders as they passed through. An

On March 22, 1982, police used tear gas to disperse people staging a sit-in on the road leading into Precision Park. Images of the event spread across the country and overseas, leaving lasting, indelible images of a long and bitter strike at Brown & Sharpe. Picketing dragged on into 1985, and court cases generated by the strike were not resolved until 1997 (courtesy Duane Clinker).

unmarked police car pulled up to the door, and out stepped Mary Elizabeth Sharpe, Henry Jr.'s mother. At 97 years old, she was less than eight years removed from her most recent triumph—convincing politicians, neighborhood groups, and a scrap-yard dealer to move a mountain of rusted scrap iron away from Providence's India Point in order to create an urban, waterfront park. She had spent a dozen years on the project, and now her vision of a pedestrian bridge spanning Route 195 to connect a neighborhood to an urban oasis had become a reality. As she stepped out of the unmarked sedan, leafleters and stockholders alike quietly parted to make way for the dignified doyenne of Providence society. She passed through, pushing her empty wheelchair instead of riding in it.

A month later, the temporary truce wore off. The union called for another Solidarity Day on May 24, and police beefed up their manpower to 100 officers drawn from multiple departments. They arrested 19 men including the union president, Billy Martin, charged with hurling a rock at an unmarked police car. Someone picked up the rock and mounted it on a plaque that read: "Actual rock thrown by union president Billy Martin at Capt. Tommy Griffin, R.I. State Police." For years, Henry Sharpe, Jr., displayed that rock in his office as a reminder of how crazy things got: shootings, attempted arson, pepper gas, a bombing, broken windshields.

Once workers ran the picket-line gamut and arrived inside Precision Park they were locked in for the day—no offsite lunch breaks allowed. At night a small "Lock In Crew" of managers slept on cots inside the plant to protect property. After a few weeks, managers and the new workers developed a soldier-like camaraderie. "It was a little bit of a war mentality," recalled Christine Jocelyn, a human resources manager.

In order to fill orders, managers worked on the factory floor. "I was doing panel assembly," Chris Jocelyn recalled. "I was doing the union president's job out on the floor"—cutting wire, stripping the ends, putting on the prongs to plug the controls into wire harnesses. As they did the work, managers noticed that for many jobs the production quotas were way too "loose"—in some cases standards allowed five hours for work they could easily turn out in three.

The discovery of loose standards prompted the company to add a new demand to its contract package: Managers would be able to spend 20 percent of their week on the floor, doing union work. And because managers liked the replacement workers, the men and women who carpooled with them through the hostile picket lines, management decided that even if the strikers opted to return to work the replacement workers would get top priority for keeping their jobs.

The union's Negotiating Committee responded angrily to the company's new proposals covering managers doing union work and the retention of replacements. Thayer said he had "no goddamn intention of giving rights to bargaining unit work to management." And as far as replacement workers went, Thayer said there was "no way the scabs could stay." A federal mediator noted that at this point, seven months into the strike, the two sides were farther apart than they had been at the beginning.

One issue that compounded the problem, one that neither side had wholly anticipated, was the absolute collapse of the United States machine tool industry in 1982. The Reagan Administration's policy of cutting taxes while boosting defense spending was running up a budget deficit of $128 billion, triple the deficit of 1979. International markets began to notice America's problems with paying its bills. Banks in the world's leading economies like to hold a variety of currencies as a hedge against any one currency's collapse; because America's currency was becoming increasingly risky, it was losing its value relative to a healthy currency such as the Japanese yen. So the yen could buy a lot more than a dollar, making it impossible for American companies to compete against foreign competitors, particularly the Japanese, who could deliver quality machines, more quickly, for far less money.

One of the reasons the Japanese could deliver products faster was America's overreliance on the defense industry. Even in times of relative peace, defense orders comprised at least 20 percent of U.S. machine tool orders. By government policy defense contractors received top priority on their orders, forcing civilian customers to wait or look elsewhere while American firms first catered to the defense department's considerable appetite, which only grew during the Reagan defense buildup.

After the Soviet Union invaded Afghanistan in 1979, the Carter administration blocked American firms from selling machine tools to Eastern Bloc nations, an income source that had helped lift Brown & Sharpe from the doldrums of the early 1970s. Competitors in other capitalist countries picked up the orders, so the Soviet Union and its satellites still bought machine tools, but U.S. firms were shut out of the multi-million dollar market.

Global corporations wanted the flexibility to move manufacturing where it was cheapest, and the Trade Agreement Act of 1979 catered to them by making it possible for the federal government to buy products that global corporations manufactured outside of the United States, as long as the country of manufacture had negotiated free trade agreements with the United States.

There were other reasons for the collapse, besides federal policy. Items that had once been cut from metal, cash register keys for example, were now molded from plastic. Capital markets favored short-term financial thinking rather than long-term productivity commitments. And many American machine tool companies such as Brown & Sharpe offered an array of products—grinders, screw machines, and milling machines—while Japanese machine tool companies tended to specialize in one area. Okamato Machine Tool Works received 90 percent of its revenue from grinders while Okuma specialized in lathes, allowing these firms to reap advantages from greater economies of scale.[9]

Japanese companies also built simple, modular designs that reduced necessary parts and increased "design overlap," so that one design worked across several models of machines, reducing production costs. A Japanese company called FANUC developed one simple numeric control that worked across different types of machines, and it soon owned 50 percent of the global market for numeric controls. American companies, meanwhile, often built sophisticated, custom controls for a specific, often defense-related product, that had no practical commercial use.

All of these factors, combined with high interest rates for capital set by the Federal Reserve, acted in synergetic fashion to gut America's machine tool industry in 1982. That year, while Brown & Sharpe workers walked the picket line, orders for American-made machine tools fell by 80 percent, employment in the industry fell by 20,500 workers, and one-out-of-five American machine tool workers lost a job.[10]

Raul Lovett, the "Mickey Mouse" lawyer, had built the bulk of his practice on worker's compensation law; in Rhode Island legal circles he was called "the King of Worker's Comp." But Lovett had little experience in labor law, and he now found himself a primary actor in one of the longest, most visible strikes in the nation. To help lighten the workload Lovett hired a lawyer fresh out of Boston University law school, Marc Gursky, a young man from the Bronx.

With negotiations stalled, Lovett turned to litigation, papering Brown & Sharpe with a flurry of lawsuits. In the first year of the strike his firm filed 15 complaints with the regional office of the National Labor Relations Board (NLRB.) The gist of the union's argument was that Brown & Sharpe did not have a legitimate economic interest in the machine preference and transfer language. The company only made these strike issues because it knew they would provoke the union to strike, giving the company a chance to put the union out on the street.

William Powers, the company's high-powered labor lawyer with a wealth of experience, countered that Brown & Sharpe had offered the union a contract extension specifically to avoid a strike while continuing negotiations. If the company had wanted to rid itself of the union it could have legally moved work to its non-union plants for far less money than the strike had cost. He also offered notes from a planning session in which Henry Sharpe Jr. specifically told company negotiators that the company had no interest in trying to break the union.

At the regional level, the union's lawsuits mostly failed—the regional director of the NLRB dismissed 12 of Lovett's complaints, the union withdrew one, and the company agreed to settle two: one by paying benefits to workers who were on disability at the time of the strike, the other by issuing striking workers their 1981 Christmas vacation pay. Lovett appealed many of the dismissed charges to the NLRB's general counsel down in Washington, D.C.

While awaiting word on appeals, the union tried another tack: in June of 1982 the national office of the machinists union hired a well-known labor strategist named Ray Rogers to put economic pressure on Brown & Sharpe's business partners. Rogers, a charismatic, 38-year-old with a full head of black hair and a mustache, touted his demonstrated record of success. Rogers's company, Corporate Campaign, Inc., had recently used unorthodox techniques to force J.P. Stevens & Co. into accepting unions at 10 of its 70 textile mills. Some of Stevens' directors held posts at major banks and insurance companies, and Rogers had put pressure on those institutions to dismiss the Stevens directors or face union boycotts.

In the summer of 1982, Rogers focused his energies on Rhode Island Hospital Trust National Bank, which had a long history with Brown & Sharpe. Henry Sharpe Jr. had recently passed his seat on the bank's board to Don Roach. Rogers began a campaign asking depositors to remove their savings if the bank did not dump Roach from its board. He staged demonstrations outside of the highly visible Hospital Trust Tower, Providence's second-tallest building, and pickets leafleted outside the bank's branches.

Rhode Island Hospital Trust was then in negotiations to buy the Columbus National Bank of Providence, and the union pressured federal banking regulators to reject the trust company's bid. Within a year, this became a moot point—the same Reagan-era banking deregulations that made it easy for Rhode Island Hospital Trust to swallow Columbus National made it easy for Bank of Boston to acquire Rhode Island Hospital Trust. The Rhode Island bank sold to Bank of Boston for $150 million, a windfall for its shareholders, including Brown & Sharpe. Bank of Boston was in turn absorbed by Fleet Financial, which eventually wound up in the belly of Bank of America.

The Rogers campaign proved ineffective, but the union's legal cases did gain some traction down in Washington, D.C.; when Lovett appealed the regional board's dismissals of his complaints to the NLRB's general counsel, the counsel issued an order in November 1982 telling the regional board to reinvestigate the complaints.

That fall, Brown & Sharpe announced that despite laying off dozens of managers and even hundreds of replacement workers, the company had lost $10.3 million through October, the current equivalent of $25 million. To save money, the firm planned to subcontract many of the products made at Precision Park, including radius gauges, straight edges, and Vernier height calipers; in addition, the company announced it could save $10,474 per screw machine if it built them at its plant in Melbourne England. The firm had given up even trying to compete in manufacturing machining centers and partial turning centers.

In November, Jocelyn told union leaders that Brown & Sharpe was closing its Greystone plant, where the company made cutting tools. McDonnell-Douglas had pulled out a major order, resulting in a monthly loss of $300,000. According to the company's bargaining notes, Jocelyn said that he thought Brown & Sharpe "would survive in Rhode Island, but not as it had before the strike." Small tools, screw machines, and cutting tools were moving elsewhere.

In June 1983, more than two-and-a-half years into the strike, the two sides made a last-ditch effort to resolve the strike. Jocelyn lead off by telling the union:

> First, we think we have won the strike. We took everything you threw at us—violence, vandalism, character assassination, and untrue unfair labor practice charges.
>
> We have a work force who works with us. We have more people who want to come to work. The strike taught us many lessons. We learned the value of flexibility.
>
> We discovered many wasteful work practices which must be corrected. We find people will work hard for a cause they believe in. Management needs the ability to direct the work force. We must adapt to meet the challenge of the competition.

He outlined the costs of the strike—training a new work force; $600,000 in property damage; millions in security; "astronomical" legal fees. "People bled and suffered," he said. "Now! I have returned to negotiations—with a proposal—at your request. We have taken all we have learned and have attempted to anticipate the needs of the future. There will be many of our proposals which you will not like."

"If you want a contract, you will have to address our concerns."

He then laid on the table a contract that he knew the union would hate—it contained the company's language on machine seniority and transfers, and rolled back an agreement that allowed early retirees to buy health insurance through the company. "Something very strange happened at that point," recalled Powers, the company lawyer. "The union took the proposal with restraint, with courtesy, at an open meeting between the parties. There was no posturing. Bob Thayer did not, as I had predicted, take the agreement and fling it on the floor."

By now, the national office of the IAM felt the need to settle—it had spent millions in striker benefits, and it hoped to get workers back into the plant, paying dues. A straight-shooting business agent from the union's New England office, John Capobianco, took over negotiations from Thayer. Capobianco told Thayer, the local business agent, "Bob, one of these days you're gonna have to wake up. Bob, we lost it."

"What the hell can you do?" Thayer said. "I know it. How do you get out of it?"[11]

In December 1983, the NLRB's general counsel gave Thayer that out—he overruled the regional office and determined that Brown & Sharpe had engaged in illegal, surface bargaining designed to cause a strike to rid itself of the union. Brown & Sharpe appealed to the full National Labor Relations Board, and the case became something like the Jarndyce and Jarndyce case of Dickens' *Bleak House*—complex, convoluted, and interminably long.

For years the long, bitter strike caused divisiveness in Rhode Island, with union sympathizers arguing that Brown & Sharpe had wounded itself by trying to kill the union, and management supporters contending that the union had killed itself by trying to hurt the company. Ultimately after the blood, bomb, bullets, and tear gas, the strike had absolutely no effect on what happened at Brown & Sharpe. Even if the union had accepted the offer of a 33 percent pay raise over 3 years, Brown & Sharpe would have been forced to lay off most of its workers to cope with the machine tools industry's near total collapse of 1982, a collapse engineered not by union workers or company management but by federal policies on taxes, spending, interest rates, and trade. If the union had agreed to the initial contract, workers would have seen larger unemployment checks; but they still would have been out on the street, and the company still would have been edging its way out of the machine tools business.

22. Locked Out

Henry Sharpe, Jr., nursed a late-season flu, too woozy to get up out of his bed. So he summoned Don Roach, the man whom he had entrusted to run his family's business, to his bedroom. Here, in April of 1990, Henry Jr. told Roach that he was fired.[1]

Roach, the company's 59-year-old president, had recently dragged Brown & Sharpe's name into the headlines when a 30-year-old woman who managed the plant's in-house travel agency filed a sexual harassment suit against him. The woman accused Roach of relentlessly pressuring her to have a relationship, though she continuously rebuffed him. At first, Henry Jr. rallied managers to support Roach. He had placed a lot of faith in Don Roach, and he believed in what he called "airplane management"—just as an airplane needs a lot of runway to get airborne, he felt that a new management team needed time and space to get its wings. But by the end of the month a couple of other women had come forward with allegations, and Roach quietly settled the case with the travel agent out of court. Henry Jr. let Roach stay on through December, which gave the company time to search for his successor while Roach wrapped up a few projects.

Before he left Brown & Sharpe, Don Roach completed some major acquisitions: in April 1990 he negotiated a deal to acquire Leitz, the metrology division of a large German company. In November he announced an agreement with Precizika of Lithuania, in which Brown & Sharpe would build the software and electronics for metrology machines sold in the Soviet Union.

At the time of the strike in the early 1980s, machine tools comprised 75 percent of Brown & Sharpe's sales, while metrology products comprised 25 percent. By 1990 those figures were reversed—Brown & Sharpe had transitioned from an industrial company that made machine tools to a technical concern designing entire manufacturing systems. Besides making coordinate measuring machines, Brown & Sharpe also wrote the software that connected the machines to databases so manufacturers could prevent mistakes on the production line.

By 1990 the company was building sophisticated measuring machines such as the Validator Process Control Robot, a rugged machine that could stand the heat, vibrations, and oily mist of a factory floor while measuring parts within seconds. The robot projected when a machine would begin making bad parts—extrapolating, for example, how many cuts a tool could make before its edges would become too dull to cut parts within precise tolerances. Customers could still buy the ruby-tipped Renishaw

Probes for measuring, or they could buy laser and optical probes that did not even have to touch the product to make dozens of measurements every second. The probes reported the measurements to a central database, allowing manufacturers to study their processes in real time and in minute detail. By monitoring parts as machines shaped them, coordinate measuring machines eliminated scrap, products that came off the line out of tolerance.

By the early 1990s these metrology systems assisted in manufacturing everyday products—Ford, Toyota, and GM used them to align the panels of auto bodies; Pratt & Whitney ensured that the blades in its jet engines did not snap off by measuring their connecting slots to within 50 millionths of an inch; a manufacturer of soft drink cans measured rims for uniform thickness so the top of a can would not break off with the pull of a tab. Computerized coordinate measuring machines shaped the daily experience of modern living in innumerable ways, standardizing everything from the weights of candy bars to huge sections of airliner fuselages.

In December, Brown & Sharpe announced Roach's replacement: Dr. Frederick Stuber, a managing director at the company's largest European subsidiary, Tesa S.A. of Switzerland. Besides making precision tools and gauges, Tesa manufactured some statistical process control devices such as probes and data logs, so Stuber was up on the company's metrology plans. He was a brilliant electrical engineer whom NASA had singled out for his work transmitting color TV pictures from the Apollo moon landing. Stuber was from Switzerland, and he planned to continue living there while making periodic visits to the company's titular headquarters in Rhode Island.

Stuber's promotion to the presidency came at an inopportune time: the recession of 1990–1991, a brief but intense recession. A major stock market crash and the collapse of the Savings & Loan industry in the late 1980s had staggered the U.S. economy. Then in the summer of 1990, Iraq invaded oil-rich Kuwait, driving up oil prices at the same time the Federal Reserve was tightening the money supply to dampen inflation.

A computerized coordinate measuring machine (CMM) developed by Brown & Sharpe to gather measurement data about products as they are being made, no matter how complex their shapes. Today CMMs shape the daily experience of modern living in innumerable ways, standardizing everything from every day products like candy bars and aluminum cans to car bodies and airliner fuselages (courtesy Hexagon Manufacturing Intelligence).

As a result the global economy tanked precisely as Stuber was taking the helm at Brown & Sharpe, a company that still depended upon a robust capital goods market to sell its manufacturing systems.

In the first quarter of 1991, the firm lost $14.6 million. In March, just five months into his tenure, Stuber announced that Brown & Sharpe was quitting the machine tool business, 139 years after Joseph Brown first sold turret lathes and a universal miller to Providence Tool. The company laid off 140 of its U.S. workers, mostly in Rhode Island. It sold the last of its machine tool inventory through a factory auction.

With the historic machine tool line gone, Precision Park became a white elephant, overly large and expensive to maintain. The company made plans to lease 70 percent of the plant to various users while contracting its workforce to a third of the space. Employment there fell to 650, just a fraction of the company's 2,400 workers worldwide. Then in July 1992, Standard & Poor's dropped Brown & Sharpe from its index of 500 stocks because of the low trading volume of its shares. A lot of investors liked to buy a mix of stocks that were on the index; when the company stock dropped from the index, many investors sold it, lowering its value.

Whenever he flew into Precision Park from Switzerland, Stuber liked to hold "Coffee Meetings" with a couple of dozen workers randomly selected to talk with the boss over coffee. He found morale was terrible. At a Coffee Meeting held days after the stock dropped from the index people complained, "There is never any positive news, just rumors and they are always negative." And, "We never hear anything new."

One worker asked, "Why don't we sell sewing machines again?"

"That's an interesting question for me," Stuber said. "Being relatively new at Brown & Sharpe, I didn't know that we ever made sewing machines."[2]

With Brown & Sharpe hemorrhaging money in 1992, the board of directors opted to give their new president some help on the business end by hiring a chief financial officer, Charles Junkunc, a former CFO of a company that made computer printers. Junkunc's first order of business was raising capital by selling the pump and hydraulic businesses, and by issuing a secondary stock offering. Then with cash in hand, Brown & Sharpe continued with its plan to grow its metrology business through acquisition. In 1994, Brown & Sharpe bought two metrology companies in Germany, Roch SA and Mauser; it bought another one in the United Kingdom, Metronic; and before the year was out, Brown & Sharpe announced it had acquired a key competitor in the metrology field—DEA of Turin, Italy.

"Brown & Sharpe decided that there was huge potential to partner with DEA," recalled Henry Sharpe III, who had joined his father for a term on the board of directors. "I think they just wanted to have a single brand to offer anything you wanted for measuring product. DEA was a big bite at the apple—about the same size Brown & Sharpe was at the time."[3] To make sense of all these acquisitions Brown & Sharpe created three divisions, the largest of which remained in Rhode Island:

- The Measuring Systems group. This largest group made the process control robots and the coordinate measuring machines, which ranged in price from $10,000 to more than $3 million for a high-speed machine with integrated

software systems. DEA made some of these products in Italy, Leitz made some in Germany, and Brown & Sharpe made the rest at Precision Park.
- The Precision Measuring Instruments group manufactured the smaller, handheld tools, calipers, micrometers and the like, at Tesa in Switzerland, and at manufacturing plants in France and in Poughkeepsie, New York.
- The Custom Metrology group in Telford, England. This group generated just 5 percent of the company's sales, designing special systems for package and can manufacturers, car companies, oil drillers, and aerospace companies.

Stock pickers liked Brown & Sharpe's acquisitions strategy; by buying up competitors, the company was in a position to reduce costs by merging duplicative assets. But cutting costs through standardizing designs proved to be a lot more difficult than anyone realized. When Henry Sharpe, Jr., set out to grow Brown & Sharpe as a machine tool company in the 1960s, he acquired United States companies that made pumps, hydraulics, and gauges. What Brown & Sharpe was trying to do now was an order of magnitude more difficult. The metrology technology was much more sophisticated than hydraulics and pumps; and the companies dealt in multiple currencies, were spread widely across time zones, workers spoke different languages, and culturally were worlds apart.

Henry Sharpe III set out to visit some of the European acquisitions in 1995. In Lithuania, recently out from under Soviet rule, managers had no understanding of how products were developed and sold in a market economy—they had never had to consider markets, pricing, and distribution. Managers there had been marketing machines at 10 percent above cost. "When I explained that most Western companies wouldn't sell a product unless it could be sold for three to five times cost, they were dumbfounded," Henry III reported.[4]

The Leitz plant in Germany was already in poor repair—the roof leaked, and a lot of old machinery and metal stock was piled up without any inventory system. German managers confided that they were reluctant to suggest streamlining measures for fear that their divisions would be the ones cut. "At an engineering, marketing or sales level, there doesn't seem to be much team work," Henry III wrote.

Melding technologies, currencies, and cultures of French, Italian, Swiss, German, Lithuanian, English, and American assets into one shared ethic represented a real challenge. The cost savings expected from the mergers were not coming to fruition; every year since Don Roach left, the firm posted losses in the millions. The board of directors formed an auditing committee to look into the company's basic business practices, and the results were not good. In early 1995 the committee reported: "During this past year we were surprised at the level of financial control discipline and preparedness and the number of potential serious issues that surfaced. We have concluded that changes are necessary."[5]

Changes included: moving all firms to U.S.–based generally accepted accounting principles; having all plants conduct inventory so the company would know what it owned; and replacing Doctor Stuber as chief executive officer. No one doubted Stuber's integrity or his brilliance as an engineer; but in his four years at the helm the company had lost $40 million. Stuber was out, and in the interregnum, Junkunc would serve in his capacity.

In hiring a new president the board of directors reached out to an old hand in the machine tool business: Frank T. Curtin, a former vice president at Cincinnati Milacron, once one of Brown & Sharpe's biggest competitors. Curtin was now 60 years old, and in many ways his career arc paralleled Brown & Sharpe's recent path. He had left the machine tools business to serve as vice president of a Michigan consortium that researched new manufacturing technologies for car companies and other high-volume industries.

Curtin came aboard in May 1995, and a year later he and Junkunc raised $96 million through a secondary stock offering. They planned to draw on that money for research and for manufacturing, while waiting for customers to pay their bills. The banks and insurance companies that invested in the stock offering built in covenants guaranteeing financial performance before pouring their money into Brown & Sharpe. Stock pickers approved. The firm of Donaldson, Lufkin & Jenrette gave Brown & Sharpe's stock an *outperform* rating for 1996, noting: "The cost reductions the Brown & Sharpe managers have identified are substantial."

The company did make money that year, a net profit of $7.8 million. With the company back in the black, Henry Sharpe, Jr., decided that 50 years after beginning his apprenticeship on Promenade Street, the time had for him to step aside from Brown & Sharpe. In April 1996 he resigned from the board of directors, telling Curtin that he wanted no fanfare about his retirement. He got his wish.

Henry Sharpe, Jr., could hardly believe the words coming out of his own mouth. In the fall of 1996, at a special shareholders meeting of the Providence Journal Company, he rose to support a motion to sell the company to A.H. Belo of Dallas, Texas.

"My grandfather bought an interest in this paper and became one of its senior executives in the late nineteenth-century," Henry Jr. said, his voice wavering with emotion.[6] He had a sentimental attachment to the Journal—his first paying job was working there as a summer copy boy in 1941, between leaving Exeter and enrolling at Brown. "My father was an intimate participant in this paper throughout his life, and he willed me his shares in the Providence Journal. In his *will* it says, 'This is to be treated as a public trust and not as a financial investment.'

"And I stand here today talking to you," he paused. "'How in the world,' you ask, 'did you search your soul and be in favor of this arrangement?'"

"The only answer to that question is that, in my judgment, this *is* an act of public trust."

The way Henry Jr. saw it, the Providence Journal's situation in 1996 was a lot like Brown & Sharpe's before the strike. Technology and government policy were about to totally transform the media business in the same way they had acted to reshape the machine tool industry in the 1980s. In many ways the company was very strong. Michael Metcalf, the late grandson of Henry Jr.'s aunt, Louisa, had proven to be an ingenious businessman. He built a media empire that included the *Providence Journal*, two smaller newspapers, and ownership or partnership in 13 television stations.

Though he could not yet envision what the new media landscape would look like, Henry Sharpe, Jr., knew that a "tremendous earthquake" would soon reshape it, and as a mid-sized player the Providence Journal Company was not big enough to survive the trembler on its own. "It makes me want to cry to think of some of the things that have happened, will happen, must happen," Henry Jr. told stockholders gathered in the flagship newspaper's auditorium. "But you must remember that we are in a tumultuous technological time.

"Newspapers are no longer solely a career for delivery boys, nor are they even printed solely on paper. We have just *begun* to see what the electronic revolution is all about. And the changes will be enormous. And there is nothing that this newspaper, as a single entity, can do to stave off the changes that will be inflicted upon us."

The company that sought to buy the Journal, A.H. Belo, published the *Dallas Morning News*, but it was primarily interested in the Providence Journal's TV stations. Television accounted for 80 percent of the Journal Company's $45.4 in pre-tax earnings; Belo's broadcasting and newspaper segments had cash flow of $117.3 million split almost evenly between print and broadcasting.

Prior to the recently passed Telecommunications Act of 1996, the merger would have been illegal. When radio and television were in their infancy, Congress viewed the airwaves as an ocean owned by everyone but leased to companies; as trustees of the public's space, media companies had to comply with regulations set by the Federal Communications Commission regarding fairness, tastefulness, and the number of broadcasting companies that one company could control. In the 1950s, the FCC set a limit of seven television stations per broadcaster. President Reagan's FCC eliminated the Fairness Doctrine in the mid–1980s, and expanded the multiple ownership rule to 12 stations. The Telecommunications Act of 1996 removed all limits on ownership, and increased the allowable percentage of household reach in a local market.[7] The Act's passage spawned a frenzy of broadcasting deals, including Belo's bid to buy the Providence Journal Company.

Henry Sharpe, Jr., had seen this before. Almost every legacy seat he had inherited on a corporation's board from his father and his grandfather, had vanished before the era's waves of deregulation. He had seen bigger banks subsume Rhode Island Hospital Trust after banking deregulation until it wound up in the belly of Bank America; following deregulation in the utilities industry the Providence Gas Co., where he and his fathers had served on the board, became a subsidiary of National Grid PLC of London. Henry Jr. told stockholders that a merger of the Providence Journal Company was inevitable. If they acted now they could at least choose their suitor. "That is why I am responding to the public trust as I see it. I hope you will vote with me." The stockholders did vote with him, opting to sell the company for $1.5 billion.

In December 1997 the world's longest strike came to an end, 16 years after it started. In a convoluted ruling befitting of a tangled case, the U.S. District Court of Appeals ruled that Brown & Sharpe had cooperated with an official investigation of the causes of the strike, so the National Labor Relations Board's general counsel had erred when he re-issued unfair labor practice complaints against the company seven months after an arm of the board had dismissed them. For those complaints to stick, the general counsel would have had to reinstate them within six months, not seven—unless Brown & Sharpe had tried to conceal evidence. Ultimately, the court ruled in this final case, there was no concealment, thus the general counsel had indeed made a mistake in November of 1981 by reinstating dismissed charges. Thus the charges of bad faith bargaining remained dismissed. What it all meant was: after 16 years, Brown & Sharpe had "won" the strike.

Had the company lost it might have owed the union $60 million in back wages, more than its entire net worth; but management had long stopped worrying that it would lose in court. Investors, too, had shown no concern about the company losing that case, but they had been waiting for a few years to see Brown & Sharpe make significant cost-cutting moves by streamlining its operations. By 1997 they were running out of patience. Donaldson, Lufkin & Jenrette backed off from its bullish appraisal of the company's stock, telling investors that year: "there is little upside in the shares." Credit Suisse recommended the stock to long-term investors but with caveats: internal communications seemed confused—customers did not know whom to call for service—and the company's management may have been "spread too thin, reducing focus."

Henry Sharpe, Jr., in 1994, two years before he rose at a meeting of Providence Journal Company stockholders to advocate selling the company to A.H. Belo of Dallas, Texas. "It makes me want to cry to think of some of the things that have happened, will happen, must happen," he told stockholders gathered in the newspaper's auditorium (courtesy Peggy Sharpe).

One of the ways the company satisfied investor demand to cut costs was by outsourcing the manufacture of small, precision tools to a factory in China. Henry Sharpe III decided to buy one of these tools after recalling "one of the best presents I ever got," a Brown & Sharpe combination square that his father had given him for his 18th birthday. "Dad gave this to me," Henry III said while fingering the ruled portion of the decades-old combination square made in Providence. "I like the way it feels in my hand. They really paid attention to the surface quality" by etching in easy-to-read numbers. Henry III also owned the combination square that his father bought as an apprentice, and it was so similar to the one that he had received as a present a generation later that the parts of one fit snugly with the parts of the other.

Henry III decided to surprise his son with a Brown & Sharpe combination square manufactured in China. When he tried to lock the ruled blade into the tool's pivoting head it wobbled. "It's good enough," Henry III said of the tool's performance. "But in comparison, it's junk. There's a fit and a feel that's not very nice. The magic leaked out of the tool." He never did give it to his son. After buying that tool, Henry III said, "I could see things going south. ... Brown & Sharpe set out from its earliest days to build the best product possible. I think the world no longer wanted the best. They wanted good enough."⁸

Some man down in New York City had been asking Henry Sharpe, Jr.'s friends for his contact information, so Henry Jr. decided to call him first. On an unusually warm day in late April 1998, he dialed the 212 exchange and reached one Bob McCabe.

McCabe spoke familiarly. He tried to refresh Henry Jr.'s memory of an earlier meeting back in 1968, when he had been an assistant to a broker at Lehman Brothers who helped Brown & Sharpe go public on the New York Stock Exchange. Of course there was more to this conversation than revisiting days of auld lang syne. McCabe now served as a director of the Thermo-Electron Corporation, which made industrial and laboratory equipment. Its headquarters were nearby in Waltham, Massachusetts, but it was a big company with workers in 22 countries and $3 billion worth of annual sales. In a memo to Frank Curtin about the conversation, Henry Jr. wrote:

> He recalled our contact of 30 years ago and felt impelled to call me for old time's-sake, "In view of the fact that there will be a B&S annual meeting tomorrow and it has to be made public that Thermo is now a more than 5% shareholder in B&S."
> He told me that Thermo had an interest of "getting together" with B&S.
> I told him that I knew absolutely nothing of Thermo, its interests or the man whom, he said, heads it, one Mr. Hatsopolous, and that I certainly held nothing against them, but that, speaking now as a private individual, he should realize that I had consistently advised against the company's joining forces with other entities throughout my 50 year career at B&S.

McCabe spent a few minutes telling Henry Jr. about what a good company Thermo was, and about the way they structured their widespread businesses to make them efficient. Henry Jr. thanked him for calling, said goodbye, and immediately informed Frank Curtin, Brown & Sharpe's president, who was busily preparing for the next day's annual meeting.

Sure enough, Thermo Electron, now controlling five percent of the shares of Brown & Sharpe, announced at the meeting that it would like to acquire the company for $13.50 a share. Curtin responded that offer was "grossly inadequate" and "we do not believe any further communication on this subject is necessary."

But Thermo Electron was persistent. In late May of 1998 it offered to buy Brown & Sharpe at $15.50 a share. The answer was the same: Brown & Sharpe was not for sale. So Thermo Electron dumped its shares in the company, and went away. And then shares of Brown & Sharpe stock began a free fall. Charles Junkunc bailed out at the end the year, citing "management errors."

In 1999 the company lost $42.9 million. Streamlining operations to cut costs required initial cash outlays—it was expensive to lay people off, particularly in European countries, and it cost money to move operations from one plant to another. In September the company announced an $11 million restructuring cost; by October its stock was trading at $2.81 a share. Henry Jr. acknowledged that he was concerned about a hostile takeover. "When the stock gets as low as it is, naturally that is a worry," he told the Texas-owned *Providence Journal*.

A new year brought a new chief executive officer—Curtin was out, and Kenneth Kermes came in. Kermes had worked in a private equity investment business, and as the vice president of business and finance at the University of Rhode Island. He started work in April, and after two months of assessing the company's finances he concluded that it was time to consider selling Brown & Sharpe. The company had failed to keep the covenants on its bond and bank loans; though Brown & Sharpe had not missed a payment, it was in "technical default" brought on by poor sales, exchange rate problems, investments it had made in new technologies, writing down inventory of discontinued products, and restructuring charges. Brown & Sharpe was saddled with the complexity of a global Fortune 500 business, but did not have the corresponding cash flow to ride out a dip in sales or unfavorable exchange rates. As Kermes explained it to the *Providence Journal*: "We attempted to do too many things, pursue too many different kinds of investments over the past few years, with too few resources."

On November 17, 2000, the company's board of directors agreed in principle to sell Brown & Sharpe to Hexagon A.B., a metrology company based in Stockholm, Sweden. The price was $180 million, which eventually came to about $4 per share. Before the sale could take place there was one formality left: approval by the stockholders at the annual meeting on April 27, 2001. Henry Sharpe, Jr., now a 78-year-old veteran of World War II, arrived at Precision Park wearing his trademark bowtie, his posture habitually erect. The lone item on the agenda was whether the company should sell "substantially all the assets" of Brown & Sharpe to Hexagon, ending the company's 168-year run.

Standing outside the plant that he had built, Henry Sharpe, Jr., fished through his wallet for the worn pass card that he had long used to scan himself in. As he slid his card into the scanner it cracked into two, brittle pieces, a pair of plastic fragments useless in the palm of his hand. He peered through the dark glass door then rapped with his knuckles, locked on the outside, hoping that somebody might let him in.

Chapter Notes

Preface

1. Shannon Lee Dawdy, *Building the Devil's Empire* (University of Chicago Press, 2009.) Dawdy coined the term "rogue colonialism" in showing the common threads between the Atlantic Revolutions of the mid-eighteen century.

2. Dennis Conrad, ed., *The Papers of Nathanael Greene,* Volume XIII (Chapel Hill: University of North Carolina Press, 2006), 231. As the South Carolina legislature convened in 1784, lawyer Nathaniel Pendleton wrote to Nathanael Greene: "What they may do is uncertain, but as Major [Pierce] Butler is at the head of the Committee of ways and means; and as he has been for some time studying the *Wealth of Nations,* we may be expect the budget will produce money; but whether it be for this state, the United States, or individuals [I leave it for you] to determine from your knowledge of the people in general & that gentleman in particular." Butler later served at the Constitutional Convention.

3. Joyce Appleby, *Inheriting the Revolution: The First Generation of Americans* (Cambridge: Harvard University Press: 2000), 4. "Facing a dramatically different challenge from that of their parents, the men and women born between Independence and 1800 worked out the social forms for the new nation. They—some enthusiastically, others reluctantly—took on the self-conscious task of elaborating the meanings of the American Revolution."

4. Seth Rockman, *Scraping By* (Baltimore: The Johns Hopkins University Press, 2009), 10.

5. William Kleinknecht, *The Man Who Sold the World: Ronald Reagan and the Betrayal of Main Street America* (New York: Nation Books, 2009), xix. "America's prosperity had reached its highest levels when government activism—the legacy of Progressivism and the New Deal—was at its peak. ... And yet it was also a period of high capital formation, rising profits, rising productivity, and increasing living standards for even the poor and the middle class. In 1957, even left-leaning social critic Max Lerner could call it 'a people's capitalism.'"

Chapter 1

1. Richard Slaney, a tool historian, writes that Joseph R. Brown's understanding of gearing greatly influenced later gear theory as it evolved in America: "As a young man working with his father in a small watch and clock making shop in Pawtucket, Joseph Brown had learned the basics of gearing, an empirical knowledge gained by making gears on the workshop floor. The subject must have interested Brown because in the 1840s he set out to master a correct mathematical understanding of the theory of gears. He acquired a copy of the book *Principles of Mechanism* by Robert Willis (published London: 1841), the first text in the English language to present a unified theory of gearing. Brown used his understanding of Willis' book to offer his customers gears that were designed following the Willis system. In 1852 Brown felt accomplished enough in gear theory to publish his own short "Memorandum on Gears." He was one of the first in America to specialize in gearing and was rewarded by strong customer support, especially in the 1840s and 1850s before other mechanics took up this study. When Joseph R. Brown died in 1876, his work in gearing was carried on at B&S by Oscar Beale, another giant in the field." Author correspondence with Richard Slaney, 2016.

2. Thomas McFarlane, "Reminiscences. 1856–1877," Rhode Island Historical Society (RIHS) 826, Series I, subseries 3, Volume 1, pg. 26.

3. Journal of Lucian Sharpe, Vol. 6, 64. RIHS Library, New Box 29.

Chapter 2

1. For common school curriculum see Carl F. Kaestle, *Pillars of the Republic* (New York: Hill & Wang, 1983), 96.

2. Slaney writes that the earliest known Vernier caliper in North America was made by Horace W. Babbitt in 1846. "Babbitt was a watch repairer and

scientific instrument maker working in Providence, RI. When Babbitt died in March, 1850, Joseph R. Brown took the inventory of his estate which included mostly tools. The exact relationship between Babbitt and Brown is unknown, but the 24-inch bench Vernier caliper that Brown introduced in 1851 has many of the design features found in Babbitt's earlier example." Author correspondence with Slaney.

3. Lucian describes Louisa Dexter as "a very sensible and intelligent girl and a pleasant talker" in his journal entry of July 6, 1851. RIHS Library, New Box 29.

4. Michael Zakim, "Book-keeping as Ideology: Capitalist Knowledge in Nineteenth-Century America." *Common-place* Vol. 6 No. 3 April 2006.

5. David A. Hounshell. *From the American System to Mass Production: 1800–1932* (The Johns Hopkins University Press, 1984), 4: "The makers of machine tools worked with manufacturers in various industries as they encountered and overcame production problems relating to the cutting, planing, boring, and shaping of metal parts. As each problem was solved, new knowledge went back into the machine tool firms, which then could be used for solving production problems in other industries."

6. Bradley Stoughton, *History of the Tools Division War Production Board* (McGraw Hill, 1949), 79.

7. Hounshell, *From the American System to Mass Production*, 17–19.

8. Joseph Wickham Roe, *English and American Tool Builders: The men who created machine tools* (New York: McGraw-Hill, 1916), 205.

9. "Chronological Records of Brown & Sharpe Business and of Persons Connected Therewith." RIHS Library, MSS 826 SF ss4 B19 Vol. 1.

10. Alex I. Askaroff, http://www.sewalot.com/willcox_gibbs.htm Askaroff, who has written extensively about sewing machines, has researched and written a concise Gibbs biography on his sewalot.com website.

11. Roe, *English and American Tool Builders*, 206.

12. Amy Breakwell, "A Nation in Extremity: Sewing Machines and the American Civil War," *Textile History and the Military*, May 2010.

13. Lucian's analysis was spot on. While many European nations had banned slavery in their territories, the United States Congress and the Supreme Court had strengthened slavery's iron grip at almost every turn: from the Fugitive Slave Act of 1850, that allowed enslaved people to be captured in free states, to popular sovereignty that allowed citizens of new territories to expand slavery by popular vote beyond a previously established westward line, through the Dred Scott decision of 1857 in which the Supreme Court ruled that blacks "had no rights that the white man was bound to respect."

14. Lucian Sharpe Personal Letter Book #2, 9/20/1858-4/12/1866 BROWN & SHARPE RIHS/826/I/3/New Box 29/vol. 3, letter 198.

15. Lucian Sharpe Personal Letter Book #2, 9/20/1858-4/12/1866 BROWN & SHARPE RIHS/826/I/3/New Box 29/vol. 3, letter 196.

Chapter 3

1. Brown & Sharpe sold the clock and watch repair business to H.W. Pray on Aug. 10, 1860. "Chronological Record of Brown & Sharpe."

2. Max Holland, *When the Machine Stopped* (Boston: Harvard Business School Press, 1989,) 13.

3. L.T.C. Rolt, *A Short History of Machine Tools* (Cambridge, Mass.: The M.I.T. Press, 1965), 171–172.

4. Nathan Rosenberg, *Perspectives on Technology* (Cambridge University Press: 1976), 23. "The Brown & Sharpe sales records show, furthermore, that the largest single group of buyers of their universal milling machine was other machine tool producers. Thus, the creation of a new machine tool to solve technical problems in the production of a final product resulted in a significant source of increased productive efficiency in the machine tool industry itself."

5. Throughout that summer monthly sales plummeted from a record high of 410 in February, to just 12 that October. The next year's early returns were better, 153 in February, but still less than half the previous February's total. Duncan McDougall. "Machine Tool Output, 1861–1910," Dorothy S. Brady, ed. *Output, Employment, and Productivity in the United States after 1800* (National Bureau of Economic Research: 1966).

6. RIHS: Brown & Sharpe New Box 33 RIHS/826/L/11/folder 2.

7. *Providence Daily Journal*, December 19, 1863.

8. Amy Breakwell, "A Nation in Extremity: Sewing Machines and the American Civil War," *Textile History*, 41, May 2010, 98–107.

9. Grace Rogers Cooper, *The Sewing Machine: Its Invention and Development*, Smithsonian Institution Press (Washington, D.C.: 1976), 58.

10. Mark R. Wilson. "The Extensive Side of Nineteenth-Century Military Economy: The Tent Industry in the Northern United States During the Civil War," *Enterprise and Society 2* (June 2001): 297–337.

11. *Ibid.*, 312.

12. *Scientific American, Vol. 14* (old series), November 28, 1863.

13. Duncan McDougall, "Machine Tool Output, 1861–1910," Dorothy S. Brady, ed. *Output, Employment, and Productivity in the United States after 1800* (National Bureau of Economic Research: 1966), 505.

14. Rolt, in *Short History of Machine Tools*, employs similar language in discussing Brown's universal milling machine, 172, and his formed milling cutter, 172.

15. BROWN & SHARPE RIHS/826/I/3/New

Box 29/vol. 3 Lucian Sharpe Personal Letter Book #2, 9/20/1858-4/12/1866. Lucian forwarded the letter to E.P. Hatch at Willcox & Gibbs and included a terse note: "Gibbs has turned up—I enclose a copy of his letter which was rec'd on the 3rd [—] I shall write him soon though it is doubtful if a letter will reach him now."

16. Askaroff, http://www.sewalot.com/willcox_gibbs.htm.

Chapter 4

1. "Typed Copies, Letters from Lucian Sharpe to Louisa Sharpe, Dated 1865." Private collection. This chapter is largely reconstructed from 17 letters that Lucian wrote to Louisa while Lucian toured Europe.

2. David Wilkinson of Pawtucket invented an industrial screw-cutting lathe at about the same time as Maudslay. See Anthony Connors. *Ingenious Machinists* (Albany: Excelsior Editions/State University of New York: 2014).

3. L.T.C. Rolt. *A Short History of Machine Tools* (Cambridge, Mass. The M.I.T. Press: 1965),151.

Chapter 5

1. William A, Viall, "Reminiscence," circa 1920. RIHS/826/New Box 23 /G/1 Folders 1 & 2 of Viall, W.A.

2. Duncan M. McDougall, "Machine Tool Output, 1861–1910."

3. Philip S. Foner, *The Great Labor Uprising of 1877* (New York: Pathfinder), 23. Also: Robert Cook, *Civil War America: Making a Nation, 1848–1877* (Longman: 2003), 304.

4. Brown & Sharpe Catalogue from 1868 in *A Brown & Sharpe Catalogue collection*, 189, with introduction by Kenneth L. Cope. (The Astragal Press: Mendham, New Jersey: 1997.).

5. *Bangor Daily Whig and Courier*, April 4, 1844.

6. Henry Dexter Sharpe, "Joseph R. Brown, Mechanic, Newcomen Speech," 21.

7. B&S: Darling Medals RIHS/826/Museum Box 2/Series J/subseries 2/Bag A.

8. BROWN & SHARPE RIHS/826/I/3/New Box 29/vol. 3 Lucian Sharpe Personal Letter Book #2, 9/20/1858-4/12/1866, letter 474. March 17, 1866.

9. Henry Dexter Sharpe, "Joseph R. Brown, Mechanic, Newcomen Speech," 21.

10. Lucian Sharpe's letter to *American Machinist*, December 15, 1892, 9 and 10.

11. BROWN & SHARPE RIHS/826/new box 29/I (letter 'I')/3/volume 4 LS Personal Letter Book #3: 4/15/1866-3/1/1872 Lucian Sharpe letters to Darling.

12. Thomas McFarlane, "Reminiscenses, 1856–1877," 30, RIHS 826 Series I, subseries 3, volume 1.

13. Sewing machine sales totaled $194,387, about $3.3 million in current value, with more than a thousand machines sold every month; precision tools exceeded the 30-percent margin they had met in Lucian's first report.

14. BROWN & SHARPE SERIHS/826/new box 29/I (letter 'I')/3/volume 4 LS personal letter book #3 4/15/1866-3/1/1872.

15. Lucian Sharpe Letters to Darling.

16. William A Viall, "Reminiscences," 24. RIHS/826/newbox 23/G/1. Later photographs of Darling's grinding machines confirm Viall's observation of six-foot diameter grindstones on the machines.

17. RIHS/226/new box 22/G/1 Folder: Phillips, C.H.

18. BROWN & SHARPE New Box 22 RIHS/826/G/1/folder 1 of 1–26 Luther Burlingame Interview with Robert Dunlap.

Chapter 6

1. Robert S. Woodbury, "History of the Grinding Machine" in *Studies in the History of Machine Tools* (Cambridge, Mass: 1961 The M.I.T. Press), second section, 62.

2. BROWN & SHARPE RIHS/826/new box 29/I (letter 'I')/3/volume 4 LS personal letter book #3 4/15/1866-3/1/1872.

3. Lucian Sharpe to J.R. Brown in the letter of June 11, 1868, referenced above.

4. RIHS/826/New Box 22/G/1, folder: Oscar Beale. Unpublished manuscript, Oscar J. Beale. "Technical Contributions of Oscar J. Beale," 3.

5. RIHS 826 Series I Subseries 3, Volume 1. Unpublished manuscript, Thomas McFarlane. "Reminiscences, 1856–1877," 43.

6. Before his death in 1867, Howe controlled much of the sewing machine business; every manufacturer, including Willcox & Gibbs, had been required to pay Howe and three other companies for use of their patents on sewing machine components.

7. *New York Times*, March 6, 1884.

8. Nathan Rosenberg. *Perspectives on Technology* (Cambridge University Press: 1976), 25. "The requirements of bicycle production played a crucial role in the development of effective techniques for making ball bearings which, in turn, had an incalculable impact through reducing the effects of wear and friction on all machine processes. The highly exacting requirements of the ball bearing, as well as of the hardened cup and cone on which bicycle parts roll, necessitated grinding operations of great precision. In some cases the grinding machines which had been designed by Brown & Sharpe for grinding sewing machine needle bars were adapted for this use. The eventual solution to the grinding machine problems involved, however, as they pertained both to the bicycle and automobile, relied heavily upon the improvements in grinding from the work of men like Charles H. Norton and Edward G. Acheson,

who revolutionized grinding operations through the introduction of artificial abrasives on the grinding wheel itself."

Chapter 7

1. *Providence Journal*, October 23, 1876.
2. Candace Millard, *Destiny of the Republic* (Doubleday: 2011).
3. Duncan McDougall, "Machine Tool Output, 1861–1910."
4. McFarlane, "Reminiscences."
5. RIHS/826/newbox 23/G/1 folders 1 and 2. Unpublished manuscript, William A. Viall, "Reminiscences."
6. David A. Hounshell, *From the American System to Mass Production, 1800–1932* (Baltimore: The Johns Hopkins University Press, 1984), 81.
7. Viall, "Reminiscences."
8. RIHS/826/new box 29/I (letter "I"/volume 5 Lucian Sharpe personal letter book #4 3/1/1872-1/1/1886.
9. *Ibid.*
10. RIHS Library/New Box 23/G.
11. W.H. Van Der Voort, *Modern Machine Shop Tools: A Practical Treatise* (New York: The Norman Henley Publishing Co., 1904), 440–449.
12. Mary Sharpe, Nancy Chafee Brien, ed., "The Sharpe Family Grand Tour of Europe," Sharpe family collection.
13. Donald L. Smith, *Zechariah Chafee, Jr., Defender of Liberty and the Law* (Cambridge: Harvard University Press, 1986), 61 and *n*298. Also William G. McLoughlin, *Rhode Island: A History* (W.W. Norton, for the American Association of State and Local History, 1978), 167–168.
14. Hounshell, *From the American System.*
15. *Ibid.*, 85.

Chapter 8

1. Robert S. Woodbury, "History of the Grinding Machine," 33.
2. Welcome Arnold Greene, *The Providence Plantations for 250 Years* (Providence: J.A. & R.A. Reid, 1886), 262.
3. Re: Lucian's investment in *The Providence Journal*, Charles H. Spilman and Garrett D. Byrnes. *The Providence Journal: 150 Years*. Providence: The Providence Journal Co., 1980), 226, 228.
4. David A. Hounshell, *From the American System to Mass Production, 1800–1932* (Baltimore: The Johns Hopkins University Press, 1984), 224.
5. Jeff McVey. "A History of the Leland and Faulconer Manufacturing Company," (2004) http://vintagemachinery.org/mfgindex/detail.aspx?id=146.
6. Robert W. Rydell, "World's Columbian Exposition," Encyclopedia of Chicago. http://www.encyclopedia.chicagohistory.org/pages/1386.html.
7. "The Starrett Story," (Athol, MA: The L.S. Starrett Company, 2012), 7.
8. Reminiscences of William Viall. Viall must have heard this story second hand, as he came to work at Brown & Sharpe in 1890, 13 years after it occurred.

Chapter 9

1. Jermone L. Sternstein, "Corruption in the Gilded Age Senate: Nelson W. Aldrich and the Sugar Trust," *Capitol Studies: A Journal of the Capitol and Congress*, Volume 6, Number 1 (Spring 1978).
2. Sternstein, "Corruption," citing February 2, 1893 contract in the Aldrich Papers, RIHS Library.
3. William G. McLoughlin, *Rhode Island: A History*, 179.
4. Scott Reynolds Nelson, *A Nation of Deadbeats: An Uncommon History of America's Financial Disasters* (New York: Alfred A. Knopf, 2012), 180–207.
5. William A. Viall, "Reminiscence," 13, RIHS/826/ ne box 23/G/1/ Folders 1 and 2.
6. Patrick Meehan, "The Big Wheel," originally published in the *UBC (University of British Columbia) Engineer*, 1964, and republished in the *Hyde Park Historical Society Newsletter*, Spring 2000.
7. Viall, "Reminiscences," 19.
8. Scott Nelson, *Nation of Deadbeats*, 183. "Indeed from 1871 until World War 1, America's most important and fastest-growing export was manufactured food, mostly canned, bottled, and preserved. This export's value far exceeded finished manufactures until 1900."
9. Duncan McDougall, "Machine Tool Output, 1861–1910," 505. Sales to bicycle builders were likely higher; McDougall could not determine the eventual user of Brown & Sharpe machine tools in 37.8 percent of sales through 1904, because sales agents acted as middlemen in the transactions.
10. Spilman and Byrnes, *The Providence Journal: 150 Years*, 344. The demonstration by "the cycling army of Providence" prompted "probably the first editorial cartoon" drawn by a Journal employee, William H. Loomis, on May 13, 1897.
11. Hounshell, *From the American System*, 190.
12. *Ibid.*, 202.
13. *Ibid.*, 201 and *n*367. Hounshell notes that "Production figures for the bicycle vary greatly," with bicycling trade journals claiming much higher numbers.
14. *Ibid.*, 214, quoting from Maxim's reminiscences *Horseless Carriage Days*.
15. Susan Hale, Caroline P. Atkinson, ed., with an introduction by Edward E. Hale, *Letters of Susan Hale* (Boston: Marshall Jones Company, 1918), 329–330.
16. Woodbury, "Grinding," 97–107.
17. Daniel Nelson, *Managers and Workers: Origins of the New Factory System in the United States*,

1880–1920 (The University of Wisconsin Press, 1975), 163–164.

18. Oscar J. Beale, ed. *A Handbook for Apprenticed Machinists, Third Edition* (Providence: Brown & Sharpe Mfg. Co.: 1901), Introduction.

19. *Ibid.*, 14.

20. BROWN & SHARPE RIHS/826/new box 33/M/1/folder 14 Minutes of the meetings of the B&S Apprentice Assoc., Sept. 14, 1894.

Chapter 10

1. *The Providence Journal*, September 24, 1893, 3.

2. William Viall, letter to his son, Richmond, January 14, 1920, 13. RIHS/826/new box 23/G/1, folders 1 and 2.

3. Henry Sharpe, Lucian Jr.'s brother, told his brother about this codicil years later, when the two were locked in a dispute about Lucian's role in the company. "In father's closing days it was [Viall's] earnest expressions of confidence in you, added to Mother's entreaties and hopes on your behalf,— that induced Father to withdraw from the envelope in which his will was placed, a memorandum which was intended to seriously curtail your activity in the business."

4. *Iron Molders' Journal, Vol. 36*, 1900, 279.

5. Eight months after Lucian sailed, bales of cotton ignited on the Hoboken Piers, a conflagration that swept through four docked ships including the Staale. Crewmembers trapped below decks struggled to squirm through small port holes but could not slip through, resulting in the deaths of 391 crew. As a result of the disaster, new laws required that port holes be of sufficient diameter to allow the passage of an average-sized man.

6. *The Providence Journal*, November 21, 1909.

7. Woodbury, *Grinding Machines*, 101–102.

8. Duncan M. McDougall, "Machine Tool Output."

9. Black Metal Box with "Lucian Sharpe" painted on the side. Personal property of the Sharpe family.

10. Garrett D. Byrnes, and Charles H. Spillman, *The Providence Journal: 150 Years* (Providence, RI: The Providence Journal Co., 1980), 234–241.

Chapter 11

1. Sheila Gamble Cook, ed., *Dear Miss Hyde: A Collection of Letters of the Chafee, Sharpe, & Gamble Families 1898 to 1932* (Cambridge, MA: 1932).

2. Quotes in this and paragraphs in Chapter 11 are drawn from Sheila Gamble Cook, ed., *Dear Miss Hyde: A Collection of Letters of the Chafee, Sharpe, & Gamble Families 1898 to 1932.* (Self-published, Cambridge, Mass.: 1993.) Description of the School of Matrimonial Bliss is drawn from the same work.

3. "Chronological Record of Brown & Sharpe Business And of Persons Connected Therewith. RIHA 826 SF SS4 B19 Vol. 1.

4. Edward Shorter, "The History of Lithium Therapy," PubMed Central, U.S. National Library of Medicine, National Institutes of Mental Health. Available online at http://www.ncbi.nlm.nih.gov/pmc/articles/PMC3712976/.

5. Correspondence drawn from black metal box with "Lucian Sharpe" painted on the side. Personal property of the Sharpe family.

6. Slack work: W.A. Viall "Strike of 1915," RIHS Brown & Sharpe 826/nb4/C/1/volume 1.

7. Evelyn Savidge Sterne, *Ballots and Bibles: Ethnic Politics and the Catholic Church in Providence* (Cornell University Press: 2004), 105–107.

8. "Secretary's Book, Vol. 1, Jan. 26 1915 to April 30 1919." RIHS/826/new box 15/E/1.

9. Reminiscences of William A. Viall.

10. "Proceedings at the Strikers' Meeting, Kingsley Avenue, Providence, R.I., October 30, 1915." RIHS 826/new box 5/C/1/ folder 4.

Chapter 12

1. The first section of this chapter draws heavily from oral history interviews conducted with Mary Elizabeth Evans Sharpe by Jeannette B. Cheek in 1979 for the Schlesinger Library's "History of Women in America" project.

2. Mary Elizabeth Evans Sharpe sometimes remembered that her father died in 1892, when she was seven, but this is probably the time when he absconded from his in-laws' house for New York City. In a 1979 interview she recalled her mother's anxiety upon learning that William Evans planned to join the Klondike Gold Rush, which occurred in 1896, when Mary Elizabeth turned 12.

3. Seventy-eight years after receiving this notice, Mary Elizabeth still held onto it, reading directly from it to Jeannette Cheek in her oral history interview.

4. Unfortunately, Mary Elizabeth Evans Sharpe did not mention Harriet's last name.

5. *Post Standard*, Syracuse, August 9, 1916, quoting the *Chicago Herald*. Available at http://freepages.genealogy.rootsweb.ancestry.com/~taughannock/reigle/maryelizabethnews.html.

6. Evelyn Savidge Sterne. *Ballots and Bibles: Ethnic Politics and the Catholic Church in Providence* (Cornell University Press: 2004), 158.

7. *Memorial to the Employees of the Brown & Sharpe Mfg. Co. Who Served at Home and Abroad in the Great Work War* (Providence: circa 1919), 92.

8. John Ihlder, Madge Headley, Udetta D. Brown. *The houses of Providence, a study of present conditions and tendencies, with notes on the surrounding communities and some mill villages* (Providence: Snow & Farnham Co.: 1916), 20.

9. *The Houses of Providence*, 48.

10. Sterne, *Ballots and Bibles*, 156.

11. Erik Oberg, ed. *Modern Apprenticeships and Training Methods.* Luther D. Burlingame, "Experience of the Brown & Sharpe Mfg. Co. during the war" (New York: The Industrial Press, 1921), 103–109.

12. Mrs. Wilfred C. Leland, with Minnie Dubbs Millbrook, *Master of Precision: Henry M. Leland* (Detroit: Wayne State University Press: 1966), 171–190.

13. *Memorial to the Employees of the Brown & Sharpe Mfg. Co. Who Served at Home and Abroad in the Great Work War,* 78.

14. Robert H. Zieger, *America's Great War: World War I and the American Experience* (Rowan & Littlefield, 2000), 77.

15. *Memorial to the Employees of the Brown & Sharpe Mfg. Co.,* 92–93.

16. Jane Drummond, *Beyond the Neverland* (self-published: 2006), 55.

17. RIHS/826/New Box 23/G/Folders 1 "Viall, W.A."

18. Sheila Gamble Cook, ed., *Dear Miss Hyde: A Collection of Letters of the Chafee, Sharpe, & Gamble Families 1898 to 1932* (Cambridge, MA: 1932), 208.

Chapter 13

1. Cook, ed., *Dear Miss Hyde,* 208, 209.
2. Scott Nelson, *A Nation of Deadbeats* (New York: Alfred A. Knopf, 2012), 209.
3. Hounshell, *From the American System,* 9–10.
4. Ibid., 276.
5. Ibid.
6. Ibid., quoting *New York Times,* December 26, 1926.
7. Nelson, *A Nation of deadbeats,* 218.
8. Ibid., 222.

Chapter 14

1. A 1931 survey issued by the Massachusetts Institute of Technology asked Henry: "To what extent do you think that the stock market distracted attention from business and thus brought on the recession of retarded recovery?" He replied: "Think this had little to do with it." He blamed the Depression on "Cessation of consumer purchasing., following drop in commodity prices, partly through fear caused by the stock market conditions."

2. Jeannette B. Cheek, "Mary Elizabeth Evans Sharpe," oral history interviews for the Schlesinger Library's "History of Women in America" project, 1979.

3. Sterne, *Ballots and Bibles,* 245.
4. Gordon E. Ellis, RIHS 826/nb33/M/1 Folder 15.
5. The rest of this section is paraphrased or drawn directly from an essay "A Son's Eye View of Henry D. Sharpe, Jr.," by Henry D. Sharpe, Jr., unpublished, 1998.

Chapter 15

1. Bradley Stoughton, *A History of the Tool Division War Production Board* (New York: McGraw-Hill, 1949), 106.

2. Of 6,201 qualified voters in the machinists unit, 5,909 cast ballots with 4,586 in favor of the union, 1,272 voting against. The foundry workers unit was similarly one-sided: of 709 eligible voters, 618 cast ballots with 455 in favor and 161 against. Figures drawn from: *In the Matter of Brown and Sharpe Manufacturing company and International Association of Machinists, Affiliated with the American Federation of Labor.*

3. *The Providence Journal-Bulletin,* April 18, 1943.
4. 2015 interview with Velia Constantino.

Chapter 16

1. Dean Edgar Lanpher did not hold a grudge on this score. He and Henry Jr. corresponded frequently and in friendly fashion while Henry Jr. was in the South Pacific with the U.S. Navy.

2. Victor J. Logan, *Innovation in American Industry—Manufacturing Engineering: Evolving As Innovators During Brown & Sharpe's Second Renaissance* (Glenn Ellen, IL: Logan Consultants, 1985), 39.

3. In a coincidence worthy of Dickens, the Carbone in this scene was Aldo Carbone of Waltham, Mass., the authors' uncle. Carbone came aboard the LSM 312 at Pearl Harbor, as a motor machinist's mate, second class.

4. Max Holland. *When the Machine Stopped: A Cautionary Tale from Industrial America.* (Boston: Harvard Business School Press, 1989), 14 and 24.

5. Using a Boston post office box as a return address, CIA agents Harrison Reynolds and Lawrence Foster wrote to Brown & Sharpe seeking information about inquiries for B&S machines from countries under Russian influence.

6. Holland, *When the Machine Stopped,* 17.
7. *Review of Voluntary Agreements Program Under the Defense Production Act: Expansion of Machine Tool Industry, Report Dated February 10, 1958 by the Attorney General,* 85th Cong., 2d sess., 1958, 6. Cited by Holland. *When the Machine Stopped,* 18.
8. Holland, *When the Machine Stopped,* 23.

Chapter 17

1. *Brown & Sharpe News*/Jan 1983, Vol. 4 no. 3 RIHS/nb6/C/1/folder 31b.

2. In the 1930s Brown & Sharpe and five other machine tool companies formed a holding company called Associated Patents, Inc., which they used as a holding company to pool their patents, allowing the member companies to decide which competitors they would license to use their patents. Member firms also agreed not to directly compete in each other's specific fields. On June 30, 1955, Judge

Thornton of the U.S. District Court in Detroit ruled: "Since August of 1933, the defendants have been engaged in a combination and conspiracy in unreasonable restraint of interstate trade and commerce in machine tools in violation of Section 1 of the Sherman Act, the purpose and effect of which has been to confine the manufacture of machine tools by each of them to fields of specialization that were not competitive with each other."

3. The source of this following scene is: *The Providence Sunday Journal*, September 23, 1951; S1, 1 "A Union Member Looks at the B&S Strike."

Chapter 18

1. William F. Patterson, and M.F. Hedges, *Educating for Industry* (New York: Prentice-Hall, 1946).

2. Alfred D. Chandler, Jr., *The Visible Hand* (Cambridge and London: The Belknap Press of Harvard University Press, 1978), 1.

3. Max Holland, *When the Machine Stopped*, 38.

4. "Japan Makes Its Bid for World Machine Tool Markets," *American Machinist*, June 1, 1959; also quoted in Holland, *When the Machine Stopped*, 116.

Chapter 19

1. Advertising man Arthur Bartlett won $25 for submitting the name Precision Center. Henry Jr. changed it to Precision Park. See: John Ward. *Providence Sunday Journal*, June 5, 1966.

2. 2015 interview with Bruce MacGunnigle, East Greenwich Town Historian.

3. 2016 interview with Ed Green.

4. Holland, *When the Machine Stopped*, 84.

Chapter 20

1. Donald Roach, "Statement of Operating Strategy 1973," RIHS/820/nb1-B/ Director's Preview, February 23, 1973.

2. James H. Rigney, "The Longest Strike: A Research Study of the Brown & Sharpe-I.A.M. Strike," unpublished master's thesis, 1992.

3. *Ibid.*, 22.

4. 2013 Interview with Henry Sharpe III.

5. Bulletin Board memo, written by Rigney: "For years, conditions seemed to permit, as a practical matter, the luxury of a relatively relaxed level of plant security. Circumstances, however, are undeniably changing, and with developments of recent times we have now concluded that we must significantly improve security measures at Precision Park, in the interests of safeguarding both company and employee property." BROWN & SHARPE RIHS/826/nb6/C/1/Volume 3 "Bulletin Board Notices 1966–1994 (Incomplete.)."

6. Dan Clawson, and Mary Ann Clawson, "Foundations of the New Conservatism," published in Michael Schwartz, ed., *The Structure of Power In America: The Corporate Elite as a Ruling Class* (Holmes & Meier: New York, 1987), 205–211.

7. Interview with Henry Sharpe III.

Chapter 21

1. Duane Clinker, *Standing Together: Union Busting at Brown & Sharpe, A Photo Essay With Strike Documents*.

2. Rigney, 26. In his paper, Rigney cites "interview with Roach" as the source for information regarding the breach between Roach and Thayer.

3. This paragraph is reprinted verbatim from Powers, Whitney, & Kinder Inc. "NLRB Trial Materials (1982–1991) RIHS/826/nb5/C/1/folder.

4. "NLRB Trial Materials (1982–1991).

5. *The Providence Journal*, November 6, 1981.

6. Interview with Ed Green, 2015.

7. Interview with Chris Jocelyn, 2015.

8. Interview with Merrick Leach, 2014.

9. For causes of the collapse of the American machine tools industry see: Arthur J. Alexander. "Adaptation to Change in the U.S. Machine Tool Industry and the Effects of Government Policy." (The RAND Corporation, 1979.) Also: Artemis March. "The U.S. Machine Tool Industry and its Foreign Competitors" published in Michael L. Dertouzos, Richard K. Lester, Robert M. Solow. *Made in America: Regaining the Productive Edge* (Cambridge: MIT Press, 1989) And: Max Holland. *When the Machine Stopped: A Cautionary Tale from Industrial America*. (Boston: Harvard Business School Press, 1989).

10. "In 1982, employment in the machine tool industry fell by 20,500 workers, 21 percent of the industry's total labor force." Alexandar, RAND Report, 30.

11. "Telephone Conversation Between Dick Jocelyn and John Capobianco, September 14, 1983," transcript, NLRB Trial Book, November 1987. RIHS/826/nb5/c/1.

Chapter 22

1. Interview with Henry Sharpe, Jr., November 2013.

2. "President's Coffee Meeting No. 1, July 16, 1992," RIHS/826/nb7a/c/a/folder 48.

3. Interview with Henry Sharpe III, 2014.

4. Henry Sharpe III, "Notes on visits by H. Sharpe III to: Renens, Rolle, Leitz, Teford, Lithuania."

5. "Confidential Memo," Brown & Sharpe RIHS/826/nbA-1/A/1/folder 2a.

6. Henry Sharpe, Jr.,'s speech is preserved on a CD and transcript in the Sharpe family's possession.

7. William Kleinknecht, *The Man Who Sold the World: Ronald Reagan and the Betrayal of Main Street* (New York: Nation Books, 2009), 132–134.

8. Interview with Henry Sharpe III, 2014.

Bibliography

American Machinist Magazine, multiple issues.

Appleby, Joyce. *Inheriting the Revolution: The First Generation of Americans.* Cambridge: Harvard University Press, 2000.

Ashton, Thomas Southcliff. *Iron and Steel in the Industrial Revolution.* Manchester, England: The University Press, 1924.

Askaroff, Axel I., sewalot.com.

Atkinson, Caroline P., ed. *Letters of Susan Hale.* Boston: Marshall Jones Company, 1918.

Bangor Daily Whig and Courier, April 4, 1844.

Beale, Oscar. *A Treatise on the Construction and Use of Grinding Machines for Cylindrical and Conical Surfaces.* Providence: Brown & Sharpe Manufacturing, 1885.

Beale, Oscar. *The Brown & Sharpe Handbook: A Guide for Young Machinists.* Providence: Brown & Sharpe Manufacturing, 1945 (reprint).

Block, Fred, and Matthew R. Keller, eds. *State of Innovations: The U.S. Government's Role in Technology Development.* Boulder, CO.: Paradigm, 2011.

Boggs, Carl. *Phantom Democracy: Corporate Interests and Political Power in America.* New York: Palgrave McMillan, 2011.

Breakwell, Amy. "A Nation in Extremity: Sewing Machines and the American Civil War," *Textile History and the Military*, May 2010.

Brown & Sharpe News, multiple issues.

Buhle, Paul, and Scott Molloy, and Gail Sansbury, eds. *A History of Rhode Island Working People.* Kingston, RI: URI Printing Services, 1983.

Byrnes, Garrett D., and Charles H. Spilman, *The Providence Journal, 150 Years.* The Providence Journal Company, 1980.

Calvert, Monte. *The Mechanical Engineer in America: Professional Cultures in Conflict.* Baltimore: The Johns Hopkins University Press, 1967.

Chandler, Alfred. *The Visible Hand.* Cambridge, MA: The Belknap Press, 1977.

Cheek, Jeannette B. "Mary Elizabeth Evans Sharpe," oral history interviews for the Schlesinger Library's "History of Women in America" project, 1979.

Conrad, Dennis, ed. *The Papers of Nathanael Greene*, Vol. XIII. Chapel Hill: University of North Carolina Press, 2006.

Cook, Sheila Gamble, ed. *Dear Miss Hyde: A Collection of Letters of the Chafee, Sharpe, & Gamble Families 1898 to 1932.* Cambridge, MA: Self-published, 1996.

Cooper, Grace Rogers. *The Sewing Machine: Its Invention and Development.* Washington, D.C.: The Smithsonian Institution Press for the National Museum of History and Technology, 1976.

Cope, Kenneth L. *A Brown & Sharpe Catalogue Collection.* Mendham, NJ: The Astragal Press, 1997.

Dawdy, Shannon Lee. *Building the Devil's Empire.* University of Chicago Press, 2009.

Foner, Philip S. *The Great Labor Uprising of 1877.* New York: Pathfinder, 1977.

Friedman, Lawrence M. *A History of American Law, Second Edition.* New York: A Touchstone Book, 1985.

Gomberg, William. *A Labor Manual on Job Evaluation.* Chicago: The Labor Education Division Roosevelt College, 1947.

Greene, Welcome Arnold. *The Providence Plantations for Two Hundred and Fifty Years: The People and Their Neighbors, Their Pursuits and Progress, 1636–1886.* Providence: J.A. & R.A. Reid, 1886.

Harris, Howell John. *The Right to Manage: Industrial Relations Policies of American Business in the 1940s.* University of Wisconsin Press, 1942.

Hindle, Brook, and Steve Lubar. *Engines of Change: The American Industrial Revolution, 1790–1860.* Washington, D.C., and London: Smithsonian Institution Press, 1986.

Holland, Max. *When the Machine Stopped: A Cautionary Tale from Industrial America.* Boston: Harvard Business School Press, 1989.

Hounshell, David A. *From the American System to Mass Production 1800–1932.* Baltimore and London: The Johns Hopkins University Press, 1984.

Ihlder, John, and Madge Headley, and Udetta D. Brown. *The Houses of Providence: A Study of Present Conditions and Tendencies, with Notes on the Surrounding Communities and Some Mill Villages.* Providence, RI: Snow & Farnham, 1916.

"Interchange: The History of Capitalism." *The Journal of American History,* September 2014.

Iron Molders' Journal, Vol. 36, 1900.

Kaestle, Carl F. *Pillars of the Reublic.* New York: Hill & Wang, 1983.

Kidwell, Claudia B., and Margaret C. Christman. *Suiting Everyone: The Democratization of Clothing in America.* Washington, D.C.: The Smithsonian Institution Press for the National Museum of History and Technology, 1974.

Kleinknecht, William. *The Man Who Sold the World: Ronald Reagan and the Betrayal of Main Street America.* New York: Nation Books, 2009.

Koistinen, Paul A.C. *Arsenal of World War II: The Political Economy of American Warfare, 1940–1945.* University Press of Kansas, 2004.

Lamoreaux, Naomi R., and Damiel M.G. Raff. "Beyond Markets and Hierarchies: Toward a New Synthesis of American Business History." *The American Historical Review,* Vol. 108, No. 2, Apr. 2003.

Leland, Ottilie M., with Minnie Dubbs Millbrook. *Master of Precision: Henry M. Leland.* Detroit: Wayne State University Press, 1966.

Lipartito, Kenneth. "Reassembling the Economic: New Departures in Historical Materialism." Oxford University Press on behalf of the American Historical Association.

Logan, Victor J. *Innovation in American Industry—Manufacturing Engineering: Evolving As Innovators During Brown & Sharpe's Second Renaissance.* Glenn Ellen, IL: Logan Consultants, 1985.

Lubrano, Annteresa. *The Telegraph: How Technology Innovation Caused Social Change.* New York & London: Garland, 1997.

McDougall, Duncan. "Machine Tool Output, 1861–1910," Dorothy S. Brady, ed. *Output, Employment, and Productivity in the United States after 1800.* National Bureau of Economic Research: 1966.

McGaw, Judith A. *Early American Technology: Making and Doing Things from the Colonial Era to 1850.* Chapel Hill and London: University of North Carolina Press, for the Institute of Early American History and Culture, Williamsburg, VA, 1994.

McLoughlin, William G. *Rhode Island: A History.* New York, London: W.W. Norton & Co. for the American Association for State and Local History, 1986.

McVey, Jeff. "A History of the Leland and Faulconer Manufacturing Company." vintagemachinery.org, 2004.

Meehan, Patrick. "The Big Wheel," originally published in the *UBC (University of British Columbia) Engineer,* 1964, and republished in the *Hyde Park Historical Society Newsletter,* Spring 2000.

Memorial to the Employees of the Brown & Sharpe Mfg. Co. Who Served at Home and Abroad in the Great Work War. Providence, RI: circa 1919.

Meyer, David R. *Networked Machinists: High Technology Industries on Antebellum America.* Baltimore: The John Hopkins University Press, 2006.

Millard, Candace. *Destiny of the Republic.* Doubleday, 2011.

Neal, Larry, and Jeffrey G. Williamson. *The Cambridge History of Capitalism, Volumes I and II.* Cambridge University Press, 2014.

Nelson, Daniel. *Managers and Workers: Origins of the New Factory System in the United States.* University of Wisconsin Press, 1975.

Nelson, Donald M. *Arsenal of Democracy: The Story of American War Production.* New York: Harcourt, Brace & Co., 1946.

Nelson, Scott Reynolds. *A Nation of Deadbeats.* New York: Alfred A. Knopf, 2012.

New York Times, March 6, 1884.

Noble, David F. *America by Design: Science, Technology, and the Rise of Corporate Capitalism.* New York: Alfred A. Konpf, 1977.

Oberg, Erik, ed. *Modern Apprenticeships and Training Methods.* Luther D. Burlingame, "Experience of the Brown & Sharpe Mfg. Co. during the war." New York: The Industrial Press, 1921.

Patterson, William F., and M.H. Hedges. *Educating for Industry.* New York: Prentice Hall, 1946.

Pawtucket Times, multiple issues.

Providence Journal, multiple issues.

Rhode Island Historical Preservatioon Commission. *Warwick, Rhode Island: Statewide Historical Preservatioon Report K-W-1.* Self-published, 981.

Rigney, James H. "The Longest Strike: A Research Study of the Brown & Sharpe-I.A.M. Strike." Unpublished master's thesis, 1992.

Rockman, Seth. *Scraping By.* Baltimore: The Johns Hopkins University Press, 2009.

Roberts, Paul Craig. *The Supply-Side Revolution: An Insider's Account of Policymaking in Washington.* Cambridge and London: Harvard University Press, 1984.

Roe, Joseph Wickham. *English and American Tool Builders.* New York: McGraw-Hill, 1916.

Roediger, David R., and Elizabeth D. Esch. *The Production of Difference: Race and the Management of Labor in U.S. History.* New York: Oxford University Press, 2012.

Rolt, L.T.C. *A Short History of Machine Tools.* Cambridge, MA: MIT Press, 1965.

Rorabaugh, W.J. *The Craft Apprentice.* New York: Oxford University Press, 1986.

Rosenberg, Nathan. *Inside the Back Box: Technology and Economics.* Cambridge University Press, 1982.

Rosenberg, Nathan. *Perspectives on Technology.* Cambridge University Press, 1976.

Rydell, Robert W. "World's Columbian Exposition," *Encyclopedia of Chicago.* encyclopedia.chicagohistory.org.

Sawyer, James E. *Why Reaganomics and Keynesian Economics Failed.* New York: St. Martin's Press, 1987.

Schrank, Robert. *Ten Thousand Working Days.* Cambridge, MA: MIT Press, 1978.

Schwartz, Michael, ed. *The Structure of Power In America: The Corporate Elite as a Ruling Class.* New York, London: Holmes & Meier, 1987.

Scientific American, Vol. 14 (old series), November 28, 1863.

Scranton, Philip. *Endless Novelty: Specialty Production and American Industrialization, 1865–1925.* Princeton University Press, 1997.

Shop Theory Department, Henry Ford Trade School. *Shop Theory.* New York, London: McGraw-Hill, 1942.

Shorter, Edward. "The History of Lithium Therapy," PubMed Central, U.S. National Library of Medicine, National Institutes of Mental Health.

Sloan, John W. *The Reagan Effect: Economics and Presidential Leadership.* University Press of Kansas, 1999.

Smith, Donald L. *Zechariah Chafee, Jr., Defender of Liberty and the Law.* Cambridge: Harvard University Press, 1986.

"The Starrett Story." Athol, MA: The L.S. Starrett Company, 2012.

Steeds, W. *A History of Machine Tools 1700–1910.* Oxford at the Clarendon Press, 1969.

Sterne, Evenlyn Savidge. *Ballots and Bibles: Ethnic Politics and the Catholic Church in Providence.* Ithaca and London: Cornell University Press, 2004.

Sternstein, Jerome L. "Corruption in the Gilded Age Senate: Nelson W. Aldrich and the Sugar Trust." *Capitol Studies: A Journal of the Capitol and Congress,* Volume 6, Number 1, Spring 1978.

Stevens, Edward W., Jr. *The Grammar of the Machine: Technical Literacy and Early Industrial Expansion in the United States.* New Haven and London: Yale University Press, 1995.

Stoughton, Bradley. *History of the Tools Division War Production Board.* New York: McGraw-Hill Book Company, Inc., 1949.

Stutz, C.C. *Formulas in Gearing, with Practical Suggestions.* Providence: Brown & Sharpe Manufacturing Company, 1882.

Van Der Voort, W.H. *Modern Machine Shop Tools: A Practical Treatise.* New York: The Norman Henley Publishing Co., 1904.

Wilson, Mark R. "The Extensive Side of Nineteenth-Century Military Economy: The Tent Industry in the Northern United States During the Civil War," *Enterprise and Society* 2, June 2001.

Wittner, Lawrence S. *Cold War America: From Hiroshima to Watergate.* New York, Washington: Praeger, 1974.

Woodbury, Robert S. *Studies in the History of Machine Tools.* Cambridge, MA, and London: The MIT Press, 1972.

The Wonders of Machinery Hall: World's Columbian Exposition—Chicago 1893. Reprinted by Lindsay Publications, 2012.

Zakim, Michael. "Book-keeping as Ideology: Capitalist Knowledge in Nineteenth-Century America." *Common-place,* Vol. 6 No. 3, April 2006.

Zieger, Robert H. *America's Great War: World War I and the American Experience.* Rowman & Littlefield, 2000.

Index

Page numbers in **_bold italics_** indicate pages with illustrations.

A. & W. Sprague Company 56, 72
aerospace industry 210–11, 224, 225, 230, 247; *see also* airplanes/airplane manufacturing
AFL-CIO 105, 119, 228
African Americans: as barbers 68; as businesswomen 123, 124; as male factory workers 7, 94–95, 110–11, 143; slavery and 25–26, 254*ch*2*n*13; union desegregation and 95; *see also* white male factory workers
agricultural equipment manufacturing 140, 152
A.H. Belo company 248, 249, 250
airplanes/airplane manufacturing: flying rule for B&S business and 183, 185; Johannson Blocks and 180; Leland and 131; machine tools manufacturing for 131, 159, 180, 201, 226, 242; metrology/metrology products for 245; retooling and 140, 201; Sharpes as pilots and 144, 185; supersonic transport planes and 210–11, 215; turret drills and 200–201
Alcott, Mr. **_197_**
Aldrich, Nelson 85, 86–87, 106
Allen and Thurber Pepperbox Pistols 18, 19
American Civil War 21, 25, 26, 27, 30, 31, 32–33, 97
American Federation of Labor 105, 119, 228
American Federation of Technical Engineers 119, 221
American Industrial Revolution 17, 109–10
American Revolution 5, 253*prefacen*1–3

American (Armory) System *see* Armory (American) System
American (Standard) Wire Gauge (Brown & Sharpe Standard) 19, **_20_**, 78–79
Americanization 126, 127; *see also* citizenship; loyalty, to U.S.
Angell, William 22, 27
Anocut Engineering 210–11
anti-trust laws 152, 184, 208–9, 258*ch*17*n*2
Appleby, Joyce 5, 253*prefacen*3
apprenticeships: African Americans and 94, 143; Apprenticeship Association and 95–96, 97; executives and 216; federal programs for 160, 196, 197; HS, Jr. and 178–79; JRB and 13, 15, 94; LS, Sr. and 11, 12, 13, 14, 16, 17, 29, 94; managers and 216; standardization of 94; unions and 196, 197; for white male factory workers 94–96, 142–44, **_197_**; *see also* mentorships; vocational schools
arbitration 141, 142, 187, 216, 219; *see also* unions/union strikes
Armistice Day 134, 154–55, 187, 189
Armory (American) System: in arms manufacturing 18–19, 36, 168; assembly lines and 138, 139, **_195_**, 198–99, 217–18; B&S and 23, 24, 34, 49, 168, 193; Ford Motor Company and 139; in Great Britain 19, 36; gun bolts and 168; LS, Sr. and 23, 24; Singer Sewing Machine and 74; *see also* machine tools manufacturing and sales

arms manufacturing: Apprenticeship Association lectures on 97; Armory/American System in 18–19, 36, 168; Europe and 19, 31, 35, 36, 37, 79, 81; FWH on improvements in 28, 29–30; gear cases for anti-aircraft guns and 179, 184; LS, Sr. and 27–29, 30, 31; machine tool companies and 19, 29–30; milling machines for 28, 30, 32, 35, 168; screw machines for 81; for World War I 81, 117
Army, U.S. 29, 84, 168, 221
Asia exchange rates in 239; Korean War in 180–81, 184, 185, 186, 190, 194, 196, 208, 211–12, 215; management/systems improvements in 228; manufacturing competition in 201–2, 213, 215, 240; marketing and sales in 137, 138; outsourcing in 250; salaries for factory workers in 232; Soviet influence in 180, 190; Vietnam War in 208, 211
assembly lines 138, 139, **_195_**, 198–99, 217–18; *see also* Armory (American) System
Austin, Frederick, Jr. 184, 203, **_209_**
Australia 138, 159
automatic screw machines 84, 88, 155, 193–94, 195–96, 202, 212; *see also* screw machines
automobile manufacturing: assembly lines and 138, 139; bicycle technology and 89, 91; cash versus credit sales in 140; development of 91–92; fuel efficient engines for 224,

263

226; gear-cutting machines for 77; Johannson Blocks and 180; machine tools manufacturing for 125, 140, 201; metrology/metrology products for 245; steam engines and 91; universal grinding machine in 62, 63, 93, 255ch6n8; Wankel engine and 215, 223–24; *see also specific automobile companies*
awards and medals, for manufacturing companies 42, 45

Babbitt, Horace W. 253ch2n2
Bainton, Arthur H. 165
Bainton, Wallace 168, 173
Bank of England 136
Baptists 172, 189
barbershop hair clippers 67–68, 100, 193, 196
Beale, George 87, 88
Beale, Oscar: biographical information about 52, 76, **76**, 87; gear-cutting machines and 76, 77, 103; gear theory and 253ch1n1; handbooks/blue books as written by 76, 77–78, 95; inventions of 76, 77; machinists/mechanics and 52; metrology and 78; shop rules and 76–77, 95
belt-driven machine tools 112–13, **112–13**
Bessemer, Henry 69
bicycling/bicycle technology: automobile manufacturing and 89, 91; Centennial Exposition in 1876 and 90; description of 89, 91, 92, 256ch9n13; machine tools manufacturing in B&S and 89, 90, 256ch9n9; metalsworking manufacturing for 92; in Providence, RI 89, 256ch9n10; universal grinding machine in 63, 91, 255ch6n8
Biltmore Hotel 138
blacks *see* African Americans
blue books/handbooks 76, 77–78, 92, 95
bond market 151–52
bookkeeping systems 14, 17, 169
Boulware, Lemuel R. and "Boulwarism" 234
Boyd, Peggy 6, 189–90, **191**, 203, 206, 226
Bradley, Jesse 7
Bridgeport Brass Company 44–45
Brown, Arnold K. (AKB) 178, 181

Brown, Caroline 17, 61
Brown, David 5, 11, 15, 43–44, 52
Brown, James 35, 37
Brown, Jane Francis (JFB): estate of, JRB and 61, 65; Jane Brown Hospital and 61, 203; marriage between, JRB and 25, 49; as stockholder in B&S 66, 85, 102, 148–49; on succession in B&S 103; telephones and 68
Brown, Joseph D. 75–76
Brown, Joseph Rogers (JRB): apprenticeships and 13, 15, 94; biographical information/characteristics of 5, 9, 10, **10**, 14–15, 43–44, 45–46, 49, 53, **55**, 60–61, 63; bookkeeping system and 169; as B&S cofounder 2, 6, 7; children of 46, 49, 61, 66–67, 102, 103, 128; clock manufacturing/repairs and 5, 9, 254ch3n1; dividends paid and 46, 56; father of 5, 11, 15, 43–44, 52; gear-cutting machines and 19–20, 22, 31; gear theory and 9, 253ch1n1; graduating machines/linear graduating machines by 9–10, **14**, 14–16, 46–47; J.R. Brown, L. Sharpe, Clock and Watchmakers and Manufacturers of Small Machinery and 14; J.R. Brown Company and 11, 12–13, 14, 16–17; lathe grinders and 47, 51; legacy and donations by 61; Leland on inventions of 51, 62, 63; LS, Sr. and 2–3, 6, 7, 13, 14, 16, 45–46, 49, 94; marriages of 17, 25, 61; mathematical and nautical instruments and 5; McFarlane on 9, 53; metrology and 41, 211; and micrometer, Pocket Sheet Metal Gauge 1, 2, **2**, 44–45, **45**, 63, 78, 79; repairmen/repairs and 9, 22; rulers manufacturing and 9–10, 14, **15**, 16, 211; sewing machine manufacturing and 30; tooth (milling) cutter machine (milling cutter machine) and 31, 32, 77; turret lathes and 29–30, 193–94; universal grinding machine and 52, 60, **62**, 63, 83, 91, 161; universal milling machine and 30, **31**, 41, 60; Vernier system for manufacture of calipers and 16, 222–23, 253ch2n2; watch manufacturing/repairs and 5

Brown, Walter 61
Brown & Sharpe (B&S): awards/medals for 42, 45; Brown & Sharpe Ltd. and 198, 241; Brown & Sharpe Manufacturing Co. incorporation and 46, 158–59, 208, 209; *Brown & Sharpe News* and 228, 229; catalogs for 2, 20, 45, 78–79, 81, 83–84; competitors/mergers with 41, 42, 43; as founded in 1833 2, 6, 7; legacy of 53; as sold in 2001 6, 7, 252; *see also* Darling, Brown & Sharpe (DB&S); executive board and stockholders for B&S; finances for B&S; inventions; machine tools manufacturing and sales; male factory workers; manufacturing/manufacturing machines; partnerships; plants/factory space, for B&S; plants/factory space, in Providence, RI; Precision Park plant, in North Kingstown, RI; women; *and specific founders, executives, managers, and employees*
Brown & Sharpe Ltd. 198, 241
Brown & Sharpe Manufacturing Co. 46, 158–59, 208, 209
Brown & Sharpe News 228, 229
Brown & Sharpe Standard (Standard [American] Wire Gauge) 19, **20**, 78–79
Brown University: HS, Jr. and 172–73, 177, 184; HS, Sr. and 84, 188, 190; landscape design for 188–89; LS, Jr. donations and 105; Sharpes' home as donated to 105; sit-ins by students from 237
Builders Iron Foundry 72, 83, 109, 142
Bull, Thomas 47, 53
Burlingame, Luther D. 131, **146**
business acquisitions/subsidiaries, for B&S 198, 206, 210–11, 213, 244, 245, 247; *see also* finances for B&S
Butler, Pierce 253prefacen2

Cadillac Motor Company 51, **62**, 63, 83, 131, 146
calipers: Bridgeport Brass Company 44–45; DigitCal Micrometer 222–23, **223**; Palmer's "screw caliper" 44, **44**; Vernier system for manufacture of 16, 222–23, **223**, 253ch2n2
can manufacturing 245, 247

Index

capitalism: American Revolution cause and 5, 253*prefacen*2; double entry bookkeeping and 17; government-regulated 125–26, 128, 134, 142, 152, 154, 158, 162, 167; industrial capitalism and 6, 7–8, 17, 111; Industrial Revolution and 17, 109–10; metrology and 6; "people's capitalism" and 7–8, 253*prefacen*5; during post–American Revolution era 6; standardization and 6; time-keeping and 6; *see also* economy
Carbone, Aldo 178, 258*ch*16*n*3
Carleton, Cyrus 67, 68
Carter, Jimmy 224, 239
catalogs for machine tools, B&S 2, 20, 45, 78–79, 81, 83–84
Catholicism 76, 126, 127, 154
Cedasize grinding machine 199, 202
Centennial Exposition in 1876 64, 73–74, 75, 90
Central Intelligence Agency (CIA) 179, 258*ch*16*n*5
Chafee, Elisabeth 109, 110, 132, 135
Chafee, Henry 110
Chafee, John 131–32
Chafee, John, Jr. **205** 206
Chafee, Mary Sharpe (MSC) 72–73, 103, 104, 109–10, 114, 158
Chafee, Zechariah (II) 72, 75, 83, 85, 109, 149
Chafee, Zechariah (III) 109, 110, 131–32
Chafee, Zechariah (Sr.) 72
Chaffee, Sally 39, 43
Chamber of Commerce in Washington, D.C. 142
Chevrolet 139, 140
chief executive officers/presidents for B&S *see* presidents/chief executive officers for B&S
China 137, 180, 250; *see also* Asia
CIA (Central Intelligence Agency) 179, 258*ch*16*n*5
citizenship 126, 127, 132; *see also* loyalty, to U.S.
Civil War, American 21, 25, 26, 27, 30, 31, 32–33, 97
Civilian Conservation Corps 154
classes, social *see* social classes; social equality/inequality
clerical workers 129, 221
Clev-Co. Jig Boring Company 210, 211

Cleveland Grinding Machine Company 210, 211
Cleveland Precision Instruments 210, 211
clock manufacturing and repairs 14, 15, 19, 21, 28, 254*ch*3*n*1
CMMs (coordinate measuring machines) 210, 211, 222, 226, 244, 245, **245**, 246–47
CNC (computer numerical control) machines 212, 215, 223
Cobham, Alan 144
Code of Fair Competition 152
Colt, Samuel (Jr.) 106
Colt, Samuel (Sr.) 19
Columbian Exposition in 1893 75, 83–84, 87–88, 116
combination square 84, 250–51, 256*ch*7*n*8
computer numerical control (CNC) machines 212, 215, 223
computers CNC or computer numerical control machines and 212, 215, 223; HS, Jr. and 200, 204; software for metrology products and 244, 246–47
conglomerates and mergers 43, 208–9, 210, 247, 249
consumer price index 102, 209, 212
Cook, Osceola 68
coordinate measuring machines (CMMs) 210, 211, 222, 226, 244, 245, **245**, 246–47
Corliss, George 17, 23, 24, 64
Corliss Steam Engine Company 111
Corliss Steam Engine Works of Providence 18, 29
Corp, Henry 60–61, **62**, 67, 75–76, 79
corporate laws 152, 154, 184, 208–9
corporate taxes 161, 209
Costantino, Jovanna "Jenny" 170
Costantino, Velia 166–69, 170, 179
court cases and laws *see* laws and court cases; *and specific bills and laws*
Curtin, Frank T. 248, 251, 252
Cutting Tools Division, and B&S 198, 242

Daimler, Gottlieb 91
D'ambra, Mike 207
Darby, Abraham 38
Darling, Brown & Sharpe (DB&S): employment statistics for 52; finance, profits/losses/sales for 43, 46, 49, 56; grinding machine and 47, **48**; machinists/mechanics and 47–48; partnerships and 2, 42, 43, 53, 66, 85; plant/factory space for 47, 50; precision tools and instruments and 43, 48, 49, 84–85; rulers manufacturing and 42, 47, **48**; universal grinding machine and 63
Darling, Samuel: biographical information/characteristics of 41–43, **42**; B&S and 2, 41, 42, 43, 53, 66, 85; combination square and 84, 256*ch*7*n*8; Darling and Schwarz company and 41, 42; graduating machine/linear graduating machine invention of 2, 42, 43, 46–47; as grinding machine inventor 47, **48**; LS, Sr. and 43, 53, 85; machinists/mechanics and 52–53; McFarlane on 41; metrology and 41; *see also* Darling, Brown & Sharpe (DB&S)
Darling and Schwarz company 41, 42
Davenport, William 194
DEA company 246–47
Defense Department, U.S. 180, 197, 201, 221, 222, 224, 226, 239, 240
defense products 161, 180, 186–87, 197, 199, 201, 221, 222, 224, 226, 239, 240
depressions, economic *see* economy
Detroit, Michigan 83, 92
Dexter, Louis 54, 56
Dexter, Louisa *see* Sharpe, Louisa Dexter
dividends paid to B&S stockholders: FWH and 46, 59, 65; HS, Jr. and 200; HS, Sr. and 66, 105, 108, 138, 149, 160, 171; JRB and 46, 56; LS, Jr. and 66, 103, 105, 109, 144; LS, Sr. and 46, 56, 59, 65, 66–67, 85, 102, 103; *see also* executive board and stockholders for B&S; finances for B&S; profits/losses and sales for B&S
Double A Products Company 198, 206, 213, 215, 216
double entry bookkeeping 14, 17
Drury, J. Henry 67–68, 87, 88
Dumba, Konstantin 119
Dunlap, Robert 50, 59
Durant, Will 131

Eagle Screw (Machine) Company 17, 22
Eastern Bloc nations 139, 258*ch*16*n*5
economy: consumer price index and 102, 209, 212; corporate taxes and 161, 209; Depression of 1921 144; exchange rates and 198, 209, 213, 239, 252; Federal Reserve and 136, 140, 145, 147, 151–52, 209, 240, 245; Great Depression of 1929 to 1939 147, 149, 150, 152, 153, 155, 156, 158, 160, 163, 165, 258*ch*14*n*1; income taxes and 132, 136, 169, 170, 211–12, 213; inflation rate in U.S. and 102, 116, 170, 179, 208, 209, 212, 224, 232, 245; oil/energy crises and 224, 245; Panic of 1837 11, 56, 68; Panic of 1857 19, 22, 56, 68; Panic of 1873 56, 57–58, 57–59, 60, 68, 149; Panic of 1893 87, 88–89, 90, 93, 116, 121; during post–American Revolution era 5, 6, 253*prefacen*3; recession of 1958 199, 201; recession of 1990–1991 245; savings and loan crises in 1980s and 245; in U.S. 5–6; *see also* capitalism; dividends paid to B&S stockholders; finances for B&S; global economy; male factory workers; New York Stock Exchange (stock market); profits/losses and sales for B&S; salaries; women
Eiffel, Gustave 81
Eiffel Tower 81–82
eight-hour day/40-hour week for factory workers 118–19, 120, 128, 162, 179
Eisenhower, Dwight 180, 197, 203
electronics 179–80, 211, 215, 223
Ellsworth, Mr. **197**
employment/employees *see* male factory workers; manager/s (foremen/shop superintendents); women
England *see* Great Britain
entreprenuers, women as 122–24, 125, 127, 134
estrangement between Sharpe family, and LS, Jr. 101–2, 108, 114–16, 144, 149
ethnic immigrants: citizenship/Americanization for 126, 127; factory workers in B&S and 130, 158; in Providence, RI 110–11, 126, 204, 235; voting rights for 116, 117
Europe: arms manufacturing in 19, 31, 35, 36, 37, 79, 81; bicycling/bicycle technology in 73, 90, 91; business acquisitions/subsidiaries in 244, 245, 246–47; economies of 213; and loans from U.S. banks 140, 144, 151; LS, Jr. trips to 79, 80–82, 99–100, 127; machine tools manufacturing in 201; machine tools sales for B&S in 35, 39, 40, 49, 89, 125, 138, 159, 160; manufacturing companies in 34, 35, 73; markets for B&S in 35, 39, 40, 89, 117, 125, 138, 156, 158, 159, 179, 258*ch*16*n*5; metals-working manufacturing in 38, 88, 89, 201; telegraph communication with 40; watch manufacturing and repairs in 39; World War II in 159; World War I in 116, 119; *see also specific countries*
Evans, Fannie Riegel 121, 122
Evans, Henry 122, 123–24
Evans, Lizzie 122
Evans, Mary Elizabeth *see* Sharpe, Mary Elizabeth Evans (MEES)
Evans, William 121–22, 257*ch*12*n*2
Evening-Bulletin board 156
Excess Profits Tax 162, 171
exchange rates 198, 209, 213, 239, 252
executive board and stockholders for B&S: A.H. Belo purchase of Providence Journal Company and 248, 249, 250; apprenticeships and 216; centralization of B&S and 138; CFO or chief financial officer and 246, 248, 251; FWH and 46, 59, 65; headquarters relocation issue and 231; HS, III and 246; HS, Sr. and 66; incorporation/public stock and 46, 158–59, 208, 209, 251; JFB and 66, 85, 102, 148–49; JRB and 46, 56; LDS and 98, 102, 103; LS, Jr. and 66, 103, 105, 109, 144, 148, 149; LS, Sr. and 46, 56, 66–67, 85, 102; management-union relationship and 215, 218–19, 225, 228; mentorships for 184; MSC and 104, 114, 158; Nickerson and 102, 103, 128; professional managers and 215, 216, 218, 219; reorganization of B&S and 198; RV and 102, 104, 111, 165; sale of B&S and 6, 7, 252; secondary stock offerings and 246, 248; and secretaries, executive 102, 105, 137; sexual harassment incident and 244; social classes and 167, 185; stock ratings and 246, 248; stock ratings/value and 246, 248, 251; stock restructuring and 210, 251; stock valuation and 145, 147; succession to LS, Sr. and 79, 80, 81, 97–98, 100–104, 105, 257*ch*10*n*3; takeover of B&S and 251–52; Thermo-Electron Corporation and 251; treasurers and 102, 103–4, 111, 114, 169, 184; *see also* finances for B&S; manager/s (foremen/shop superintendents); presidents/chief executive officers; profits/losses and sales for B&S; vice presidents for B&S
exhibitions for machine tools manufacturing and sales: Centennial Exposition in 1876 64, 73–74, 90; Columbian Exposition in 1893 75, 83–84, 87–88, 116; HS, Sr. on 155; International Machine Tool Exposition in 1960 201–2; International Machine Tool Exposition in 1976 223; National Machine Tool Builders' Association show 155; Paris Exposition in 1889 79, 80, 81; Paris International Exhibition in 1867 44

factory space, for DB&S 47, 50; *see also* plants/factory space, in Providence, RI
factory workers *see* male factory workers; manager/s (foremen/shop superintendents); women
Fairness Doctrine 249
Faulconer, Robert C. 82, 83, 92
FCC (Federal Communications Commission) 249
Federal Communications Commission (FCC) 249
federal government: apprenticeship program offered by 160, 196, 197; economic policies in 7, 86, 212–13, 225, 231, 239, 241, 243, 249; financial support from B&S

and 133–34; New Deal and 154, 155, 169, 253*prefacen*5; overseas plants programs and 197–98; pool order system during wartime and 161–62, 180; regulation of capitalism by 125–26, 128, 134, 142, 152, 154, 158, 162, 167; steel manufacturing priorities in 160–61; *see also* United States; *and specific legislation*
Federal Housing Authority (FHA) 154
Federal Reserve 136, 140, 145, 147, 151–52, 209, 240, 245
Federal-Aid Highway Act of 1956 201
FHA (Federal Housing Authority) 154
finances for B&S: business acquisitions/subsidiaries and 198, 206, 210–11, 213, 244, 245, 247; Custom Metrology group and 247; economic support for U.S. and 133–34; in global economy 137, 215, 216, 228, 232, 246; Hydraulics Division and 198, 206, 215, 246; incorporation of Brown & Sharpe Manufacturing Co. and 46, 158–59, 208, 209, 251; Measuring Systems group and 246–47; Panic of 1857 and 19, 68; Panic of 1873 and 56, 57–58, 60, 68, 149; Panic of 1893 and 88–89, 90, 93, 116, 121; Precision Measuring Instruments group and 247; professional managers and 215, 216, 218, 219; Pumps Group and 198, 206, 246; recession of 1958 and 199, 201; Roach and 215, 216, 218, 219, 223, 224, 225–26, 228; secondary stock offerings and 246, 248; stock ratings and 246, 248; stock restructuring and 210, 251; stock valuation and 145, 147; *see also* dividends paid to B&S stockholders; economy; executive board and stockholders for B&S; profits/losses and sales for B&S; salaries
finances, profits/losses/sales for DB&S 43, 46, 49, 56
Finnell, George 120
flying rule, for B&S business 183, 185; *see also* airplanes/airplane manufacturing
food market, manufactured 89, 256*ch9n*8

Ford, Henry 93, 138–40, 193
Ford Motor Company: Armory/American System and 139; assembly lines and 138, 139; cash versus credit sales and 140; Ford Model A and 139–40; Ford Model T and 93, 138–39; Johannson Blocks and 180; machine tools manufacturing for 140, 201; metrology/metrology products for 245
foremen/shop superintendents (manager/s) *see* manager/s (foremen/shop superintendents)
foster families' support 144
foundry, in Providence, RI 7, 68, 69, **70**, **71**, 71–72, 78, 163, 173–74, 179, **194**, 220, 231; *see also* plants/factory space, in Providence, RI
France: arms manufacturing in 79, 81; bicycling/bicycle technology in 73; B&S machinists in 117, 159; B&S plants in 247; Eiffel Tower in 81–82; JRB's trip to 44, 49; LS, Sr.'s trips to 38–39, 44, 49, 79, 80–82; machine tools sales for B&S in 49; manufacturing companies in 39, 73, 79, 81; markets for B&S in 89, 117, 156; Paris Exposition in 1889 in 79, 80, 81; Paris International Exhibition in 1867 in 44; supersonic transport planes and 211; World War I and 116; World War II in 159
Francis, Jane *see* Brown, Jane Francis
Fraser, Maybury 187–88, 189
Freer, Robert 103, 107

Gamble, Elisabeth Chafee 109, 110, 132, 135
Gamble, Jim 132, 136
gear theory 9, 253*ch1n*1
gear-cutting machines 19–20, 22, 31, 76, 77, 103
General Motors (GM) 131, 139, 140, 245
George-Dean Act of 1937 196
Germany: automobile manufacturing and 89, 91; business acquisitions/subsidiaries in 244, 245, 247; in global economy 213, 215; and loans from U.S. banks 140, 151; LS, Sr.'s trips to 98–100; machine tools manufacturing in West 201, 213, 218; markets for

B&S in 89, 117; metals-working manufacturing and 88, 89; ships/ship manufacturing in 102, 257*ch10n*3; World War I and 116; World War II and 159
Geyer, Howard 230
Gibbs, James E.A. **21**, 21–22, 24, 25, 26, 32–33, 53, 60
global economy: description of 7, 8, 137; finances for B&S and 137, 215, 216, 228, 232, 246; Japan in 213, 215; NC or numerical control machines in 240; oil/energy crises and 224, 245; Trade Agreement Act of 1979 and 240; West Germany in 213, 215; *see also* economy
GM (General Motors) 131, 139, 140, 245
Goodrum, Thomas ("the Silk Hat machinist") 51, 52, 53, 55, **55**
Gordon, John 229, 230, 231
Goss, J.E. 142–43
government-regulated capitalism 125–26, 128, 134, 142, 152, 154, 158, 162; *see also* capitalism; federal government; United States
graduating machines/linear graduating machines: Darling's invention of 2, 42, 43, 46–47; JRB's invention of 9–10, **14**, 14–16; for ruler manufacture 9–10, 14, **15**, 16, 52–53, 211; steam engines and 22
Gray, Rockwell 169
Great Britain: American Civil War and 36; Armory/American System in 19, 36; arms manufacturing and 19, 31, 36, 37; Bank of England in 136; bicycling/bicycle technology in 90, 91; Brown & Sharpe Ltd. plant in 198, 241; B&S's plants in 198, 241; Custom Metrology group in 247; exchange rates for pound and 198; iron manufacturing in 38; LS, Jr.'s trips to 99–100; LS, Sr.'s trip to 34–38, 72–73; manufacturing companies in 19, 31, 36, 37, 38, 73; markets for B&S in 89, 117, 125, 156; metals-working manufacturing in 38; metrology/metrology products and 245; Panic of 1873 in 60; sewing machine manufacturing in 37; social classes in 37; steel manufacturing in 38;

supersonic transport planes and 211; telegraph communication with 40; World War I and 116, 136; World War II in 159
Great War *see* World War I
Green, Ed 207–8, 212, 233
Green, Theodore Francis 154–55
Greene, Nathanael 5, 253-*prefacen*2
Greene, Welcome Arnold 78
grinding machine/s: Cedasize grinding machine and 199, 202; Darling as inventor of 47, **48**; FWH as inventor of 49; JRB as inventor of lathe grinder and 47, 51; universal grinding machine and 52, 60, **62**, 63, 69, 83, 91, 92, 93, **161**, 255*ch*6*n*8

hair clippers, barbershop 67–68, 100, 193, 196
Hall, John 47, 60
handbooks/blue books 76, 77–78, 92, 95
Harriet (African American business woman) 123, 124
Harvard University 14–15, 132, 174, 215, 216
Hexagon A.B. 252
Hill, Thomas, and machine shops 11, 12, 17
holidays, for factory workers 134, 154–55, 170, 187, 189
Holland, Max 208–9
Hook Richard 216
Hoppin, William A. 106–7
horology, and capitalism 6, 16
horse hair clippers 50, 67
Houghton, Donald M. 211–12
Hounshell, David A. 254*ch*2-*n*5, 256*ch*9*n*13
household management, and women 109–10
Howe, Elias 56, 255*n*6
Howe, Frederick W. (FWH): arms manufacturing improvements and 28, 29–30; biographical information/characteristics of 28–29, 46, 54, 59; dividends received by 46, 59, 65; grinding machine invention and 49; LS, Sr. and 9, 46, 59, 60, 65; machinists/mechanics and 52–53, 59, plant/factory space design/construction by 54–55, 194; Providence Tool Company and 29, 46; sewing machine improvements by 28; turret lathe and 29, 193–94

Howe Machine Company of Bridgeport 56–57, 58–59, 65
Hydraulics Division 198, 206, 215, 246

IAM (International Association of Machinists) 162–64, 218, 221, 225, 230; *see also* machinists' strike of 1981–1997
immigrants, ethnic *see* ethnic immigrants
income taxes 132, 136, 169, 170, 211–12, 213
industrial capitalism 6, 7–8, 17, 111; *see also* capitalism
Industrial Products Division 198, 208, 230
Industrial Relations department 218, 224, 228, 229, 230
Industrial Revolution, American 17, 109–10; *see also* capitalism; labor history
inflation rate, in U.S. 102, 116, 170, 179, 208, 209, 212, 224, 232, 245
insurance plans for factory workers 115, 128–29, 135
interior design 145, 151, 189–90, 206
International Association of Machinists (IAM) 7, 95, 119–20, 162–64, 218, 221, 225, 230; *see also* machinists' strike of 1981–1997
International Association of Machinists and Aerospace Workers 225, 230; *see also* International Association of Machinists; machinists' strike of 1981–1997
International Cooperation Administration 197–98
International Machine Tool Exposition in 1960 201–2
International Machine Tool Exposition in 1976 223
internships *see* apprenticeships; federal government; vocational schools
inventions: American (Standard) Wire Gauge 19, **20**, 78–79; B&S's legacy and 53; Cedasize grinding machine 199, 202; combination square 84, 256*ch*7*n*8; gear-cutting machines 19–20, 22, 31; gear theory and 9, 253*ch*1*n*1; grinding machines 47, **48**, 49; lathe grinders 47, 51; micrometer, Pocket Sheet Metal Gauge 1, 2, **2**, 44–45, **45**, 63, 78, 79; rulers manufacturing and 9–10, 14, **15**, 16, 211;

tooth (milling) cutter machine (milling cutter machine) 31, 32, 77; turret lathes 29–30, 193–94; universal grinding machine 52, 60, **62**, 63, 83, 91, 161; universal milling machine 30, **31**, 41, 60; Vernier system for manufacture of calipers 16, 222–23, **223**, 253*ch*2*n*2; *see also* graduating machines/linear graduating machines; metrology/metrology products; *and specific inventors*
investment casting 198–99
Iranian Revolution in 1979 224
Iraq invasion in Kuwait 245
iron manufacturing 38, 46, 56, 59, 72, 83, 109, 142
Italy 156, 245, 246–47

Jane Brown Hospital 61, 203; *see also* Brown, Jane Francis
Japan: 201–2, 213, 215, 228, 232, 239, 240; *see also* Asia
Jocelyn, Christine 235, 239
Jocelyn, Dick (DJ): as contract negotiator during strike of 1981–1997 229, 230, 231, 232, 233, 236, 242; on global economy 232; Industrial Relations department and 218; on profits/losses and sales for B&S 232, 242; on replacement workers 235; on violence during union strike 234, 237
Johannson Blocks 180
J.R. Brown & Sharpe (JRB&S) *see* Brown, Joseph Rogers (JRB); Brown & Sharpe (B&S); Sharpe, Lucian, Sr. (LS, Sr.)
J.R. Brown Company 11, 12–13, 14, 16–17; *see also* Brown, Joseph Rogers (JRB)
J.R. Brown, L. Sharpe, Clock and Watchmakers and Manufacturers of Small Machinery 14
Junkunc, Charles 246, 248, 251

Kane, Joseph 185, 218
Kermes, Kenneth 252
Keynes, John Maynard 154
Koehler, Florence 145
Korean War 180–81, 184, 185, 186, 190, 194, 196, 208, 211–12, 215; *see also* Asia

labor history 7, 17, 109–10; *see also* laws; machinists' strike of 1981–1997; male factory

workers; unions/union strikes; women
Labor Reform Bill 225
landscape designs 187–89, 205, 238
Lanpher, Edgar J. 173, 258-*ch*16*n*1
lathes: lathe grinders 47, 51, 60; screw-cutting lathe 37, 255*ch*4*n*2; turret lathes 29–30, 193–94, 195
laws and court cases: anti-trust 152, 184, 208–9; Armistice Day holiday and 154–55, 187, 189; corporate 152, 154, 155, 184, 208–9; labor 125–26, 128, 134, 142; machinists' strike of 1981–1997 lawsuits and 238, 240–41, 249–50; *see also specific bills and laws*
Leach, Merrick 236–37
Leitz 244, 247
Leland, Henry M.: airplane manufacturing and 131; as Cadillac Motor Company founder 51, **62**, 63, 131, 146; on, JRB 51, 62; as manager/foreman/shop superintendent 63, 66, 76; metalsworking manufacturing and 83, 92; pay rates set by 66; photograph of **146**; piecework/scientific management and 66, 82, 93, 126, 217; production systems and 82; shop rules and religion and 75–76; on universal grinding machine 63; World War I hawk-dove disputes and 131
Lerner, Max 253*prefacen*5
Lincoln, Abraham 25, 26, 30
Lincoln Motor Company 131
Lindbergh baby's kidnapping 150–51
linear graduating machines/graduating machines *see* graduating machines/linear graduating machines
Lithuania 244, 247
Logan, Victor J. 173, 200
Longstreth, Morris 113, 115
Lovett, Raul 236, 240, 241
loyalty, to U.S. 127, 132; *see also* Americanization; citizenship
Lyon, Elijah 69

Macaroni Riots in 1908 116, 117
Machine Tools Division of B&S 198, 208, 215, 230, 246
machine tools manufacturing and sales: Armory/American System and 23, 24, 34, 49, 168, 193; assembly lines and **195**, 195–96, 198–99, 217; catalogs for 2, 20, 45, 78–79, 81, 83–84; collapse in U.S. of 239–40, 243, 259*ch*21*n*10; computers and 200, 204; corporate tax deductions for 161; Excess Profits Tax and 162, 171; factory conditions for 141, **141**; foundry equipment and 71, 79; gear theory and 9, 253*ch*1*n*1; oil/energy crises and 224, 245; plant/factory space for 50–51, 100, 107–8, 138, 144; pool order system and 161–62, 180; repairmen/repairs and 19, 82, 183; retooling and 140, 193, 195, 196, 201; salesmen for 82, 87, 88, 199–200; scrap metal collection and **163**; specialization versus diversification in 240; subcontracts for 241; surplus during postwar years for 137, 179, 197; technological convergence and 17–18, 29, 30, 254*ch*3*n*4, 254*ch*2*n*5; technological innovations in 11, 179–80, 212, 215; *see also* executive board and stockholders for B&S; exhibitions for machine tools manufacturing and sales; finances for B&S; global economy; male factory workers; manager/s (foremen/shop superintendents); manufacturing/manufacturing machines; marketing and sales; metrology/metrology products; plants/factory space, for B&S; plants/factory space, in Providence, RI; Precision Park plant, in North Kingstown, RI; precision tools and instruments; profits/losses and sales for B&S; women; *and specific countries, industries, and machine tools*
machinists/mechanics: DB&S and 47–48; federal apprenticeship program for 160; in France 117, 159; in labor history 7; for machine tools 52–53; photograph of **161**; salaries for 51, 108, 224; women as **130**; *see also* machine tools manufacturing and sales; male factory workers; unions/union strikes; *and specific unions*
machinists' strike of 1981–1997: B&S Negotiating Committee and 229, 231, 232, 233; B&S's direct communication with factory workers and 232, 233–35; contract negotiations with 229, 230–35; DJ as contract negotiator for B&S during 229, 230, 231, 232, 233, 236, 242; economic pressure strategy of union during 241; Greystone plant and 232, 237, 242; headquarters relocation issue and 231; job preference/seniority issue and 229–30, 231, 232, 233, 235, 242; job transfers issue and 230, 231, 232, 233, 234–35, 242; lawsuits/court cases and 238, 240–41, 249–50; lock-in strategy for B&S during 238–39; managers as union factory workers during 239; mass media campaign during 235; NLRB and 230, 234, 240, 241, 242, 249; pension plans and 232–33; picketers/picket lines during 234, 236–38, 239; profits/losses for B&S and 236, 241; replacement workers/scabs and 235–36; salaries and 232–33; settlement agreement for 242; sit-in during 237, **238**; terms and settlement strategies for B&S during 235–36; Thayer and 232, 233, 234, 236, 239, 242; unemployment compensation during 234; violence during 234, 236–37, 238, **238**; *see also* machinists/mechanics; male factory workers; unions/union strikes
Machon, John 196–97
male factory workers: African Americans and 7, 94–95, 110–11, 143; American Federation of Labor and 105; apprenticeship programs for 94–96, 97, 142–44, **197**; Armistice Day holiday and 134, 154–55, 187, 189; citizenship/Americanization and 126; closed shop clause and 164–65; contract negotiations and 164, 165, 184; corporate laws and 152, 154, 162; eight-hour day/40-hour week and 118–19, 120, 128, 162, 179; eight-hour day/40-hour week for 118–19, 128, 162, 179; employment statistics for 41, 52,

107, 113, 128, 155, 160, 161, 184, 190, 194, 196, 246; ethnicities and 130; flying rule for B&S business and 183; in foundry 173–74; furloughs/shut-downs and 219–20; handbooks/blue books and 76, 77–78, 92, 95; health benefits for 170, 186, 225, 242; insurance plans and 115, 128–29, 135; insurance plans for 128–29, 135, 141; labor laws and 125–26, 128, 134, 142, 162; layoffs for 116, 138, 149, 152, 190, 196–97, 211, 212, 220; loyalty for U.S. citizens and 132–33, *133*; machinist strike of 1981–1997 and 231, 235, 238, 240–41; management/systems improvements for 93, 94; mentorships for 11, 12, 16–17, 111, 116; molders' strike and 99; pension plans and 135, 141–42, 170, 220, 225; progressivism and 128, 134; salaries for 51, 66, 99, 108, 140, 155, 165–66, 169, 170, 179, 187, 208, 220, 221; sexual harassment incident and 107; shop rules for 75–76, 95, 183; social class and 120, 126; unemployment compensation for 150, 155, 169, 234; unions/union strikes and 118–20, 134, 142, 155, 162–64, 258*ch14n2*; War Bonds and 170, 171; whites as 7, 8, 94, 143, 204; workplace shooting among 107; World War I recruiters and 117–18; *see also* finances for B&S; machinists' strike of 1981–1997; manager/s (foremen/shop superintendents); unions/union strikes; women
manager/s (foremen/shop superintendents): apprenticeships and 216; Leland as 63, 66, 76; LS, Sr. on 66; professional managers and 215, 216, 218, 219; RV as 53, 65, 69, 75, 111, 116; social classes and 167, 185; as union factory workers during machinists' strike of 1981–1997 239; union relationship with 215, 218–19, 225, 228; WV as 129; *see also* executive board and stockholders for B&S; male factory workers; production systems; women
Manchaug Mill Co. 56

manufactured food market 89, 256*ch9n8*
manufacturing companies *see specific companies, and countries*
manufacturing/manufacturing machines: during American Civil War era 28, 30; apprenticeships and 11, 12, 13, 14, 16, 17, 29, 94; awards/medals for 42, 45; bookkeeping systems and 14, 17, 169; development of 19; management and systems for 93–94; molders and 99; piecework/scientific management and 66, 82, 126, 128, 167, 217–18, 230; repairmen/repairs and 9, 19, 22; technological convergence and 17–18, 29, 30, 254*ch3n4*, 254*ch2n5*; *see also* Armory (American) System; machine tools manufacturing and sales; metrology/metrology products; *and specific companies, countries, industries, inventions, and machines*
Marine Corps, U.S. 127, 172
marketing and sales: in Asia 138, 228; cash versus credit sales and 140; in Europe 35, 39, 40, 49, 89, 117, 125, 138, 156, 158, 159, 160, 179, 258-*ch16n5*; for Singer Sewing Machine versus Willcox & Gibbs 73–74; *see also* machine tools manufacturing and sales
Martin, Billy 232, 234, 238
Mason, Owen 11, 12, 16–17
mass media 235, 248–49; *see also* print media
mass production, and assembly lines 138, 139, *195*, 198–99, 217–18; *see also* Armory (American) System
Masser, William 230
Maudslay, Henry 37, 255*ch4n2*
Mauser company 245
Maxim, Hiram Percy 91
McCabe, Bob 251
McDougall, Duncan 256*ch9n9*
McElroy, Ed 228
McFarlane, Thomas: biographical information/characteristics of 46, 52, 65; B&S and 46, 52, 56, 65; on CW 53; on Darling 41; on FWH 59; on Gibbs 53; on, JRB 9, 53; LS, Sr. and 9, 65; on plant/factory space design/construction 54; on Stockwell 56, 65
McMurtry, David 222, 226

Measuring Systems group 246–47
medals and awards, for manufacturing companies 42, 45
media, mass 235, 248–49; *see also* print media
men, and workers at B&S *see* male factory workers; women
mentorships 11, 12, 16–17, 111, 116, 184; *see also* apprenticeships
mergers and conglomerates 43, 208–9, 210, 247, 249; *see also* partnership/s:
metals-working manufacturing: bicycling/bicycle technology and 92; in B&S 52, 88, 97, 99; Builders Iron Foundry 72, 83, 109, 142; in Detroit, Michigan 83, 92; in Europe 38, 88, 89; iron manufacturing and 38, 46, 56, 59, 72, 83, 109, 142; scrap metal collection for *163*; steel manufacturing and 38, 69, 71–72, 74, 160–61
Metcalf, Jesse 106, 109, 169
Metcalf, Louisa Sharpe 106, 109, 145, 169
Metcalf, Michael 248
Metcalf, Stephen O. 106–7
methodology and resources, for history of B&S 6, 7
metrology/metrology products: for airplanes/airplane manufacturing 245; for automobile manufacturing 245; Beale and 78; B&S and 211, 244, 246–47; capitalism and 6; definition of 16; Germany and 244, 245; JRB and 41, 211; profits/losses and sales of 244; software for 244; standardization across technologies/currencies/cultures and 247; systems for 244–45; technological development of 244, 245, 246, 247; Validator Process Control Robot and 211, 244
Metronic company 245
micrometer/s: DigitCal Micrometer and 222–23, *223*; Pocket Sheet Metal Gauge and 1, 2, *2*, 44–45, *45*, 63, 78, 79, 186; and Pocket Sheet Metal Gauge micrometer 1, 2, *2*, 44–45, *45*, 63, 79, 186
milling machines: for arms manufacturing 28, 30, 32, 35, 168; assembly lines and 217; motor-driven versus belt-driven *112*–*13*; Numericam,

tape-controlled cam milling machine and 200, 202; redesign of 111; tooth (milling) cutter machine (milling cutter machine) and 31, 32, 77, 111; women factory workers and *130*; *see also* universal milling machines
missile manufacturing 199, 202, 204
molders 99
motor-driven machine tools 112–13, *112–13*
Mushet, Robert 69

National Cash Register Company 208, 212
National Industrial Recovery Act 152, 154
National Labor Relations Board (NLRB) 154, 163–64, 230, 234, 240, 241, 242, 249
National Metal Trades Association 120
Navy, U.S. 171, 172, 173, 174, 175, 176–78, 184
NC (numerical control machines) 200, 202, 204, 212, 240
Nelson, Scott 256ch9n8
New Deal 154, 155, 169, 253prefacen5
New England Butt Company 22, 24, 50, 52, 55–56
New York Stock Exchange (stock market): corporate taxes/tax-free stock swaps and 209; crash in 1929 and 149, 151, 258ch14n1; crash in 1980s and 245; incorporation of Brown & Sharpe Manufacturing Co. and 158–59, 208, 209, 251; stock valuation and 145, 147
Nickerson, Lyra 61, 66–67, 85, 102, 103, 128
Nixon, Richard M. 211, 212–13, 225
NLRB (National Labor Relations Board) 154, 163–64, 230, 234, 240, 241, 242, 249
North Carolina plant/factory space 231
Norton, Charles H. 82–83, 92, 93, 255ch6n8
Norton, Franklin 93
Norton Grinding Company 93
nuclear power industry 206, 211
numerical control machines (NC) 200, 202, 204, 212, 240
Numericam, tape-controlled cam milling machine 200, 202, 223

oil/energy crises 224, 245
Onorato, Roland 203
outsourcing and subcontracts 241, 250–51
Owen, George, Jr. 97

Palmer, Jean Laurent 44
panics, economic *see* economy
Parks, E.H. 52, 83
Parmly, J. 57, 58
Parsons, John C. 200
partnership/s: Darling and B&S 2, 42, 43, 53, 66, 85; LS, Sr. and, JRB 2–3, 6, 7, 13, 14, 16; mergers or conglomerates and 43, 208–9, 210, 247, 249
pension plans 135, 141–42, 170, 225
"people's capitalism" 7–8, 253prefacen5; *see also* machinists' strike of 1981–1997
Perreault, Edmond J. 206
Perry, Albert Q. 185–87
Perry, Marsden 86, 87, 106
Peters, Amey Sharpe 73, 109
Peters, William 109
Phillips, C.H. 49, 50
piecework/scientific management 66, 82, 93, 111, 126, 128, 167, 217–18, 230
plant/factory space, for DB&S 47, 50
plants/factory space, for B&S: in North Carolina 231; overseas 198, 231, 241, 244, 245, 247; in Poughkeepsie, New York 231; *see also* plants/factory space, in Providence, RI; Precision Park plant, in North Kingstown, RI
plants/factory space, in Providence, RI: design and construction of 54–55, *71*, 194, *194*; foundry and 7, 68, 69, *70*, *71*, 71–72, 78, 163, 173–74, 179, *194*, 220, 231; as gas-lit crowded shop 14, 30, 41, 47; Greystone plant and 183, 184, 186, 232, 237, 242; LS, Sr. and 50–51, 53–58, *55*, *71*, 78, 93, 99–100; for machine tools manufacturing and sales 50–51, 99–100; RV on 53–54, 100; sale of 206; *see also* Precision Park plant, in North Kingstown, RI
Pleasant Valley Land Company 138
pool order system 161–62, 180
Pope, Albert 90, 91

Poughkeepsie, New York 231
Powers, William 240–41
Pratt & Whitney 91, 142, 180, 186, 245
Preble, O.L. 119
Precision Measuring Instruments group 247
Precision Park plant, in North Kingstown, RI: groundbreaking ceremony for 204–5, *205*; leasing of 246; Measuring Systems group and 246–47; move to 204–7; naming of 204, 259ch19n1; security measures for 221–22, 252, 259ch20n5; *see also* plants/factory space, in Providence, RI
precision tools and instruments: aerospace industry and 211; automobile manufacturing and 211; barbershop hair clippers and 67–68, 100, 193, 196; combination square and 84, 250–51, 256-ch7n8; Darling and Schwarz company and 41, 42; DB&S and 43, 48, 49, 84–85; electronics and 211, 215, 223; horse hair clippers and 50, 67; LS, Sr. and 41, 43, 68; metrology and 211; and micrometer, Pocket Sheet Metal Gauge 1, 2, *2*, 44–45, *45*, 63, 78, 79, 186; nuclear power industry and 206, 211; outsourcing manufacture of 250–51; universal grinding machine for 63; Validator CMM for 211; *see also* machine tools manufacturing and sales
Precizika, Lithuania 244, 247
presidents/chief executive officers for B&S: Curtin 252; DAO 222, 227; HS, Jr. 181, 222, 248; HS, Sr. 105, 108, 140; Kermes 252; Roach 222, 227; salaries for 108, 140; Stuber 245, 246, 247
print media *Brown & Sharpe News* and 228, 229; *Evening-Bulletin* board and 156; Providence Journal Company and 79, 102, 105–6, 105–7, 108, 109, 156, 190, 248, 249, 250
production companies *see specific companies*
production systems: assembly lines and 138, 139, *195*, 198–99, 217–18 (*see also* Armory (American) System); improvements in 28, 29–30,

93–94, 228; piecework/scientific management and 66, 82, 93, 126, 128, 167, 217–18, 230
professional managers 215, 216, 218, 219; *see also* manager/s (foremen/shop superintendents);
profits/losses and sales for B&S: business acquisitions/subsidiaries and 247; DJ on 232; HS, Jr. and 187, 197, 208, 211–12, 223, 224, 225–26; HS, Sr. and 105, 116, 149, 152, 153, 155, 156, 160, 162, 171, 179; LS, Sr. and 27–28, 30, 32, 43, 45, 46, 49, 56, 59, 60, 65, 67, 68, 79, 103, 255-*ch5n*13; machinists' strike of 1981–1997 and 236; metrology products and 244; restructuring and 251; Stuber and 247–48; *see also* dividends paid to B&S stockholders; finances for B&S
Progressive era 110, 134
progressivism, and B&S 115, 128, 134
Providence, RI: bicycling in 89, 256*ch9n*10; cost of living in 170; ethnic immigrants in 110–11, 126, 204, 235; history of 5, 11, 50; landscape design in 238; population statistics for 50; technological convergence in 17–18, 111, 254*ch2n*5
Providence Gas Company 102, 108, 190, 249
Providence Journal Company 79, 102, 105–6, 105–7, 108, 109, 156, 190, 248, 249, 250
Providence Machine 11, 29
Providence Public Library 61, 128
Providence Tool Company 29, 30, 31, 46, 249
Pumps Group, and B&S 198, 206, 246

railroads 41, 49, 56, 138, 140
Rathom, John 106
Reagan, Ronald 7, 225, 236, 239
Red Cross 132, 133, 134, 144–45
religious beliefs: Baptists and 172, 189; Catholicism and 76, 126, 127, 154; Jews and 158; Second Adventists 76; shop rules and 75–76; Swedenborgians and 43; Unitarians and 39, 102, 206

Renishaw Probe 222, 226, 244–45
repairmen/repairs: clock manufacturing and 14, 15, 19, 21, 28, 254*ch3n*1; JRB and 9, 22; LS, Sr. and 19; machine tools manufacturing and sales and 19, 82, 183; watch manufacturing and 5, 14, 19, 21, 28, 39, 254*ch3n*1; *see also* clock manufacturing and repairs; watch manufacturing and repairs
Rhode Island Historical Society 6
Rhode Island Hospital Trust Co./National Bank 102, 108, 128, 137–38, 190, 241, 248, 249
Rhode Island School of Design 65, 128, 160, 189, 190, 206
Richards, Charles **146**
Rider, Sidney 34, 35
Riegel, Fannie 121, 122
Riegel, Henry 121, 123
Riegel, Polly 122
rifles, Springfield 29, 97, 168
Rigney, James H.: on furlough/shut down of plant 220; Industrial Relations department and 218, 224, 228, 229; management-union relationship and 225, 228; on picketers shed 237; on professional managers 219; on security measures 221, 259*ch20n*5; union-management dispute settlements and 218, 219
Roach, Donald A.: business acquisitions/subsidiaries and 244, 247; characteristics of 215; CMMs or coordinate measuring machines and 222; DigitCal Micrometer and 222–23; Double A Products Company and 215, 216; educational achievement of 215, 216; finances for B&S and 215, 216, 218, 219, 223, 224, 225–26, 228; global economy and 215, 228; Hydraulics Division and 215; machinist strike of 1981–1997 and 231, 235; management-union relationship and 215, 218–19, 225, 228; metrology products and 244, 245; military service in Navy during Korean War and 215; as president/chief executive officer 222, 227, 244; professional managers and 215, 216, 218, 219; Renishaw Probe and

222, 226; Rhode Island Hospital Trust National Bank board and 241; sexual harassment suit against 244; subcontracts for 241; technological innovations in 212, 215; Thayer and 228; on unions/union strikes 213, 225; as vice-president 213, 215
Roch S. A. 245
Rockman, Seth 7
Rogers, Ray 241
"rogue colonialists" 5, 253*prefacen*1
Rolt, L.T.C. 27, 30, 37
Roosevelt, Franklin Delano crash in 1929 and 152; government-regulated capitalism and 162; New Deal and 154, 155, 169, 253*prefacen*5; on union desegregation 95; World War II and 159
Rosenberg, Nathan 17–18, 254*ch3n*4, 255*ch6n*8
ROTC, Naval 173, 174
rulers: Darling's manufacture of 42, 47, **48**; graduating machines/linear graduating machines for marking 9–10, 14, **15**, 16, 52–53, 211; handmade 9, 16
Russia 87, 89, 116, 139, 212, 258*ch16n*5; *see also* Soviet Union

salaries: for executive board members 104, 108, 140; for factory workers 51, 56, 66, 99, 108, 140, 155, 165–66, 167, 169, 170, 179, 187, 224; unions/union strikes and 186, 187, 213, 220, 221, 224
salesmen, for machine tools 82, 87, 88, 199–200
Schwarz, Michael 41, 42
screw machines: automatic screw machines and 84, 88, 155, 193–94, 195–96, 202, 212; electronics and 223; LS, Sr. and 74, 81, 91; turret screw machines 32, 41, 45, 194
screw-cutting lathe 37, 255*ch4n*2
Second Adventists 76
secretaries, executive 102, 105, 137; *see also* clerical workers
security measures, for Precision Park plant, in North Kingstown, RI 221–22, 252, 259*ch20n*5
sewing machine manufacturing: during American Civil

War 30–32; B&S and 23, *23*, 24–25, 26, 27, 30–32, 43, 254*ch*2*n*5; in Europe 37, 39; EW and 56, 255n6; FWH's improvements and 28; Gibbs and 21–22; Singer Sewing Machine and 21, 24, 39, 73–74, 79, 90; Willcox & Gibbs sewing machine and 22, *23*, 24–25, 30–32, 41, 43, 57, 58–59, 60, 255*ch*6*n*6
sexual harassment 107, 244
Shah of Iran 224
Sharpe, Amey 73, 109
Sharpe, Ellen 114, 115
Sharpe, Henry, Jr. (HS, Jr.): aerospace industry and 210–11; A.H. Belo purchase of Providence Journal Company and 248, 249, 250; airplane manufacturing and 226; apprenticeships and 178–79; assembly lines and *195*, 195–96, 198–99, 217; automatic screw machines and 193–94, 195–96, 202, 212; automobile manufacturing and 201, 204, 211, 226; biographical information/characteristics of 138, *139*, *139*, 150–51, 156–57, 159, *159*, 174–76, *177*, *184*, 189–90, 191, 193, *205*, 225, *250*, 252; Brown & Sharpe Ltd. and 198; Brown University and 172–73, 177, 184; business interests and support from 248, 249; Cedasize grinding machine and 199, 202; CMMs or coordinate measuring machines and 210, 211, 222; CNC or computer numerical control machines and 212, 215; computers and 200, 204; Defense Department procurement and 226; DigitCal Micrometer and 222–23, *223*; dividends paid and 200; educational achievement of 157, 167, 172–73, 177, 184; electronics and 211, 215, 223; employment statistics and 184, 190, 196; family relations and 157, 159, 174, 175–76; foundry and 179, 220, 231; furloughs/shut-downs and 219–20; gear cases for anti-aircraft guns and 184; on global economy 215, 216; and hair clippers, barbershop 196; headquarters relocation issue and 231; health and accidents and 202–3; homes/homelife of 156–57, 159; incorporation/public stock and 208, 209; investment casting and 198–99; layoffs and 190, 196–97, 211, 212, 220; machinist strike of 1981–1997 and 231, 235, 238, 240–41; metrology and 211; military service in Navy during World War II by 171, 172, 175, 176–78, 184, 215; missile manufacturing and 199, 202, 204; NC or numerical control machines and 200, 202, 204, 212, 215; nuclear power industry and 206, 211; numeric controlled turret drills and 204; oil/energy crisis and 224; overseas plants in Europe and 197–98; politics and 156; Precision Park plant and 204–7, *205*, *205*, 206, *207*, 259*ch*19*n*1; as president/chief executive officer 181, 222, 248; professional managers and 215, 216; profits/losses and sales for B&S and 187, 197, 208, 211–12, 223, 224, 225–26; recession of 1958 and 199, 201; Renishaw Probe and 222, 226; reorganization of B&S and 198; retooling and 193, 195, 196, 201; salaries for factory workers and 179, 187, 208; salesmen and 199–200; stock restructuring and 210; surplus during post-war years for 197; technological innovations and 212, 215; trips to Europe for 179; turret drills and 200–201; unions/union strikes and 184–87, 213, 217–18, 219–21, 225; Validator CMM and 211; Wankel engine and 215, 223–24
Sharpe, Henry, Sr. (HS, Sr.): agricultural equipment manufacturing and 140, 152; airplane manufacturing and 131, 159, 162; apprenticeship programs and 142–44; arbitration and 141, *142*; automatic screw machines and 155; automobile manufacturing and 125, 140, 141, *141*, 156; biographical information/characteristics of 56, *80*, *98*, 125, 129, 135, *139*, 157, 159, *159*, *191*, 192; Brown University and 84, 188, 190; business interests and support from 105–7, 108, 109, 128, 135, 137–38, 142, 144, 156, *156*, 190–91, 248, 249; business skills of 2–3; centralization of B&S and 138; children of 138, *139*; citizenship/Americanization and 126; closed shop clause and 164–65; contract negotiations and 164, 165, 184; corporate laws and 152, 154, 162; corporate tax deductions and 161; courtship and marriage of 125, 127, 134–35, 136; Depression of 1921 and 144; dividends paid and 66, 105, 108, 138, 149, 160, 171; economic support for U.S. and 133–34; educational achievement of 68, 84; eight-hour day/40-hour week and 118–19, 128, 162, 179; electronic gauges for measuring pieces and 179–80; employment statistics and 107, 128, 155, 160; European markets for machine tools and 125, 138, 156; Excess Profits Tax and 162, 171; exhibitions for machine tools and 155; factory conditions and 141, *141*; gear cases for anti-aircraft guns and 179; global economy and 137; government-regulated capitalism and 125–26, 128, 134, 142, 152, 154, 162; Great Depression of 1929 to 1939 and 144, 149, 153, 258*ch*14*n*1; gun bolts and 167–68; homes/homelife of 144–45, 151, 156–57, 159; HS, Jr.'s relations with 157, 175–76; illnesses/health of 181; incorporation of Brown & Sharpe Manufacturing Co. and 158–59; insurance plans and 128–29, 135, 141; Johannson Blocks and 180; labor laws and 125–26, 128, 134, 142, 162; layoffs and 116, 138, 149, 152; legacy of 190–92; loyalty for U.S. citizens and 132–33, *133*; LS, Jr. relationship/estrangement with 101–2, 108, 114–15; machinists/mechanics and 137; metals-working manufacturing and 97; payroll systems and 169–70; pension plans and 135, 141–42, 170; plant/factory space for B&S and 100, 107–8, 138, 144; pool order system and 161–62,

180; as president 105, 108, 140; profits/losses and sales for B&S and 105, 116, 144, 149, 152, 153, 155, 156, 160, 162, 171, 179; progressivism and 128, 134; railroads and machine tools manufacturing and 138, 140; salaries for factory workers and 140, 155, 170; salary as president and 108, 140; as secretary 102, 105; social class and 120, 126; social equality/inequality in business and 79; succession to LS, Sr. and 79, 80, 81, 97–98, 100–102, 257*ch*10*n*3; surplus during post-war years for machine tools and 137; technological innovations and 179–80; trips to Europe for 73, 79, 80–82, 145; unemployment compensation and 150, 155; unions/union strikes and 118–20, 134, 142, 155, 162–64, 258*ch*14*n*2; War Bonds and 170, 171; women as factory workers and 129–30, **130**, 165–69, 170, 179; workplace shooting and 107; on World War II and U.S. 159

Sharpe, Henry III (HS, III): on business acquisitions in Europe 247; combination square and 250–51; educational achievement of 226; on executive board of B&S 246; on foundry closing and HS, Jr. 220; at Precision Park groundbreaking ceremony 204–5, **205**; product design skills of 226–27

Sharpe, Louisa (m. Jesse Metcalf) 106, 109, 145, 169

Sharpe, Louisa Dexter (LDS): biographical information/characteristics of 16, 19, 103, 254*ch*2*n*3; children of 25, 26, 34, 49, 50, 56, 72; on succession to LS, Sr. 98, 102, 103; trips to Europe by 36, 49, 99

Sharpe, Lucian, Jr. (LS, Jr.): arms manufacturing and 81; biographical information about 56, 68, **80**, **98**, 144; Brown University support from 105; dividends paid and 66, 103, 105, 109, 144, 148, 149; estrangement between Sharpe family and 101–2, 108, 114–16, 144, 149; illnesses/health of 97, 98, 103, 104, 111, 113–14, 115, 148, 149; lawsuits against 127; management and systems for machine tools manufacturing and 93; management/systems improvements and 93; plant/factory space for machine tools manufacturing and 99–100; RV and 97, 111, 114; salaries as vice president for 104; social equality/inequality in business and 79; succession/resignation/reinstatement of 79, 80, 81, 97–98, 100–102, 104, 114, 257-*ch*10*n*3; as treasurer of B&S 102, 103–4, 111, 114; trips to Europe by 79, 80–82, 99–100, 127; as vice-president of B&S 104; WV and 111, 114

Sharpe, Lucian, Sr. (LS, Sr.): advice to managers from 66; American Civil War era and 27, 28, 34; American (Standard) Wire Gauge invention by 19, **20**, 78–79; apprenticeship programs and 11, 12, 13, 14, 16, 17, 29; apprenticeships and 94; Armory/American System and 23, 24; arms manufacturing and 27–29, 30, 31, 81; automatic screw machines and 84, 88; on barbershop hair clippers 68; biographical information about **35**, **55**, 85, 101, **101**; business interests and support from 79, 102, 105–6, 241, 248, 249; business skills of 2–3, 11, 14, 17; calipers and 44–45; children of 25, 26, 34, 49, 50, 56, 72; clock manufacturing and repairs by 21, 28; Combination group of Willcox & Gibbs stockholders and 58, 85; Darling and 43, 53, 85; death and funeral of 102; double entry bookkeeping and 14, 17; educational achievement of 16–17; and Europe, manufacturing in 34, 35, 39, 40, 73; father of 5, 11–12, 94; FWH and 9, 46, 59, 60, 65; Gibbs and 21, 25; on graduating machines/linear graduating machines pirates/competition 46–47; grandfather of 5; homes of 54, 56, 68; on horse hair clippers 50, 67; on horse team grooming 50; illnesses/health of 12–13, 34–35, 37, 39, 98–99; JRB and 2–3, 6, 7, 13, 14, 16, 45–46, 49, 94; J.R. Brown Company and 11, 12–13, 14; machinists/mechanics and 7; marriage of 19; mentorships and 11, 12, 16–17, 111, 116; on mergers and conglomerates 210; mother of 39, 43; plant/factory space and 50–51; plants/factory space, in Providence, RI and 50–51, 53–58, **55**, **71**, 78, 93, 99–100; precision tools and instruments and 41, 43, 68; profits/losses/sales and 27–28, 30, 32, 43, 45, 46, 49, 56, 59, 60, 65, 67, 68, 79, 103, 255*ch*5*n*13; repairmen/repairs and 19; salesmen and 82, 87, 88; screw machines and 74, 81, 91; sewing machines for Willcox & Gibbs and 23, **23**, 24–25, 26, 32, 57, 58–59, 60, 255*ch*3*n*15; on slavery 25–26, 254*ch*2*n*13; social equality/inequality and 7; succession in B&S after death of 79, 80, 81, 97–98, 100–104, 257*ch*10*n*3; technological innovations and 11, 179; on telegraph communication with Europe 40; on telephones 68; trips to Europe by 34–40, 44, 49, 72–73, 74, 80–82, 98–100; watch manufacturing/repairs and 21, 28; *see also* Sharpe, Henry, Sr. (HS, Sr.)

Sharpe, Mary (daughter of LS, Sr.) 72–73, 103, 104, 109–10, 114, 158

Sharpe, Mary Elizabeth Evans (MEES): biographical information about 121–22, 123, **124**, 135, 136, **139**, **191**, **205**, 257*ch*12*n*2; as business woman 122–24, 125, 127, 134, 150–51, 189; children of 138, **139**, 174; courtship and marriage of 125, 127, 134–35, 136; educational achievement and honorary degree for 189; at groundbreaking ceremony for Precision Park 204–5, **205**; homes/home-life of 144–45, 151, 156–57; HS, Jr.'s relations with 174, 175–76; illnesses/health of 139, 145; interior design skills of 145, 151; landscape design skills of 187–89, 205, 238; machinists' strike of 1981–1997 and 237–38; trips to Europe for 145; World

War I and 127, 132, 134, 144–45
Sharpe, Peggy Boyd 6, 189–90, **191**, 203, 206, 226
Sharpe, Sally Chaffee 39, 43
Sharpe, Wilkes 5, 11–12, 94
Sherman, L.M. 173–74
ships/ship manufacturing 72, 102, 206, 257*ch*10*n*3
shop superintendents/foremen (manager/s) *see* manager/s (foremen/shop superintendents)
shutoff valve patent 17
Siemens, William 69
"the Silk Hat machinist" (Thomas Goodrum) 51, 52, 53, 55, **55**
Singer Sewing Machine 21, 24, 39, 73–74, 79, 90
Slater, Mabel Hunt 148
slavery 25–26, 254*ch*2*n*13; *see also* African Americans
Slocum, John 27, 28
Smith, Adam, *The Wealth of Nations* 5, 253*prefacen*2
Smith-Hughes Act of 1917 196
social classes: in Great Britain 37; management versus factory workers and 167, 185; unions/union strikes and 83, 96, 120
social equality/inequality: for African Americans 7, 94–95, 110–11, 143, 254*ch*2*n*13; in B&S 7; description of 7; desegregation of union and 95; voting rights and 5, 109–10, 117; white male factory workers and 7, 8, 94, 204; for women 7, 79, 109–10; *see also* social classes
Social Security Act 154, 155
soft drink can manufacturing 245, 247
software, for metrology products 244, 246–47; *see also* computers
Soviet Union 180, 239, 244; *see also* Russia
Spalding, Frank 45
Spencer, Christopher 84, 193–94
Springfield rifles 29, 97, 168
standardization 6, 19, **20**, 78–79, 247
Starrett, Leroy S. 84–85
steam engines 18, 22, 23, 37, 38, 91, 103, 107
steel manufacturing 38, 69, 71–72, 74, 160–61
stock market (New York Stock Exchange) *see* New York Stock Exchange (stock market)
stockholders and executive board for B&S *see* executive board and stockholders for B&S
stockholders, and Willcox & Gibbs 57, 58, 59, 60, 65, 66, 85
Stockwell, Alden B. 56–57, 58–59, 60, 65, 66
Stuber, Frederick 245, 246, 247–48
subcontracts and outsourcing 241, 250–51
Sugar Trust (American Sugar Refining Company) 86–87
supersonic transport planes 210–11, 215; *see also* aerospace industry; airplanes/airplane manufacturing
Supreme Court, U.S. 154, 254*ch*2*n*13
Swanson, Mr. **197**
Sweden 6, 7, 252
Swedenborgians 43
Switzerland 39, 210, 211, 245, 247

taxes: corporate 161, 209; Excess Profits Tax and 162, 171; income 132, 136, 169, 170, 211–12, 213
Taylor, Fredrick 126
technology, history of: mass media and 248–49; technological convergence and 17–18, 29, 30, 111, 254*ch*2*n*5; *see also* machine tools manufacturing and sales; metrology/metrology products; *and specific manufactured products*
Telecommunications Act of 1996 249
telegraph communications and wires 11, 17, 19, **20**, 40, 78–79, 111
telephones 68, 78, 111
television stations 248, 249
Tesa, S.A. 210, 211, 245, 247
Thayer, Robert: as business agent for machinist union and strikes 225, 229, 230; machinists' strike of 1981–1997 and 232, 233, 234, 236, 239, 242; management-union relationship and 218–19, 228; as president of machinist union and strikes 218, 220, 221; Roach and 228
Thermo-Electron Corporation 251
Thomas, Sidney Gilchrist 69
time-keeping, and capitalism 6, 16; *see also* clock manufacturing and repairs; "watch clocks"; watch manufacturing and repairs
Tingley's Marble Works (Tingley Marble Company) 30, 47
tooth (milling) cutter machine (milling cutter machine) 31, 32, 77
Toyota 245
Trade Agreement Act of 1979 240
Truman, Harry S 180, 185
Turbitt, J.J. 218
turret drills 200–201
turret lathes 29–30, 193–94, 195
turret screw machines 32, 41, 45, 193–94
Twain, Mark 64
typewriter manufacturing 64, 168

unemployment compensation 150, 155, 169, 234
unions/union strikes: American Federation of Labor and 105, 119; American Federation of Technical Engineers and 221; apprenticeship programs and 196, 197; arbitration and 141, 142, 187, 216, 219; assembly lines and 217–18; on B&S's direct communication with factory workers 228, 233–34; clerical workers and 221; closed shop clause and 164–65; contract negotiations between HS, Jr. and 164, 165, 184; desegregation of 95; eight-hour day/40-hour week and 118–19, 163; Greystone plant and 186; health benefits and 170, 186, 225, 242; holidays and 170, 187, 189; HS, Jr. and 184–87, 213, 217–18, 219–21, 225; HS, Sr. and 118–20, 134, 142, 155, 162–64, 258*ch*14*n*2; IAM and 7, 95, 119–20, 162–63, 164, 218, 221, 224; Industrial Relations department and 218, 224, 228, 229, 230; management-union relationship and 215, 218–19, 225, 228; membership in 162; National Industrial Recovery Act and 152, 154; National Metal Trades Association and 120; NLRB and 163–64,

230, 234, 240, 242, 249; pension plans and 170, 220, 225; piecework/scientific management and 217–18; railroad buyout bill and 138; salaries and 186, 187, 213, 220, 221, 224; social classes and 83, 96, 120, 185; unemployment compensation paid by 234; unemployment compensation paid by unions and 234; Wagner Act and 154, 162; WV and 118–19; *see also* machinists' strike of 1981–1997; male factory workers; *and specific unions*
Unitarians 39, 102, 206
United States: American Civil War in 21, 25, 26, 27, 30, 31, 32–33, 97; American Revolution in 5, 253*prefacen*1–3; Army of 29, 84, 168, 221; consumer price index in 102, 209, 212; Defense Department of 180, 197, 201, 221, 222, 224, 239; economy of 5–6; and European loans 140, 144, 151; exchange rates for dollars and 198, 209, 213, 239, 252; Federal Reserve in 136, 140, 145, 147, 151–52, 209, 240, 245; import/exports statistics for 201, 209; Industrial Revolution in 17, 109–10; inflation rate in 102, 116, 170, 179, 208, 209, 212, 224, 232, 245; interstate highway building in 201; Iranian Revolution in 1979 and 224; machine tools manufacturing and sales collapse in 239–40, 243, 259*ch*21*n*10; Marine Corps of 127, 172; national budget of 132–33; National Industrial Recovery Act and 152, 154; Navy of 171, 172, 173, 174, 175, 176–78, 184; Supreme Court of 154, 254*ch*2*n*13; surplus of machine tools during postwar years in 137, 179, 197; Vietnam/Vietnam War and 190, 208, 211; in World War I 127, 131–32, 133–34; in World War II 159, 161; *see also* African Americans; economy; federal government; labor history; New York Stock Exchange (stock market); *specific legislation*
United Traction and Electric Company (UTE) 87, 89
United Way 135, 190–91

universal grinding machine 52, 60, **62**, 63, 69, 83, 91, 92, 93, **161**, 255*ch*6*n*8; *see also* grinding machines
universal milling machines: for arms manufacturing in Europe 35; B&S and 30, **31**, 32, 41, 60, 75, 88, 254*ch*3*n*4; description and history of 27; steel manufacturing and 69; universal grinding machines and 63; *see also* milling machines
UTE (United Traction and Electric Company) 87, 89

Validator CMM 211
Validator Process Control Robot 244–45
Vernier system for manufacture of calipers 16, 222–23, **223**, 253*ch*2*n*2
Viall, Maybury Fraser 187–88, 189
Viall, Richmond (RV): apprenticeship program and 97; apprenticeships and 94; barbershop hair clippers and 67; biographical information about 65–66, 75; executive board and stockholders for B&S and 102, 111, 165; LS, Jr. and 97, 111, 114; managers and 53, 65, 69, 75, 111, 116; piecework/scientific management and 66, 126, 217; plant/factory space construction/relocation and 53–54, 100; shop rules and 75–77; succession to LS, Sr. and 97–98, 100; universal grinding machine modifications and 92
Viall, Richmond (son of William Viall): 104, 132, 134, 165, 187–88
Viall, William (WV): on apprenticeship standardization 94; catalogs for machine tools and 83–84; on Darling 43, 84, 256*ch*7*n*8; on Drury 87; factory shop site and 54, 55; on FWH 54; on global economy 137; on illnesses/health of RV 104; LS, Jr. and 111, 114; management and systems for manufacturing and 93; as manager in B&S 129; on manufacturing at B&S 47–48, 60; markets for B&S in Europe and 117, 125; on Panic of 1876 57–58; as secretary of B&S 137; on

succession to LS, Sr. 97–98; unions/union strikes and 118–19; women as factory workers in B&S and 129–30, **130**
vice presidents for B&S: AKB 178, 181; Austin 203, **209**; Geyer 230; LS, Jr. 104; Masser 230; Roach 213, 215; salaries for 104
Vietnam and Vietnam War 190, 208, 211; *see also* Asia
violence during machinists' strike of 1981–1997 234, 236–37, 238, **238**; workplace shooting and 107
vocational schools 17, 142, 196; *see also* apprenticeships
voting rights 5, 109–10, 116, 117, 136

Wagner Act 154, 162
Wankel engine 215, 223–24
Wary, Mr. **197**
Washington, George 5, 36
"watch clocks" 28, 30
watch manufacturing and repairs 5, 14, 19, 21, 28, 39, 254*ch*3*n*1; *see also* clock manufacturing and repairs
Waterman, Dave 224, 225, 229, 230
Watt, James 37, 38
The Wealth of Nations (Smith) 5, 253*prefacen*2
Weld, Calvin 47, 48, 60, 93–94
Weld, Charles 48, 60, 93
Weld, George 60, 93–94
West Germany 213, 215; *see also* Germany
white male factory workers 7, 8, 94, 143, **197**, 204; *see also* African Americans; male factory workers
Whitworth, Joseph 78
Wilkinson, David 255*ch*4*n*2
Wilkinson, John 37, 38
Willard, Fitzroy 107
Willard, Walter, H. 107
Willcox, Charles (CW) 22, 26, 32, 35, 53, 58, 60
Willcox, James (JC) 22, 23, 24, 25, 26, 27, 30, 32
Willcox & Gibbs: B&S manufacturing of sewing machines for 23, **23**, 24–25, 26, 27, 30–32, 41, 43, 49, 55, 57, 58–59, 60, 73, 75, 168–69, 193, 254-*ch*2*n*5, 255*ch*5*n*13; Combination group of stockholders and 58, 65, 85; EW and 56, 255*ch*6*n*6; Gibbs and **21**, 21–22, 24, 25, 26, 32–33, 53, 60;

JC and 22, 23, 24, 25, 26, 27, 30, 32; marketing and sales for 73–74; Stockwell and 57, 58, 59, 60, 65, 66
Wilson, Woodrow 116, 126, 127
Wingo, Richard **146**
Winslow-Townsend Bill 138
wires, telegraph 11, 17, 19, **20**, 40, 78–79, 111
women: African American 123, 124; as business women 122–24, 125, 127, 134, 150–51, 189; as clerical workers 129; as entreprenuers 122–24, 125, 127, 134; as factory workers wartime 129–31, **130**, 165–69, 170, 179; household management and 109–10; pay rates for 165–66; and salaries as factory workers 165–66, 167; sewing machine as used by 31, 32; sexual harassment of 107, 244; social equality/inequality for 7, 79, 109–10; voting rights for 109–10, 136; *see also* male factory workers; manager/s (foremen/shop superintendents)
Work Projects Administration (WPA) 154
Workmen's Compensation Insurance 128
World War I: Armistice Day and 134, 154–55, 187, 189; arms manufacturing during 81; B&S and 117, 125–26, 128, 129–31, **130**, 133–34; in Europe 116, 119, 136; hawk-dove disputes during 131–32; labor laws during 125–26, 128, 134; MEES and 127, 132, 134, 144–45; military service during 132, 187; Red Cross during 132, 133, 134, 144–45; U.S. entry into 127, 131–32, 133–34; women as factory workers during 129–31, **130**
World War II: corporate tax deductions during 161; Europe and 159; military service during 171, 172, 175, 176–78, 184; pool order system during 161–62; rationing during 171, 175; scrap metal collection during **163**; steel manufacturing priorities in U.S. during 160–61; U.S. entry into 159, 161; War Bonds during 170, 171; women as factory workers during 165–69, 170, 179; *see also* World War I
Wriston, Henry 188, 189

Zakim, Michael 17

www.ingramcontent.com/pod-product-compliance
Lightning Source LLC
Chambersburg PA
CBHW081545300426
44116CB00015B/2754